Industrial Internet of Things

Advances in IoT, Robotics, and Cyber Physical Systems for Industrial Transformation

Series Editors: S. Balamurugan and Dinesh Goyal

Smart home, Smart city, and Wearable Technologies are the most exponentially growing applications of Internet of Things (IoT). Wearable Technology is considered to be highly ubiquitous and the most phenomenal IoT Application. Smart World and Wearable Technology exhibits a high potential to transform our lifestyle. Ranging from healthcare tracking applications to smart watches and smart bands for personal safety, IoT has become one of the most indispensable parts of our lives. In a study by Business Insider's premium research service, by the end of 2022, the wearable IoT market is expected to grow and reach 162.9 million units. Some of the top-notch applications of IoT include smart parking, smart wearable technologies, smart clothing, smart safety, smart farming, smart industry, robotics and so on, thereby building the next-generation smart world. An important priority for today's communication world is to invest in the development of a smart world, where intelligent "things" are connected to serve people better. With the exponential growth of Internet of Things (IoT) and its applications, the implementation of a smart world becomes much more feasible. Sensor market plays a vital role in applying IoT to build the smart world. Applications such as monitoring parking space to optimally park vehicles (Smart Parking), detecting the frequency and intensity of traffic to optimally select routes (Traffic Congestion), predicting trash levels in garbage collection containers (Smart Waste Management), managing the intensity of street lights based on weather condition and sunlight (Smart Lighting), detecting and controlling excess hazardous gases coming from industries and vehicles (Smart Air Pollution Controlling), detecting the mix of hazardous chemicals from factories in drinking water (Smart Water Pollution Controlling), efficient monitoring and management of power consumption (Smart Grid), artificial intelligence driven retail marketing (Smart Shopping), detecting health abnormalities (Smart Health) and Home Automation mechanisms fall under the category of applying IoT and Robotics to building a smart world. The purpose of designing this book series is to portray certain practical applications of IoT in building a smart world. With a wide-spread application in both smart cities and smart homes, this book series is becoming especially important.

Industrial Internet of Things
Technologies and Research Directions
Edited by Anand Sharma, Sunil Kumar Jangir, Manish Kumar,
Dilip Kumar Choubey, Tarun Shrivastava, and S. Balamurugan

For more information on this series, please visit: https://www.routledge.com/Advances-in-IoT-Robotics-and-Cyber-Physical-Systems-for-Industrial-Transformation/book-series/CRCAIRCPSIT

Industrial Internet of Things

Technologies and Research Directions

Edited by

*Anand Sharma, Sunil Kumar Jangir, Manish Kumar,
Dilip Kumar Choubey, Tarun Shrivastava,
and S. Balamurugan*

CRC Press

CRC Press
Taylor & Francis Group
Boca Raton London New York

CRC Press is an imprint of the
Taylor & Francis Group, an **informa** business

First edition published 2022
by CRC Press
6000 Broken Sound Parkway NW, Suite 300, Boca Raton, FL 33487-2742

and by CRC Press
2 Park Square, Milton Park, Abingdon, Oxon, OX14 4RN

CRC Press is an imprint of Taylor & Francis Group, LLC

Library of Congress Cataloging-in-Publication Data

Names: Sharma, Anand, editor.
Title: Industrial internet of things : technologies and research directions / edited by
Anand Sharma, Sunil Kumar Jangir, Manish Kumar, Dilip Kumar Choubey,
Tarun Shrivastava, S. Balamurugan.
Description: First edition. | Boca Raton : CRC Press, 2022. | Series:
Advances in IoT, robotics, and cyber physical systems for industrial
transformation | Includes bibliographical references and index.
Identifiers: LCCN 2021044837 (print) | LCCN 2021044838 (ebook) |
ISBN 9780367702076 (hbk) | ISBN 9780367702083 (pbk) | ISBN 9781003145004 (ebk)
Subjects: LCSH: Internet of things--Industrial applications.
Classification: LCC TK5105.8857 .I483 2022 (print) | LCC TK5105.8857 (ebook) |
DDC 004.67/8--dc23/eng/20211116
LC record available at https://lccn.loc.gov/2021044837
LC ebook record available at https://lccn.loc.gov/2021044838

ISBN: 978-0-367-70207-6 (hbk)
ISBN: 978-0-367-70208-3 (pbk)
ISBN: 978-1-003-14500-4 (ebk)

DOI: 10.1201/9781003145004

Typeset in Times
by KnowledgeWorks Global Ltd.

Contents

Preface

Many research, technical, and academic communities are dynamically pursuing research on the Internet of Things (IoT). At present, as sensing, control, and communication of data become increasingly complicated, there is a considerable overlap in these communities, despite often differing opinions. Anything that can be interconnected will be linked and communicated using sensors. Once linked, these connected "things" can send stored or generated data and interact with other things and people – all in real time.

Connectivity and interoperability are of central importance for the Fourth Industrial Revolution to connect products, machines, people, and the environment in a manufacturing system. Currently, with huge applications (e.g. smart manufacturing, industrial equipment monitoring, smart factory, industrial property management), the Industrial Internet of Things (IIoT) is attracting ever-increasing attention from research, technical, and academic communities.

The integration of connectivity, physical objects, cyber objects, and social objects in industry are the key concepts that enable the IoT and cyber-physical systems (CPS) that ultimately represent a central entity of Industry 4.0. IIoT promotes and develops intelligent behavior to better serve people in industry. For instance, smart surveillance systems could be enabled for petrochemical plants with IIoT by converging CPSs and social spaces.

The trend toward Industry 4.0 intends to populate traditional shop floors with digitalized systems that are able to share their process parameters and their operative status and express their availability for collaboration with other machines or workers. By ensuring maximum uptime of hardware and efficient and cost-effective systems, industries are able to provide customers with a better, more satisfying service. In this sense, the situational awareness of the manufacturing environment greatly relies on connectivity solutions, like IoT or CPS.

By leveraging the various technological advancements in IoT for Industry 4.0, organizations will be able to increase their revenue. This is achievable in a number of ways, such as developing custom solutions for clients with improved profit margins or improving the efficiency of various production processes to decrease costs.

Although IoT and CPS provide companies with many benefits like increased efficiency, time savings, cost savings, and enhanced industrial safety, they do not come without a cost. A few of the difficulties and challenges associated with Industry 4.0 are energy efficiency, coexistence, real-time performance, interoperability, security, and privacy. Reliable, fast, and stable communication networks are also necessary.

In particular, this book focuses on the key technologies, challenges, and research directions associated with the IIoT. To provide a basis for discussing open principles, methods, and research problems in IIoT, a vision for how IIoT could change the world in the distant future will be presented. This book also provides a systematic overview of the state-of-the-art research efforts and potential research directions to deal with IIoT challenges. The key technologies will be defined and research problems within these topics will be discussed in this book.

About the Authors

S. Balamurugan received his BTech, MTech, and PhD degrees all in the field of Information Technology from Anna University, India. He received three post-doctoral degrees – Doctor of Letters (DLitt), Doctor of Science (DSc), and Doctor of Technology (DTech). He has published 50 books, 200+ international journals/conferences, and has 60 patents to his credit. He is the Director-Research and Development, Intelligent Research Consultancy Services (iRCS), Coimbatore, Tamil Nadu, India. He serves as a research consultant to many companies, startups, small and medium enterprises (SMEs), and micro small medium enterprises (MSMEs). He is the Editor-in-Chief of the Book Series *Artificial Intelligence and Soft Computing for Industrial Transformation*, Scrivener, John Wiley; *Artificial Intelligence and Learning Techniques for Engineering*, Bentham Sciences; *Advances in Quantum Computing, Artificial Intelligence and Data Sciences for Industrial Transformation;* and *Advances in IoT, Robotics and Cyber Physical Systems for Industrial Transformation*, CRC Press, Taylor & Francis Group. He is the Editor-in-Chief of Information Science Letters, Natural Sciences Publishing, NSP, and the International Journal of Robotics and Artificial Intelligence, PRIRS, United States. He is Associate Editor of IEICE Transactions on Information and Systems, Oxford University Press; Cluster Computing, Springer; Simulation, Sage Publishing; Technology and Healthcare, IOS Press; Intelligent IoT Computing, Inderscience; World Journal of Engineering, Emerald; International Journal of Intelligent Systems, Technologies and Application, Inderscience; International Journal of Renewable Energy Technology, Inderscience; and Journal of Autonomous Intelligence, Frontier Scientific Publishing. He is also in Editorial of International Journal of Intelligent Unmanned Systems, Emerald; International Journal of Automation and Control, Inderscience; International Journal of Society, Systems and Science, Inderscience; International Journal of System Dynamics Applications (IJSDA), IGI Global Publishers; International Journal of Service Science, Management, Engineering, and Technology, IGI Global Publishers; International Journal of Knowledge and System Sciences, IGI Global Publishers; and several other journals of Taylor & Francis, Inderscience, Emerald, IET, MDPI, De Gruyter, and Lead Guest Editor of several special issues with Springer, Elsevier, and the Institute of Electrical and Electronics Engineers (IEEE). He is the recipient of the Rashtriya Vidya Gaurav Gold Medal Award and The Best Educationalist Award from Hon. Justice O.P Saxena, Supreme Court, New Delhi, and the Former Chairman of Minority Council, New Delhi, India. He is the recipient of two Lifetime Achievement Awards. He is the recipient of Dr A. P. J. Abdul Kalam Sadhbhavana

Award from Hon. Balmiki Prasad Singh, Former Governor of Sikkim, Jewel of India Award from Mr Gurpreet Singh, General Secretary, India, Star of Asia Award from Mr Korn Debbaransi, Former Dy. Prime Minister, Thailand, at an International Summit in (Bangkok) Thailand, and Pride of Asia Research Excellence Award from Hon. Anant. V. Sheth, Deputy Speaker, Goa Legislative Assembly, Best Director Award. He has received the "Active Member CSI National Award." He was awarded the Prestigious Mahatma Gandhi Leadership Award at the House of Commons, British Parliament, London, UK. The book he authored, *Machine Learning and Deep Learning Algorithms using MATLAB and PYTHON*, won the "Best MATLAB Book for Beginners" award by Book Authority. Dr S. Balamurugan won the CSI Young IT Professional Award. He is the winner of the "National CSI Youth Award 2020" by the Computer Society of India. He is also the recipient of the Best Researcher Award, Certificate of Exceptionalism, Young Scientist Award, Best Young Researcher Award, and Outstanding Scientist Award. His biography is listed in "Marquis WHO'S WHO," New Jersey, United States. His professional activities include roles as Editor-in-Chief/Associate Editor/Editorial Board Member for more than 500+ international journals/conferences of high repute and impact. He has been invited as Chief Guest/Resource Person/Keynote Plenary Speaker in many reputed universities and colleges at the national and international Levels. His research interests include Artificial Intelligence, Augmented Reality, the Internet of Things, Big Data Analytics, Cloud Computing, and Wearable Computing. He is a life member of ACM, IEEE, ISTE, and CSI.

Dilip Kumar Choubey is currently working as an Assistant Professor in the Department of Computer Science & Engineering, Indian Institute of Information Technology Bhagalpur, India. Previously, he worked as an Assistant Professor (Senior) at the School of Computer Science and Engineering, Vellore Institute of Technology, Vellore, India, and as an Assistant Professor (On Contract) at NIT Patna India. He has worked as an Assistant Professor at Lakshmi Narain College of Technology (LNCT), Bhopal, India, and Oriental College of Technology (OCT), Bhopal, India. He has more than nine years of experience in teaching and research. He received his PhD in Engineering from Birla Institute of Technology (BIT), Mesra, Ranchi, India, in 2018. He received his MTech degree in Computer Science and Engineering and BE degree in Information Technology from RGPV Bhopal, India, in 2012 and 2010, respectively. He is an author of two books and one copyrights grant and has more than 45 reputed research publications in international journals, book chapters, and conference proceedings. He also has one project as a Co-PI funded by TEQUIP Collaborative Research Scheme. He is guiding two PhD students and has guided many PG and UG projects. He has conducted several faculty development program (FDP)/workshops. He has delivered many expert talks as well as attended FDP/workshops/conferences. He has also been a member of the organizing and technical program committees of many conferences and workshops. He has reviewed many research papers in reputed journals/conferences/book chapters. He is a lifetime member of several professional bodies. His research interests include Artificial Intelligence, Machine Learning, Deep Learning, Data Science, Soft Computing, Pattern Recognition, Bioinformatics, Data Mining, Database Management Systems, and IoT.

Sunil Kumar Jangir received his BE (Information Technology) with honors from the University of Rajasthan in July 2009. He received the MTech degree with the highest distinction in Software Engineering and his PhD degree in Computer Science & Engineering. He has 12+ years of teaching experience in the field of Computer Science & Engineering. He has published more than 35 research papers in national and international conferences and journals. He has contributed more than 5 book chapters to books by reputed publishers. He is a member of IET, Professional member of ACM, and Senior member of IEEE. Dr Jangir served as a Convener of 1st and 2nd ICITDA, 1st NCITSA, and Convener of JECRC Hackathon 1.0 and had been nominated four times as Institute SPOC for Smart India Hackathon. Dr. Jangir has chaired a number of sessions in various international conferences. He is currently an Associate Professor at the Department of Computer Science & Engineering, University Institute of Engineering, Chandigarh University, Mohali, Punjab. Dr. Jangir has also served as IEEE Student Branch Faculty Coordinator of Mody University of Science & Technology, Sikar, Rajasthan and IEEE Student Branch Counselor Anand International College of Engineering, Jaipur, Rajasthan. He has been associated with CSI-India in the past. His research interests include Computer Networks, Machine Learning, Deep Learning, and Routing Protocols for ad-hoc networks. Currently, he is working in the area of IoT.

Manish Kumar received the BTech degree in Applied Electronics and Instrumentation Engineering from Biju Patnaik University, Rourkela, India, in 2010 and the MTech degree in Biomedical Engineering from the Manipal University, Udupi, India, in 2013. He received a PhD degree in Electrical and Electronics Engineering from Birla Institute of Technology, Mesra, Ranchi, India, in 2019. He worked as a Research Associate in the Indian Institute of Technology, Patna, India. He is currently an Assistant Professor at the Department of Biomedical Engineering, Mody University of Science and Technology, Sikar, Rajasthan, India. His research areas of interest are Medical Image Processing, Signal Processing, Machine Learning, and Nature-Inspired Techniques.

Anand Sharma received his PhD degree in Engineering from MUST, Lakshmangarh, MTech from ABV-IIITM, Gwalior, and BE from RGPV, Bhopal. He has been working with Mody University of Science and Technology, Lakshmangarh, for the last ten years. He has more than 14 years of experience in teaching and research. He has been invited to several reputed institutions like ISI-Kolkata, IIT-Mumbai, IIT-Jodhpur, IIT-Delhi, and RTU-Kota. He has pioneered research in areas of Information Security, IoT, WBAN, and Machine Learning. He is a member of IEEE, IET, ACM, IE(India), Life Member of CSI and ISTE. He is serving as secretary of CSI-Lakshmangarh Chapter and Student Branch Coordinator of CSI-MUST Student branch. He has more than six authored/edited books with national and international publishers. He has contributed more than ten book chapters to the books of reputed publishers. He has organized more than 15 conferences/seminars and workshops. He has chaired more than eight special sessions and delivered six keynote addresses at international conferences. He is serving in the advisory capacity in several international journals as Editorial Member and in International Conferences as a Technical Programme Committee/Organizing Committee.

Tarun Shrivastava is the Founder and CEO of Tishitu Electronics and deals with industrial, government, and commercial product development, and Tishitu Technology and Research Private Ltd. deals with educational expertise, wherein a team of 80 professionals are connected in India and abroad. Tishitu is registered with MSME iStart Rajasthan, certified with the ISO accreditation system in Scotland, UK, and a well-known name in the domain of Engineering. He has developed Hydrogen Fuel Cells for green energy fuel generation: "HHO Home gas" was his project from 2010–2016. He also worked with the company Omega Electronics Jaipur for various educational products like "Digital computer interface simulators for digital electronics" and "VLSI kit for learning VHDL and Verilog programs to embed in an environment with Xilinx for CPLD and FPGA interface."

Contributors

Amit Kumar Bairwa
Department of Computer Science and
  Engineering
Manipal University Jaipur
Jaipur, India

Vaishali Baviskar
Department of Computer Engineering
GH Raisoni College of Engineering and
  Management
Savitribai Phule Pune University
Pune, India

Manoj Kumar Bohra
Department of Computer and
  Communication Engineering
Manipal University Jaipur
Jaipur, India

Pradeep Chatterjee
Digital Transformation and Experience
  Management
TML Business Services Limited
Pune, India

Dilip Kumar Choubey
Department of Computer Science and
  Engineering
Indian Institute of Information
  Technology
Bhagalpur, India

Dilip Kumar Choudhary
Department of ETC
G. H. Raisoni College of Engineering
Nagpur, India

Shambo Roy Chowdhury
Department of Mechatronics
Manipal University Jaipur
Jaipur, India

Dharmendra Kumar Dheer
Department of Electrical Engineering
National Institute of Technology
Patna, India

Mahadev Gawas
Department of Computer Science
Government College of Arts, Science
  and Commerce
Sanquelim, India

Shankar K. Ghosh
Advanced Computing and
  Microelectronics Unit
Indian Statistical Institute
Kolkata, India

Amisha Kirti Gupta
Department of Computer and
  Communication Engineering
Manipal University Jaipur
Jaipur, India

M. Nageswara Guptha
Department of Computer Science and
  Engineering
Sri Venkateshwara College of
  Engineering
Bengaluru, India

M. S. Hema
Department of Information
  Technology
Anurag University
Hyderabad, India

Prashant Hemrajani
Department of Computer and
  Communication Engineering
Manipal University Jaipur
Jaipur, India

Nihla Iqbal
Faculty of Technology
Department of Information and
    Communication Technology
South Eastern University of Sri Lanka
Oluvil, Sri Lanka

Sunil Kumar Jangir
Department of Computer Science and
    Engineering
University Institute of Engineering,
    Chandigarh University
Mohali, India

Chhotelal Kumar
Department of Computer Science and
    Engineering
National Institute of Technology
Patna, India

Jitendra Kumar
Department of Computer Science and
    Engineering
National Institute of Technology
Patna, India

Jitendra Kumar
School of Advanced Sciences
Vellore Institute of Technology
Vellore, India

Manish Kumar
Department of Biomedical
    Engineering and Electronics and
    Communication
School of Engineering and
    Technology
Mody University of Science and
    Technology
Lakshmangarh, India

Mukesh Kumar
Department of Computer Science and
    Engineering
National Institute of Technology
Patna, India

R. Maheshprabhu
Mechanical Engineering
Aurora's Scientific and Technological
    Institute
Hyderabad, India

Sandipan Mallik
Department of Electronic and
    Communication
NIST
Berhampur, India

G. Prema Arokia Mary
Information Technology
Kumaraguru College of
    Technology
Coimbatore, India

K. Meena
Department of Computer Science and
    Engineering
Vel Tech Rangarajan Dr Sagunthala
    R&D Institute of Science and
    Technology
Morai, India

Tanishka Mohan
Department of Computer
    and Communication
    Engineering
Manipal University Jaipur
Jaipur, India

Aishwarya Parab
Department of Computer
    Science
Government College of Arts, Science
    and Commerce
Quepem, India

Akshet Bharat Patel
Department of Mechatronics
Manipal University Jaipur
Jaipur, India

Hemprasad Y. Patil
Department of Embedded Technology
SENSE, Vellore Institute of
Technology
Vellore, India

Rahul Prakash
Department of Electrical
Engineering
National Institute of Technology
Patna, India

Fanoon Raheem
Faculty of Technology
Department of Information and
Communication Technology
South Eastern University of Sri Lanka
Oluvil, Sri Lanka

R. Raja Sekar
Department of Computer Science and
Engineering
School of Computing
Kalasalingam Academy of Research
and Education
Srivilliputhur, India

Nishant Sharma
School of Computer Science and
Engineering
Vellore Institute of Technology
Vellore, India

Pranav Rajesh Sharma
Department of Mechatronics
Manipal University Jaipur
Jaipur, India

Vaibhav Shukla
Tech Mahindra
Mumbai, India

Prashant Kumar Singh
Department of Electronics and
Communication
UCET VBU
Hazaribag, India

Shashank Kumar Singh
Department of Electronics and
Communication
UCET, VBU
Hazaribag, India

Vaibhav Soni
Department of Computer Science and
Engineering
Maulana Azad National Institute of
Technology
Bhopal, India

Divya Srivastava
Bennett University
Greater Noida, India

Parveen Sultana H.
School of Computer Science and
Engineering
Vellore Institute of Technology
Vellore, India

Shipra Swati
Department of Computer Science and
Engineering
National Institute of Technology
Patna, India

Babul P. Tewari
Department of Computer Science and
Engineering
Indian Institute of Information
Technology
Bhagalpur, India

Anjini Kumar Tiwary
Department of Electronic and
Communication
Birla Institute of Technology
Mesra, India

Madhushi Verma
Department of Computer Science
Engineering
Bennett University
Greater Noida, India

1 Artificial Intelligence and Machine Learning for the Industrial Internet of Things (IIoT)

Fanoon Raheem and Nihla Iqbal
South Eastern University of Sri Lanka
Oluvil, Sri Lanka

CONTENTS

1.1 INTRODUCTION

Businesses around the world are increasingly exploiting the Internet of Things (IoT) to develop innovative creation and service networks that are opening up new marketing strategy and establishing new revenue streams. The temporal variation is rapidly transforming the way paradigm through how industries organize business activities and reach potential customers. Though the IoT connectivity ensures advanced business processes, for industries to understand the full value of empowering IoT, IoT and artificial intelligence (AI) technology must be integrated. This results in supporting smart machines that allow intellectual behavior to be replicated and well-informed decision-making practices. Continuing developments in AI would have a substantial effect throughout the near future impacting the employment, expert knowledge, and HR approaches in almost every sector of industry, stressing the fact that businesses should have the necessary experience for AI enablement. The addition of AI into IoT networks is now becoming a prerequisite to succeed in today's IoT-based artificial environments. Yet companies need to work efficiently to determine how to make it to the importance of AI and IoT integration (Chitkara, Rao, and Yaung 2017) that will result in a perfect value chain analysis to achieve the necessary core competencies along with competitive advantages over the rival industries.

1.1.1   THE CONCEPT OF IoT

The association of different physical devices and objects around the globe through the Internet is referenced as IoT. It is the physical world network that allows certain artifacts to gather and share data, including computers, instruments, cars, buildings, and other things embedded with processors, circuits, applications, detectors, and communication technologies. The IoT makes it possible to remotely track and monitor artifacts through existing network infrastructures. The concept of IoT shows it as the individually recognizable associated objects with radio frequency identification (RFID) technology. The definite concept of IoT, however, is still in the phase of creation and is subject to the viewpoints drawn. Evolving global distribution network with norms that has the ability to self-configuring technologies and networking protocols has traditionally been known as IoT (Gokhale, Bhat, and Bhat 2018). Therefore, the IoT can be seen as the IT industry's next revolution after the Internet, which can be applied anytime, wherever on-demand, to detect, track, and control anything.

The advent of the IoT is going to have a strong influence on many dimensions of the distribution of resources, such as production, sales, transportation, usage, and recycling, as well as states', companies', and individuals' actions (Khan 2019). IoT is now an evolving industrial knowledge infrastructure focused on the Internet. The importance of IoT is felt in standardizing business processes. Throughout this emerging phase, IoT and its associated support frameworks need to be quickly built and configured to satisfy requirements of the industry (Wang et al. 2015).

1.1.2   THE CONCEPT OF IIoT

The Industrial Internet of Things (IIoT) incorporates a wide range of machines linked by software for communications. The corresponding programs, and even the wireless device that compose them, will track, capture, share, interpret, and respond to information instantly to modify their actions or their environment smartly, without external influence. The architecture of the IoT framework must ensure IoT operations that connect virtual objects and physical environments. IoT framework design requires several variables, such as networking, connectivity, and procedures. Attention should be paid to the extensibility, scalability, and operability between devices. Thus, the understanding of IoT architecture is important in adoption and deployment of IIoT that can act as a framework in achieving business advancements.

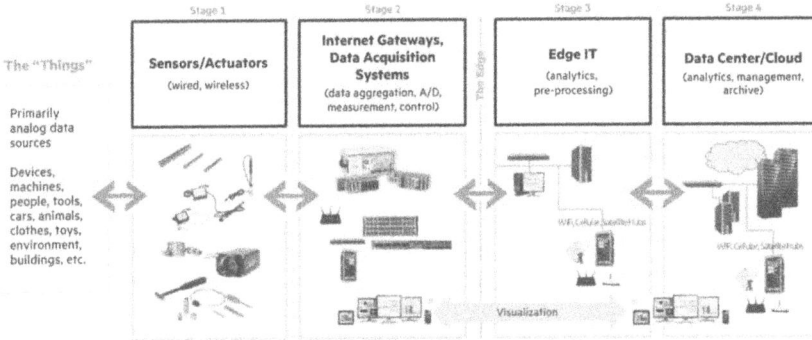

FIGURE 1.1 A four-stage IoT solutions architecture. (Boyes et al. 2018.)

For IoT systems, there is a whole range of architectures provided; however, these are general and therefore not directly relevant to IIoT implementations. Figure 1.1 shows an architecture specifically relevant to industrial applications. From this architecture, larger organizations can meet the integration of information and organizational capabilities (Boyes et al. 2018).

In simple terms IIoT can be elaborated as the one that puts a range of domains like traditional automation and machine-to-machine (M2M) communication, big data and machine learning (ML), and cyber-physical systems together to collect, analyze, and use data from industrial properties and devices, frequently in real time to improve operations. Industries embrace IoT as a means of energy optimization, optimizing wear of consumables, preventative analysis, capacity planning, standard costing, and significant trade control, among others. In keeping with the IoT model, emerging IoT technological possibilities such as new industrial sensors and sensing solutions, new factory wireless protocols, new industries based on IoT platforms, new synergistic applications for intelligent diagnoses, smart factories, smart goods, and intelligent logistics are being built for industries.

1.1.3 INDUSTRY 4.0 AND DATA

The fourth industrial revolution, described by Klaus Schwab, the founder and executive chairman of the World Economic Forum, explains the world as where people switch from digital fields to analog realities using smart devices to empower or even control their lives (Xu, David, and Kim 2018).

Thus, the Industry 4.0 introduced the latest emerging technology to make steam, equipment in factories, labor unions, and even robotics in business activities connectable to the digital and physical environment. Physical and automated interconnections contributed to the formation of volumes, culminating in new approaches. As a part of this, data has become an unavoidable aspect where digital information needs to be collected, refreshed, obtained, and distributed in various respects. Mass storage and cognitive capacity will change industry by integrating data. As already stated, the data changes the industry functionalities and demands new regulatory approaches, as given in Figure 1.2.

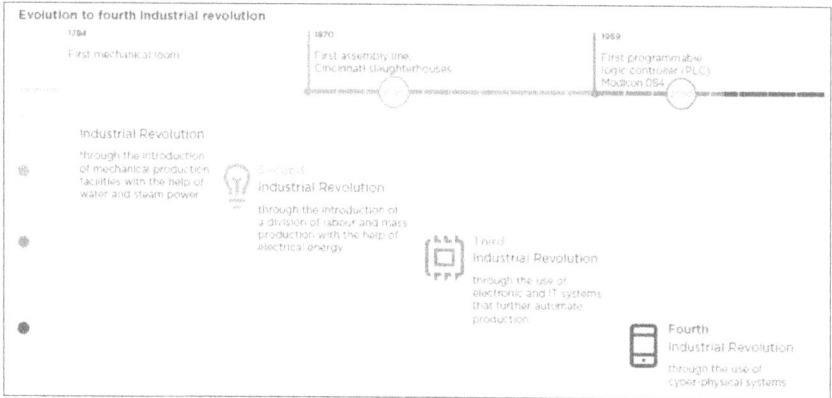

FIGURE 1.2   The evolution of Industry 4.0. (GSMA Intelligence 2017.)

Centered on IoT concepts, such as a machine, it is not smart until it is connected to another network of computers. Communicators operate together in ambient devices in IoT are smart, and the system becomes smarter with multiple cycles or iterations (Akugizibwe 2020), It is recognized that the IIoT explicitly shows and acknowledges that nearly everything in the industrial world, like production tracking, materials handling, energy management, quality control, predictive maintenance, workers safety, and smart lifecycle, must be related to everything else in order to satisfy the criteria of simulating the smart machines to showcase the intelligent behavior. Therefore, AI is required to make these things "intelligent" or "smart."

1.1.4 Artificial Intelligence and Machine Learning

AI is a technology that aims to make human reasoning accessible to computers. Through this the digital business change will be speeded up. The ML as well as data analysis (DA) is used in emulating human learning in order to make the world and its physical systems fully autonomous. ML is about designing techniques to facilitate automatic and self-sustain learning in the various components or devices of a network, while DA assesses and analyzes all data collected over time to identify trends from the past and be more influential and effective in the future. This trend has evolved and attempts to combine ML and DA in wearable technology detectors and embedded systems (Ghosh, Chakraborty, and Law 2018; Hunt 1987).

IIoT enabled by AI is far more contributing to avoiding unwanted latency. It can also help in increasing organizational performance. The capability of ML to generate fast and precise forecasts and the ability of AI technology can be used to automate a wide range of industry activities (Naganathan and Rao 2018). This is summarized in Figure 1.3.

Increasing device proliferation

Decreasing cost of megabit/sec

Increase in VC spend and investment

IoT Enabled by AI

Decreasing cost of CPU, memory, & storage

Convergence of IT & operational technology

Advent of big data & cloud/fog

FIGURE 1.3   The impact of AI in IoT/IIoT. (Chitkara, Rao, and Yaung 2017.)

AI is splashing across IoT with a boom in innovation, an introduction of advanced technologies and a growing pattern in business deployments. The AI system may need to be considered by organizations designing an IoT strategy, testing a new technical IoT proposal, or attempting to take advantage of the prevalent IoT implementation.

1.2   TACTILE INDUSTRIAL INTERNET OF THINGS (TACTILE IIoT)

The IoT connects machines or objects to maximize their utility by leveraging the opportunity for networking. The very next stage in line of such innovation is indeed the Tactile Internet. The nature of the Tactile Internet is distinguished by incredibly low latency coupled with increased accessibility, reliability, and security. This will have an immense impact on business and industry and will also provide numerous potential opportunities for the advancement of market technology and the distribution of critical services.

Under this context, it is important to understand how Tactile Internet would impact IIoT by means of industry automation. Industry automation is a rapidly developing core area of service for the Tactile Internet. Industrial robotics is predicted to provide an end-to-end latency slightly below 1 millisecond per sensor. Additional demands for large bandwidth from many other implementations are apparent. When a wireless system is used, protection is of utmost importance, because potential attacks may happen remotely. Therefore, there should be a wide variety of flexible wireless as well as wired industrial networks with varying data rates, reliability, and end-to-end latency specifications (Fettweis 2014).

The Tactile Internet is the ability to construct a view of the IoT by which perceptions or sensations are being conveyed between one particular region to another through real time. So, Tactile IoT/IIoT is the next breakthrough that lets IoT control in real time, communicating with machines by means of humans. It will use different technologies, both at the network and application level, to allow and improve cyberspace interaction.

In various fields, the Tactile Internet can have its instances, the industry being one of the most common. The Tactile IoT is used in the IIoT for medical, automobile, mining, education, autonomous driving, and many more. In any one of the Tactile IoT contact applications, the delay is calculated to be less than a millisecond.

Haptic connectivity in a Tactile Internet, as shown in Figure 1.4, can be used in different industries like if medical care is taken into example, in situations in which a patient can even be assessed digitally and handled with mobile activity is available. In transportation, particularly disabled people can be informed about their street navigation and robot searching can also be regarded as a good example. In all these, IIoT is formed through the connectivity of sensors, actuators, and detectors.

The vital industrial activities such as manufacturing or production or other mission critical services in the IIoT are handled by remotely operated machines or computers. In such scenarios, the Tactile Internet applications come into function by enabling fast data speeds with higher reliability (Aazam, Harras, and Zeadally 2019;

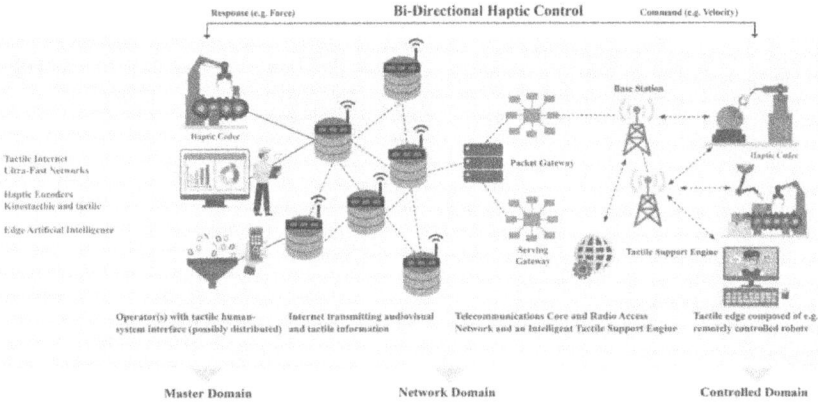

FIGURE 1.4 Tactile Internet of Things. (Vermesan, Eisenhauer, and Serrano n.d.)

Varsha and Shashikala 2017). In addition to high availabilities, stability, and security, the Tactile Internet has been considered as an enabler for mission critical IoT networks. Tactile IoT ensures that virtually all aspects of society can be moved from traditional to IoT networks (Ye et al. 2019).

Faster links to the Internet and improved capacity enhance the on-site sensor knowledge in the Industrial IoT network. A new software facilitated with lower latency and elevated reliability is needed for this. So, Tactile IoT or IIoT makes it possible to come up with such next degree hyperconnectivity. The Tactile IoT or IIoT offers the opportunity to enable remote transmission of real-time control and sensory (haptic) sensations. For example, if applied in self-car manufacturing, autonomous vehicles can identify and respond instantly to safety-critical scenarios, and in the long term the safety of their users will be ensured.

For the reason that the Tactile Internet is used for hospitals and autonomous cars, latency remains as critical as the output variables that are definitive. AI tries to overcome the latency dilemma by using predictive algorithms in implementation, providing enhanced haptic perception (Akshatha et al. 2019). Thus, intelligent networking through AI helps to make the own decisions by using AI training models to relieve the burden of collaborative teamwork and other necessities.

A variety of design criteria, including very low end-to-end latency of 1 millisecond, is required for the Tactile Internet. This primary architectural goal of the Tactile Internet can only be reached by keeping tactile apps local, close to consumers. Moreover, the Tactile Internet would lay out the challenging criteria for potential access networks in terms of bandwidth, reliability, and power. Scalable procedures for all protocol layers are considered a necessary requirement under the context of design goals (Vermesan, Eisenhauer, and Serrano n.d.).

Hence, Tactile IoT/IIoT is a change in the interactive paradigm where the network of communication modal brings a simple perspective of sensing and acting capabilities that mean people and computers no longer need to be physically interacted to each other or be near as the remote connectivity is enabled, as illustrated in Figure 1.5.

TOWARDS A WORLD OF INTELLIGENT CONNECTIVITY

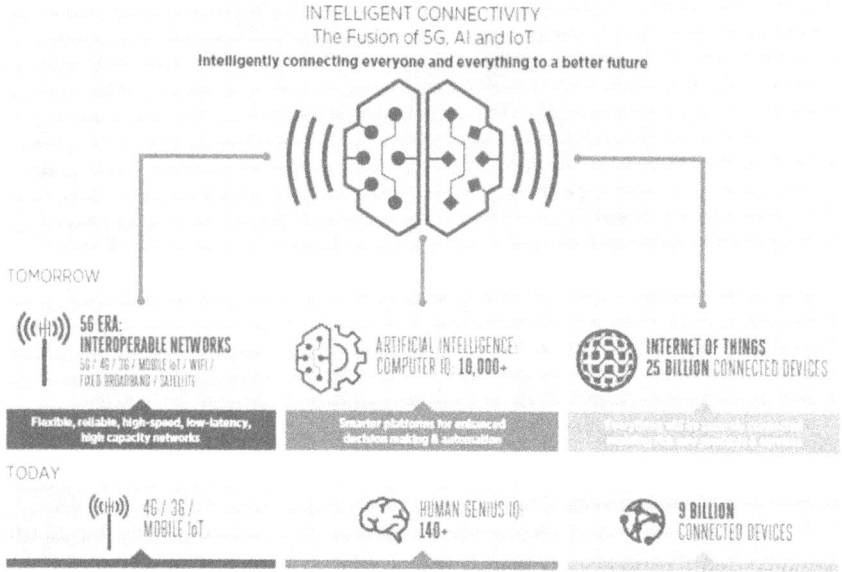

INTELLIGENT CONNECTIVITY
The Fusion of 5G, AI and IoT
Intelligently connecting everyone and everything to a better future

TOMORROW

((·|·)) 5G ERA:
INTEROPERABLE NETWORKS
5G / 4G / 3G / MOBILE IoT / WIFI /
FIXED BROADBAND / SATELLITE

ARTIFICIAL INTELLIGENCE:
COMPUTER IQ: 10,000+

INTERNET OF THINGS
25 BILLION CONNECTED DEVICES

Flexible, reliable, high-speed, low-latency,
high capacity networks

Smarter platforms for enhanced
decision making & automation

TODAY

((·|·)) 4G / 3G /
MOBILE IoT

HUMAN GENIUS IQ:
140+

9 BILLION
CONNECTED DEVICES

FIGURE 1.5 Intelligent connectivity – 5G, AI, and IoT. (GSMA 2019.)

1.3 ARTIFICIAL INTELLIGENCE AND MACHINE LEARNING

1.3.1 ARTIFICIAL INTELLIGENCE

Human muscle power has been replaced by machines with the evolution of the Industrial Revolution in England in 1760, whereas AI intends to replace human intelligence using a machine, which initiated in the early periods of the 1950s, and the word AI was introduced in 1956. AI has not a clear definition. AI and its definitions mainly focus on how the results perform to demonstrate intelligence and not on the methods applied to achieve it. But in general, AI can be defined as "The study of making computers do things that the humans needs intelligence to do" (Munakata 2008). The definition can be simplified in other words as the computer systems developed with the ability to accomplish tasks which will need human intelligence. Some of the examples include recognition of speech, visual perception, decision making, and language translation.

AI is a broader area of research that is not limited only to certain aspects. It acts as a connection media among various sectors such as computer science, mathematics, cognitive science, and many more (Villani 2018). The research studies relevant to AI are more specialized and technical. AI includes knowledge, reasoning, problem solving, perception, learning, planning, and the potential to move objects as the core problems and the primary part of AI research is knowledge engineering (Habeeb n.d.). The core scientific areas may be diversified with the detailed knowledge in the fields in which they are used, and every AI algorithm is supported by the combination of techniques, which include semantic analysis, symbolic computing, ML, exploratory analysis, deep learning, and neural networks.

Marvin Lee Minsky, one of the founding fathers of AI, interprets that the ability of a machine to imitate or outperform the capability of a human is a standard intelligent measurement of an AI machine. In contrast to his ideology, another founder of AI, John McCarthy, argues that rather than the human capability measurement, rational logic is a more standard intelligent measurement (Villani 2018).

But above all, a question "what exactly is an intelligent computation" arises in the minds of researchers. Intelligent computation can be categorized by the forms of computations that do not need intellect; for example, purely numeric computations, e.g. addition and multiplication of numbers at a rapid speed, cannot be considered as an AI. So, what exactly an intelligent computation is, a general way of characterizing intelligent computation is grounded on the vicinity of the given tasks to be performed. Inference derived through knowledge, reasoning with not certain or not complete information, diversity in perceptions, and learning are few requirements in order to solve problems that require intelligence. Apart from the appearance of the problem, another way to characterize is depending on the underlying mechanism of the process applied to derive a solution. Neural networks and genetic algorithms may be mentioned as the primary instances for the second category.

1.3.1.1 Goal of Artificial Intelligence

There is an enthusiasm among the people as to why researchers are nowadays focused on developing smart machines. The key aim of the development of these machines is that they make the lives of human beings easier by performing complex tasks and daily repetitive tasks. Examples include identifying the best route to arrive at a friend's house, selection of the utmost preferred content of a user, translation of texts, and many more. But these systems are referred to as "Weak AI" since they are capable of performing only a single and a delimited task. On the other hand, the "Strong AI" is capable of matching and also outperforming the capabilities of a human being in all aspects, but till now scientists are far from developing such a machine.

1.3.1.2 Categories of Artificial Intelligence

Approaches in the AI field are primarily categorized into two. The first one being the traditional symbolic AI, which has been leading throughout the history, is categorized according to an abstraction that is of high level and a perceptual view that is macroscopic. Systems that involve knowledge engineering and logic programming are categorized under this category. The areas of logical reasoning, systems based on knowledge, symbolic ML, natural language processing, and search techniques fall under symbolic AI.

The next approach to AI is dependent on the lower level, microscopic models on biology, which are in similar to emphasizing physiology or genetics. Key examples include neural networks and genetic algorithms. It is not necessary for these biological models to resemble the original models, but these are few significant areas where the AI approach could be practically applied in the future. In addition to these two approaches, there are also few more AI approaches such as fuzzy logic, rough set theory, and chaotic systems (Munakata 2008).

These techniques and approaches of AI should be split into definite classes and objects in order to perform tasks by the algorithms according to predefined features so that knowledge engineering could be implemented by accessing these objects, categories, properties, and relationships among them. The availability of data in abundance makes a machine to act and react like humans, but initiating common sense, problem-solving, and reasoning are very difficult to perform using a machine. Hence, ML is considered to be another core element of AI.

1.3.2 MACHINE LEARNING

Learning is a very personalized process for human beings, but it is not just limited to human beings only. Even plants do show intelligence. Learning also cannot be defined precisely, but in general it is defined using the phrases as "to gain knowledge, or understanding of, or skill in, by study, instruction or experience" (Nilsson 2005). Machines are not intelligent by nature and they are designed only to perform specific tasks, and this is the key difference between a human being and a machine. The main purpose of development of machines is to assist humans in their everyday lives; therefore, it is a very important role for the machines to reason, realize to resolve problems, and make suitable profound conclusions similar to human beings. Simply, these machines can be termed as smart machines, which in turn is a symbolic approach for the ML process.

ML has become one of the major important areas of information technology over the past two decades with the availability of huge amounts of data for smart DA along with the progress in technology. ML, being a subpart of AI, intends to make the machines accomplish tasks efficiently through the application and usage of intelligent software. The backbone of the intelligent software is the constituents of statistical learning and they require a connection with a database for ML algorithms require data to learn.

The ML process has been expanding to several broader disciplines currently. Some of the separate disciplines where the ML approach is being applied are explained briefly in the following passages below (Nilsson 2005):

1. Statistics

   The major purpose of statistical analysis is the efficient use of samples drawn from unknown probability distribution, and depending on the results, it assists in deciding from which distribution the new sample is taken. Statistical methods applied for these problems can be thought-out as ML instances since the model corpus resulting from the problem environment decides the decision and estimation rules.

2. Brain Models

   Non-linear elements having weighted inputs are simply explained as the simple models of biological neurons. The networks of the brain model elements are in the process of study from the beginning by many researchers, and brain modelers are eagerly studying the way in which these networks estimate the learning phenomena of living brains.

3. Adaptive Control Theory

   This method will have unknown parameters that are needed to be evaluated during the operational process and the role of control theorists in this field is to study to control such processes. Also, the parameters may change while the operation is on process, where the control process must keep a record of these change occurrences.

4. Psychological Methods

   The performance of humans is studied by psychologists through several forms of tasks related to the learning process.

5. Artificial Intelligence

   AI has always been interconnected with ML from the beginning. In the recent years, researchers are aimed at formulating rules suitable for expert systems. This is achieved with the application of decision-tree methods and inductive logic programming.

6. Evolutionary Models

   Though the difference between evolving and learning may be unclear in computer systems, techniques that are applied in modeling certain features of biological evolution are considered to be learning aids in improving the computer program's performance. Genetic algorithms and programming are the most prominent computational techniques for evolution.

SAS Institute Inc., North Carolina, interprets the connections of ML to other disciplines, as illustrated in Figure 1.6.

ML algorithms and techniques are very helpful in performing the learning process. The algorithms are classified as shown in Figure 1.7.

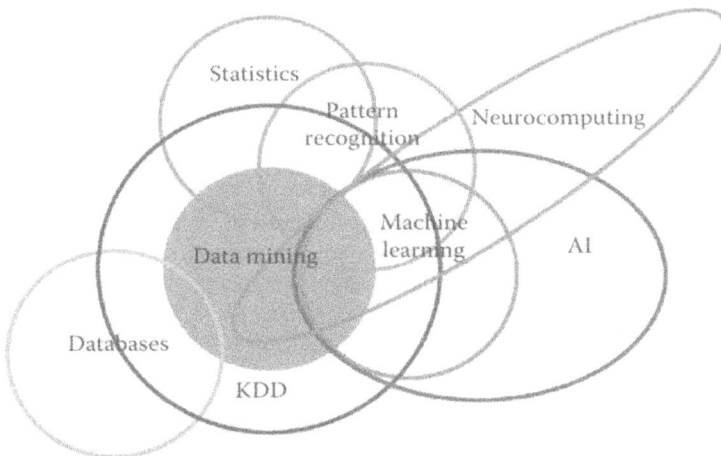

FIGURE 1.6 Knowledge and machine learning disciplines. (Nilsson 2005.)

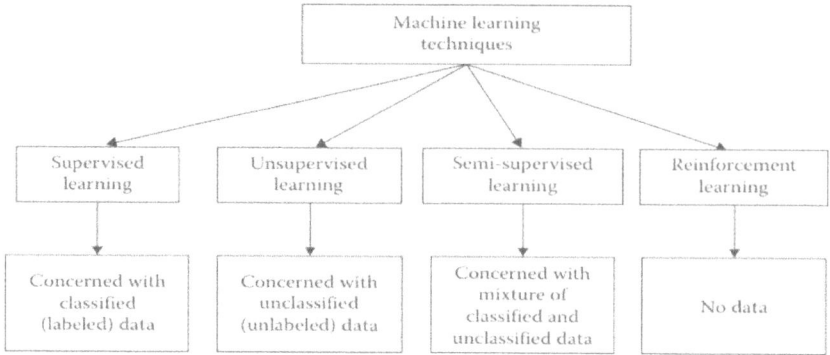

FIGURE 1.7 Machine learning techniques.

1.4 ARTIFICIAL INTELLIGENCE AND MACHINE LEARNING FOR INDUSTRIAL INTERNET OF THINGS

AI and IoT are coined terms that are pronounced every day in the media and the people are in a higher expectations and imaginations that these technologies would bring miracles and fantasy. Along with deep learning and ML approaches, industrialists expect that AI would bring more solutions to problems and requirements at present with the interconnection among components by the application of IoT techniques.

ML and deep learning algorithms, the core parts of AI, support in managing the data retrieved from IIoT by recognizing, classifying, and deriving decisions. Deep neural networks and big data analytics help in deriving eloquent and beneficial information in order to make a decision. Application of artificial neural networks (ANNs) in IoT and data analytics is critical for efficient and effective decision-making, particularly in the field of data streaming and analytics in real time related to edge computing networks.

ML and deep learning algorithms for IIoT considerably improve reliability, outcome, and satisfaction from the customer, which depend on incorporating vital technologies, devices, software, and applications. Security, interoperability, real-time response, and future readiness are major challenges faced in IIoT, and in order to overcome these challenges and enhance them, contextual analysis should be introduced along with ML algorithms.

ML is one of the major components of contextual analysis in IIoT, which helps to analyze and troubleshoot the network issues and IoT device issues, in time. The capabilities of ML are the development of prediction models, anomaly detection, and recommendation models. This can help in identifying the hidden flags and the future threats in the system. Also, it is vital to train the system to analyze the context in order to produce predictions and recommendations that are of high accuracy (Ambika 2020).

1.4.1 MACHINE LEARNING ALGORITHMS FOR IIoT

ML algorithms are used in IIoT with the purpose to save costs, time improvement, and enhancing the performance. ML algorithms are capable of handling huge and complex data from which they derive patterns and trends; for instance, anomalies. Smart machines are developed to process the information at a quicker rate of speed and form quicker decisions once a threshold is reached. The ML algorithms for a better analysis of data in the Industrial IoT devices are tabulated in Table 1.1.

TABLE 1.1
Machine Learning Algorithms for IIoT

Machine Learning Algorithm	Task
K-nearest neighbor	Classification
Naïve Bayes	Classification
Support vector machine	Classification
Linear regression	Classification/regression
Random forest	Classification/regression
K-means	Clustering
Principal component analysis	Feature extraction and reduction of dimensionality
Canonical correlation analysis	Feature extraction
Neural networks	Classification/regression

1. K-Nearest Neighbor
   K-nearest neighbor, being the most essential algorithm used for classification problems in ML, is used in classifying unseen data set with little or no knowledge about the data points. Distance metrics are used to identify the K-nearest neighbors of the given data point (Peterson 2009).
2. Naïve Bayes Algorithm
   Naïve Bayes algorithm is another technique for classification problems that is based on Bayes' theorem, which assumes independence among predictors. It is easy to build this model and it is applicable to larger data sets. Another advantage is that the Naïve Bayes theorem is capable of outperforming even for highly sophisticated classification methods (Ray 2017).
3. Support Vector Machine
   A support vector machine (SVM) is used in generalizing between two different classes when a labeled data set is input for the training set. The foremost role of the SVM is to find the separating hyperplane to distinguish between the two classes. SVMs are applied for anomaly detection, smart traffic, and smart environment traffic environment (Dwivedi 2020a).
4. Linear Regression
   Linear regression is one of the most prevalent and well-clear algorithms in both statistics and ML. It is an algorithm where the output is predicted using known parameters and correlating it with the output. The values are predicted within a given range of continuous values and not classify them. The continuous and constant slope derived from the known parameters is used to predict the values for the unknown parameters (Sarkar 2019).

5. Random Forest

Random forest is an algorithm of supervised learning, which is flexible for usage to produce outputs without hyperparameter tuning. It is much preferred due to its simplicity in usage and can perform both classification and regression tasks. The forest is built as a decision tree and the technique applied is the bagging method (Donges 2019).

6. K-Means Clustering

K-means clustering is another algorithm of unsupervised learning without labeled data and performs the separation operation of objects based on the similarity and dissimilarity of the objects in a cluster (Simplilearn 2020).

7. Principal Component Analysis

Principal Component Analysis (PCA) is a feature extraction technique that creates new features, which are independent of the old features and amalgamation of both and store only the features that can be used to predict the values. Any feature can be dropped while extraction of features depending on the importance of dependency on the target variable. That is, a less important feature may be dropped during the feature extraction process (Dwivedi 2020b).

8. Canonical Correlation Analysis

Canonical Correlation Analysis (CCA) is a technique of linear dimensionality reduction that has a relation with PCA and both differing by the number of parameters each deal with. PCA deals with one parameter, whereas CCA deals with either two or more parameters. Its purpose is to identify a consistent pair that has high cross-correlated linear subspaces so that each component will have a correlation between each and a single component from the other subspace within one of the subspaces.

9. Neural Networks

Neural networks is a technique where a network is developed composed of three layers: the input layer, the hidden layer, and the output layer. The second layer being the hidden layer has neurons that interconnect the layers. In this technique, the neural network learns from the data and fine tunes the connection weights among the neurons in the layers, thus resulting in the network to produce accurate predictions (Simpson 2018).

AI techniques and ML algorithms in Industrial IoT allow the users to gather data in a huge quantity to discover the hidden patterns within them and derive the trends to make predictions, thus benefiting industrialists in the decision-making process to manage and control their assets and hence improving the overall performance of the organization.

1.5 CHALLENGES OF IMPLEMENTING ARTIFICIAL INTELLIGENCE AND MACHINE LEARNING IN IIoT

With time changing, the physical world has become a world that is interactive. It is all about ML, AI, and robotics that led the way to increase of smart computers and innovations, which have brought together an interconnected universe. Using data exploration technologies, massive data volumes are easily handled and analyzed.

The IIoT or Industry 4.0 is one of those in which integration of M2M interaction with industry-leading data analytics will make the industries reach an exponential degree of productivity, competence, and success.

As the majority of organizations rely on IoT manufacturing firms to integrate technologies into existing sectors, with the exponential evolution of the IIoT, the rising evolution of big data processes in real time has become apparent.

However, there are problems that occur when attempting to incorporate AI and ML into IIoT. It is clear that not all organizations are able to support themselves with the advantages of IIoT enabled by AI or ML. It was observed during a survey that the health care industries lacked the ability of the facilities to consolidate and evaluate the vast amount of contradictory data available in an analysis of half-management (Sharma n.d.).

Therefore, it is necessary to have clear-cut knowledge on problems or challenges that need to be solved before AI learns automatically to optimize the processes in the factory efficiently.

In a heterogeneous and complex IIoT architecture, a number of technological problems have been developed, such as usability, security, scalability, heterogeneity, effectiveness, and management of resources. And an important issue is that it is not very straightforward to implement an efficient AI technology and an ML algorithm to produce reliable and promising results in industries. It needs an exceptional mixture of domain expertise, problem-solving skills, and comprehensive knowledge in the area considered. Such challenges are given as follows.

1. Choice of correct algorithm and training AI

   For AI and ML's implementation, there are a number of extensively used algorithms available. While algorithms will work in any normal circumstances, there are certain criteria and conditions that decide the performance of algorithms. After months of working an improperly selected algorithm, the inaccurate algorithms will not give the output or the experimental results as expected and wanted, contributing to a massive loss of effort and a garbage collection (Gupta 2017).

   After gaining enough data to begin ML, the training phase is reached as the next step. In this stage, considerations are drawn toward the particular model, data, and possible outputs. The outcome from training shows to what extent the algorithms have worked in predicting the output. Thus, more instances of machines running in real environments should be considered in order to achieve higher accuracy from ML. In that context, sensors and detectors are needed to be added in deciding the inputs to ML (Allsbrook 2017).

2. Determination of appropriate data set and understanding the data

   The correct data set collection is very essential to address the use case or problem statement considered. The competence of an ML system depends on the power, number, planning, and selection of data. Data collection can have an effect on bias. It is necessary to prevent bias in the selection process and to select the data that fully represents the cases to gain unbiased results from the data set trained and tested (Gupta 2017).

Many industries still do not comprehend how information is conveyed through their computers. To be placed in an appropriate model for training the data set, data must be prearranged, and it must be well understood. And also, without the ability to identify results, the data is useless. By waiting and observing only the output can be realized. In a point, if the outcome is not as anticipated, then it will be challenging to further continue the operation. Seeking knowledge and advice from experts based on the domain can reduce the risks arising in understanding the data (Allsbrook 2017).

3. Preprocessing of data

Missing values and outliers make the historical data set confusing. Preprocessing tasks like parsing, cleaning, and other preprocessing steps might become inefficient, which could affect the feature engineering process. To avoid certain features from dominating the whole model, feature scaling tests and procedures must be applied (Gupta 2017).

4. Labeling of data

The algorithms of supervised ML are easier and more fitting. It takes great time to discover and implement the unsupervised ML algorithms and it involves several imprecise iterations. For supervised ML algorithms, data labeling is important. Data labeling is an active operation that is done manually and should never be outsourced in a way it is wanted. For example, in the health care sector, in order to perform the predictive diagnosis, the available diagnostic data must be categorized or labeled. In such a scenario, the input needs to be constant for accurate labeling. But in the field of medicine, the decision makers consider this labeling process to be a mere waste of their assets and time (Gupta 2017).

In addition to the challenges listed above, there are many other challenges, including, while trying to adopt AI or ML in IIoT, the need for continuous networking is a must. But 100% functionality cannot be assured, particularly when using internet connectivity and it is almost impossible too. At one point, for maintenance or for some other cause, the connectivity may become broken. Thus, one of the main challenges is the connectivity outage challenge. Another challenge includes data storage issues as the predictions or forecasts are heavily contingent on previous data from storage as IIoT is occupied with a vast amount of data collected (Sharma n.d.). Due to this, efficient management of the data using proper channels and schemes is also considered a complex task since industrial sensors and actuators produce data streams of higher velocity.

To be used for ML, the retrieval of IIoT data in real time is a vital task since this information is used in on-time industrial automation operations. However, traditional database management approaches are unable to achieve the required outcomes. For robust and efficient data analytics of IIoT devices, effective big data analytics technologies are needed in real time, which makes the implementation get a bit multifaceted in terms of organization's budget and other resources (Khan et al. 2020). Consequently, there is a challenge for latency when trying to integrate IIoT applications with AI.

Using AI within the IIoT applications would cause extra communication overheads to increase the performance of IIoT systems. So, there are possibilities that the higher overhead charges are demanded or required during the implementations. And traditional or classic AI or ML strategies like ANNs are clearly inadequate if routing is taken into consideration, which makes the challenges to routing procedures and network trafficking of the systems. Also resources being consumed by AI within network equipment, such as routers and switches, cause issues for caching of the systems (Ahmad et al. 2020).

1.6　SUMMARY

IoT is considered to become an important force in order to simplify and automate the demand and growth processes of the industries. Within an industry everything is going to be smart and intelligently interconnected through the Internet. Any branch of industry will work toward providing a big value proposition by means of competencies. The industry processes like supply chain management, quality control, predictive maintenance, and energy consumption are highly empowered by AI and ML that improve accountability, speed up decision making, and provide a way to make accurate prediction. Also, the AI and ML support in boosting the response time by removing the anticipated failures.

Consequently, it is agreed that AI and ML will push industries to change and adapt product and services distributions, according to market requirements by increasing the level of intelligence in their products and services, providing opportunities for achieving business value in all industries. Increased AI adoption will accelerate the transition of the industry through competitive areas. ML and deep learning algorithms are the core parts of AI, which would support the development and advancement of the industries even more. The relationships between IoT, IIoT, AI, and ML are summarized in Figure 1.8, and the ML algorithms for IIoT are summarized in Figure 1.9.

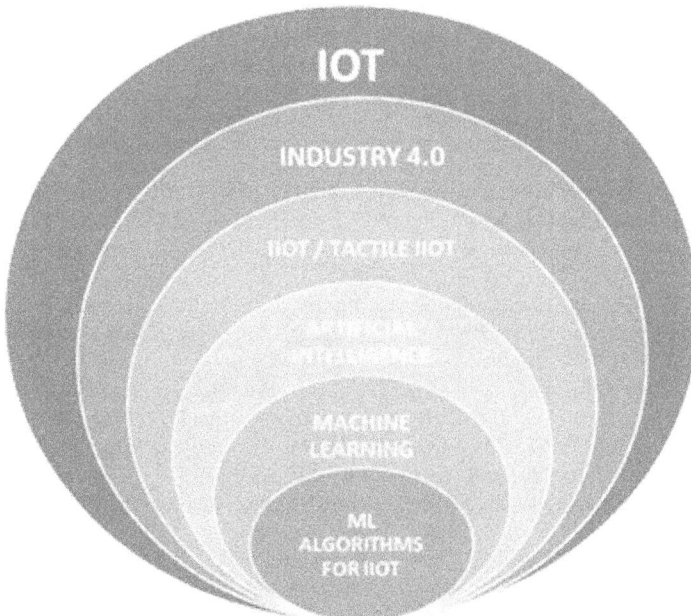

FIGURE 1.8　Relationship between IoT, IIoT, AI, and Machine Learning.

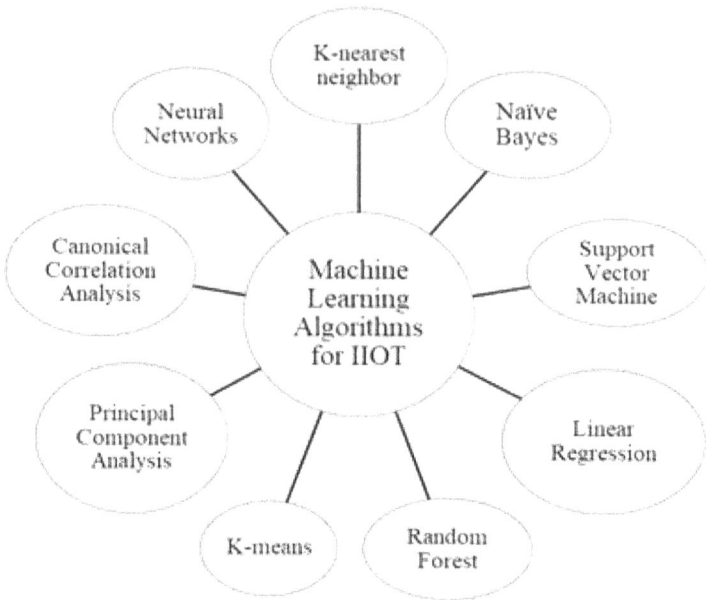

FIGURE 1.9   Machine Learning algorithms for IIoT.

Therefore, this study shows that with industries speeding up their IoT deployment, the importance of AI in addressing them has become progressively clear. IoT leading the way to IIoT is considered to be the modern master role in business, and as discussed, IIoT enabled with AI and ML will induce and sustain rapid evolutions by making ways to massive industry growths, innovations, and disruptive technologies.

REFERENCES

Aazam, Mohammad, Khaled A. Harras, and Sherali Zeadally. 2019. "Fog Computing for 5G Tactile Industrial Internet of Things: QoE-Aware Resource Allocation Model." *IEEE Transactions on Industrial Informatics* 15 (5): 3085–3092. https://doi.org/10.1109/TII.2019.2902574.

Ahmad, Ijaz, Shahriar Shahabuddin, Tanesh Kumar, Erkki Harjula, Marcus Meisel, Markku Juntti, Thilo Sauter, and Mika Ylianttila. 2020. "Challenges of AI in Wireless Networks for IoT," no. April. http://arxiv.org/abs/2007.04705.

Akshatha, N., K. Harishree J. Rai, M. K. Haritha, Rachita Ramesh, Rajeshwari Hegde, and Sharath Kumar. 2019. "Tactile Internet: Next Generation IoT." *Proceedings of the 3rd International Conference on Inventive Systems and Control, ICISC 2019*, no. ICISC: 22–26. https://doi.org/10.1109/ICISC44355.2019.9036389.

Akugizibwe, Simon Peter. 2020. "Artificial Intelligence and Internet of Things in the Development of Smart Sustainable Cities." Adoption of Circular Economies in the 4th Industrial Revolution. https://www.itu.int/en/ITU-T/Workshops-and-Seminars/gsw/201804/Documents/Simon%20Peter%20Akugizibwe.pdf

Allsbrook, Aaron. 2017. "Council Post: Four Artificial Intelligence Challenges Facing The Industrial IoT." 2017. https://www.forbes.com/sites/forbestechcouncil/2017/02/24/four-artificial-intelligence-challenges-facing-the-industrial-iot/?sh=549ea0e61640.

Ambika, P. 2020. *Machine Learning and Deep Learning Algorithms on the Industrial Internet of Things (IIoT): Advances in Computers.* 1st ed. Vol. 117. Elsevier Inc. https://doi.org/10.1016/bs.adcom.2019.10.007.

Boyes, Hugh, Bil Hallaq, Joe Cunningham, and Tim Watson. 2018. "The Industrial Internet of Things (IIoT): An Analysis Framework." *Computers in Industry* 101 (April): 1–12. https://doi.org/10.1016/j.compind.2018.04.015.

Chitkara, Raman, Anand Rao, and Devin Yaung. 2017. "Leveraging the Upcoming Disruptions from AI and IoT." *pwc*, 21.

Donges, Nicklas. 2019. "The Random Forest Algorithm: A Complete Guide | Built In." Builtin. 2019. https://builtin.com/data-science/random-forest-algorithm.

Dwivedi, Rohit. 2020a. "How Does Support Vector Machine Algorithm Works in Machine Learning? | Analytics Steps." AnalyticSteps. 2020. https://www.analyticssteps.com/blogs/how-does-support-vector-machine-algorithm-works-machine-learning.

Dwivedi, Rohit.. 2020b. "Introduction to Principal Component Analysis in Machine Learning | Analytics Steps." Analytic Steps. 2020. https://www.analyticssteps.com/blogs/introduction-principal-component-analysis-machine-learning.

Fettweis, Gerhard. 2014. "The Tactile Internet." *IEEE Vehicular Technology Magazine*, no. March: 64–70.

Ghosh, Ashish, Debasrita Chakraborty, and Anwesha Law. 2018. "Artificial Intelligence in Internet of Things." *CAAI Transactions on Intelligence Technology* 3 (4): 208–218. https://doi.org/10.1049/trit.2018.1008.

Gokhale, Pradyumna, Omkar Bhat, and Sagar Bhat. 2018. "Introduction to IoT." *International Advanced Research Journal in Science, Engineering and Technology*, no. January: 1–24. https://doi.org/10.1201/9780429399084-1.

GSMA. 2019. "The 5G Guide," 1–15. https://www.gsma.com/wp-content/uploads/2019/04/The-5G-Guide_GSMA_2019_04_29_compressed.pdf

GSMA Intelligence. 2017. "Global Mobile Trends." *Presentation*, no. September: 121. https://data.gsmaintelligence.com/api-web/v2/research-file-download?id=18809377&file=global-mobile-trends-1482139998965.pdf

Gupta, Anil. 2017. "Machine Learning Challenges in the Implementation of Industrial Internet of Things." 2017. https://iiot-world.com/artificial-intelligence-ml/artificial-intelligence/machine-learning-challenges-in-the-implementation-of-industrial-internet-of-things/.

Habeeb, Ahmed. n.d. "Artificial Intelligence." https://doi.org/10.13140/RG.2.2.25350.88645/1.

Hunt, Earl. 1987. *Machine Learning: An Artificial Intelligence Approach.* Los Alton, CA: Morgan Kaufmann 305: 299–305.

Khan, Jamil Y. 2019. "Introduction to IoT Systems." *Internet of Things (IoT)*, 1–24. https://doi.org/10.1201/9780429399084-1.

Khan, W. Z., M. H. Rehman, H. M. Zangoti, M. K. Afzal, N. Armi, and K. Salah. 2020. "Industrial Internet of Things: Recent Advances, Enabling Technologies and Open Challenges." *Computers and Electrical Engineering* 81 (March 2020). https://doi.org/10.1016/j.compeleceng.2019.106522.

Munakata, Toshinori. 2008. *Fundamentals of the New Artificial Intelligence.* Edited by Toshinori Munakata. Texts in Computer Science. London: Springer London. https://doi.org/10.1007/978-1-84628-839-5.

Naganathan, Venkatesh, and Prof K. Rajesh Rao. 2018. "The Evolution of Internet of Things: Bringing the Power of Artificial Intelligence to IoT, Its Opportunities and Challenges." *International Journal of Computer Science Trends and Technology (IJCST)* 6 (3): 94–108.

Nilsson, Nils J. 2005. "Introduction to Machine Learning an Early Draft of a Proposed Textbook Department of Computer Science." *Machine Learning* 56 (2): 387–399. http://www.ncbi.nlm.nih.gov/pubmed/21172442.

Peterson, Leif E. 2009. "K-Nearest Neighbor – Scholarpedia." Scholarpedia. 2009. http://www.scholarpedia.org/article/K-nearest_neighbor.

Ray, Sunil. 2017. "Learn Naive Bayes Algorithm | Naive Bayes Classifier Examples." Analytics Vidhya. https://www.analyticsvidhya.com/blog/2017/09/naive-bayes-explained/.

Sarkar, Priyankur. 2019. "What Is Linear Regression?" Knowledge Hut. 2019. https://www.knowledgehut.com/blog/data-science/linear-regression-for-machine-learning.

Sharma, Rita. n.d. "Top 5 Challenges for Industrial IoT (IIoT) Implementation – Finoit." Accessed November 19, 2020. https://www.finoit.com/blog/5-iiot-industrial-internet-of-things-implementation-challenges/.

Simplilearn. 2020 "K-Means Clustering Algorithm." 2020. https://www.simplilearn.com/tutorials/machine-learning-tutorial/k-means-clustering-algorithm.

Simpson, Michael. 2018. "Machine Learning Algorithms: What Is a Neural Network?" 2018. https://www.verypossible.com/insights/machine-learning-algorithms-what-is-a-neural-network.

Varsha, H. S., and K. P. Shashikala. 2017. "The Tactile Internet." *IEEE International Conference on Innovative Mechanisms for Industry Applications, ICIMIA 2017 – Proceedings*, no. December: 419–22. https://doi.org/10.1109/ICIMIA.2017.7975649.

Vermesan, Ovidiu, Markus Eisenhauer, and Martin Serrano. n.d. "The Next Generation Internet of Things – Hyperconnectivity and Embedded Intelligence at the Edge," no. Ml.

Villani C. 2018. "What Is Artificial Intelligence? Villani Mission on Artificial Intelligence," no. March. http://educationdocbox.com/Homework_and_Study_Tips/92429680-What-is-artificial-intelligence-villani-mission-on-artificial-intelligence.html

Wang, Pan, Ricardo Valerdi, Shangming Zhou, and Ling Li. 2015. "Introduction: Advances in IoT Research and Applications." *Information Systems Frontiers* 17 (2): 239–241. https://doi.org/10.1007/s10796-015-9549-2.

Xu, Min, Jeanne M. David, and Suk Hi Kim. 2018. "The Fourth Industrial Revolution: Opportunities and Challenges." *International Journal of Financial Research* 9 (2): 90–95. https://doi.org/10.5430/ijfr.v9n2p90.

Ye, Neng, Xiangming Li, Hanxiao Yu, Aihua Wang, Wenjia Liu, and Xiaolin Hou. 2019. "Deep Learning Aided Grant-Free NOMA Toward Reliable Low-Latency Access in Tactile Internet of Things." *IEEE Transactions on Industrial Informatics* 15 (5): 2995–3005. https://doi.org/10.1109/TII.2019.2895086.

2 Role of Internet of Things (IoT) in Electronic Waste Management

Shambo Roy Chowdhury, Akshet Bharat Patel, and Pranav Rajesh Sharma
Manipal University Jaipur
Jaipur, India

CONTENTS

2.1 INTRODUCTION

The Internet of Things (IoT) is devised to create a network of electro-mechanical systems capable of communicating with each other to achieve a common objective. This emerging technology is drawing attention over a vast area of application ranging from industrial management to daily household needs. With technological advancements, our day-to-day electronics are capable of doing much more than their basic applications. For example, smart home devices are able to control your electrical gadgets, play your songs, and act as personal assistance device. Incidents reported in the news confirmed smart watches were able to save the lives of individuals with their advanced technologies such as detecting health conditions, detecting accidents, or calling emergency numbers. On the other hand, Industrial Internet of Things (IIoT) enables the connection of all the equipment and assets to the business process to achieve optimized industrial operations (Sisinni, Saifullah, Han, Jennehag, & Gidlund, 2018). These applications help in scaling production, lowering latency, maintaining resources, and optimizing maintenance schedules in industries (O'Donovan, Leahy, Bruton, & O'Sullivan, 2015; Qu et al., 2016). As machine replaces human activity and has a higher intervention in industrial operation by IoT, the count of electronic devices increases exponentially. The number of electronic devices is reportedly to reach hundreds of billions and e-waste production to reach over 100 million tonnes per year by 2050. In developed countries, e-waste often constitutes 8% of the volume of total municipal waste (Widmer, Oswald-Krapf, Sinha-Khetriwal, Schnellmann, & Böni, 2005). With the advent of newer technologies more devices reach their end of "useful life" and contribute to the pile of ever-increasing electronic waste. Since current product design features and changes in technology and wireless services often make it difficult for consumers to avoid frequent replacements of the products, the technological innovations must be accessible in ways that will generate less e-waste and maximize the life expectancy of the product. Rapid changes in existing technologies can also leave many consumers helpless without support to continue to get the full functional use of the existing equipment. The contribution of a particular type of electronics device toward the annual E-waste production, A (kg/year), can be calculated from the mass of the device, M (kg), multiplied with the quantity of the device used in a year, N, and divided by its average lifespan, L (years).

$$A = \frac{M * N}{L}$$

Electronic wastes are hazardous to the environment as well as to human health. E-waste-related problems have been serious concerns for authorities of many countries and policies are developed continuously to tackle the same. Management tools and laws at the national and universal levels are introduced to alleviate the problems of e-waste. Assessments of life-cycle and estimation of e-waste generation are done to aid such policies. The ill effects of e-waste on human health, both for chronic and acute conditions, turn out to be a serious societal problem. Similar harmful effects of e-waste can also be observed toward the environment. Electronic wastes are different from other municipal or industrial waste, both in chemical and physical constituents. They contain both valuable and hazardous materials and demands special handling techniques for efficient waste management. A systematic review (Song & Li, 2014) on the effect of unregulated heavy metal extraction methods on the local environment in China presents the following conclusion:

- Risks of mortality and morbidity have a close correlation with elevated levels of total suspended particles (TSP) in the air.
- It was found that heavy metals accounted for 1% of TSP in the air around a workshop for the recycling of PCBs.
- The presence of lead (Pb) in TSP was more than 28.5 times the maximum limit of air Pb levels for non-urban European sites.
- Pollution levels inside the workshops were much more severe than that outside and are a cause for serious concerns to the workers of these workshops.
- From the records of heavy metals in dust, it was observed that concentrations of Pb and copper (Cu) in workshop dust and adjacent roadside dust were "330 and 106, and 371 and 155 times higher, respectively, than for non-e-waste sites located 8 and 30 km away."
- Similar higher concentrations of heavy metal were found in soil, sediments, and plants.

Similar results were obtained in many developing countries. Many developing countries practice smoldering plastic off cables for recovering their copper and precious metals may be extracted by acid baths: unscientific and unregulated methods of extraction, as reported in Agyei-Mensah & Oteng-Ababio (2012). Higher concentrations of toxic materials were observed in working places of e-waste management in Ghana. The lack of awareness among locals and workers on the health hazards from these materials is downplayed in these regions. Toxic materials from e-wastes cause serious health hazards in human beings. The toxic components reach the human body through breathing, water, fruits, and vegetables. Long intake of lead can cause problems to the central and peripheral neural systems (Wong et al., 2007). In addition, lead can also have adverse effects on the reproductive system of both males and females. Mercury causes damage to vital organs and cadmium can act as a long-term cumulative poison (Pinto, 2008). Cadmium also has toxic effects on the kidney and is considered carcinogenic. In addition, an unmonitored and ill-trained extraction process frequently causes cuts and burns from de-soldering methods.

2.2   GENERAL CONVENTIONS FOR ELECTRONIC WASTE MANAGEMENTS

The importance of the management of electronic waste is thus undeniable. Reduce-reuse-recycle are three prime approaches to manage electronic wastes. Reduce involves manufacturers and consumers increasing the useful life of electronic devices. This requires bringing changes in business, manufacturing, and usage practices. Equipment should be easily repaired and upgraded in contrary to becoming obsolete with rapid change in technology. Quality and safety standards must be improved for new and refurbished devices. Encourage and enable users to upgrade and repair old devices. Over 60 electronic gadgets are expected to be discarded from a regular household over a span of 20 years (Cairns, 2005). Reuse, on the other hand, allows modification or refurbishment of "obsolete" electronic gadgets for newer purposes. This encourages consumers to make creative use of old or abandoned electronic devices. Recycle is a much larger and most commonly practiced approach to handle electronic wastes. Physical materials from electronic devices are extracted to be recycled back to industries. In the recycling process, only a percentage can be reclaimed, such as metals. These are collected from separation workshops and materials with lower economic values like plastics are burned, causing harmful fumes into the environment. Due to the lack of reuse and recycle infrastructure, many consumers are storing old equipment in their homes or discarding it with their regular trash. Many countries have drafted laws against the uncontrolled disposal of electronic wastes. For proper management, every state in the world must contribute toward making and adopting policies to ban types of e-waste from landfills and encourage recycling programs. These are to be run by local municipal or public interest organizations, with recycling costs paid through a mixture of tax revenues, government and/or corporate grants, and fees charged to consumers who return equipment for recycling (Cairns, 2005). Directives drafted by the European Union Waste from Electrical and Electronic Equipment (WEEE) had been adopted by all countries in 2007. The directives provide clauses and set targets for the recovery and recycling of almost all kinds of electronic goods. Initiatives such as life cycle assessment (LCA), material flow analysis (MFA), and extended producer responsibility

(EPR) have been opted by national agencies to control electronic wastes. Here the pollution makers are included to take back or contribute toward the recycling of the devices. A UN-led initiative StEP (Solving the E-waste Problem) was organized in 2004 to generate guidelines and encourage the worldwide practice of reverse supply chain (UNEP, 2009). It provides a directive for the use of minimal toxic material and designs for the scope of easy repair and upgradations. LCA of electronic devices encourages eco-friendly designs and assesses the ecological effect of the device along with its development process (Kiddee, Naidu, & Wong, 2013). Contributing to the rapid rate of product replacement is the lack of built-in, effective tools and technical support to help consumers maintain, upgrade, and repair existing equipment. Progress is needed in designing products for easy maintenance, upgrading and repair, and finding ways to introduce new features and technological innovations in a manner that does not impede the functionality or flexibility of existing equipment. Device compatibility for upcoming or multiple technologies is the need of the hour. For example, mobile phones compatible with different technologies communication (Union, 2004) can significantly improve the life cycle period of the device. Making battery replacement easier and more affordable and providing consumers with better tools and technical support to make other repairs and upgrades are critical to extending product life and reducing waste. In recent times, authorities like the Cellular and Telecommunications Industry Association (CTIA) have also taken steps to establish guidelines and standards for recycling and reuse. MFA tracks the pathways from the initial to the final destination to assess the major recycling process of electronic devices. Developed countries like the United States, Japan, and the countries of the European Union are leading producers of e-waste. They are also the leading exporters of electronic waste. These countries are estimated to export 4.49 million tons of household e-waste in 2010. Developing countries like China, India, and West Africa are major importers of electronic waste (Zoeteman, Krikke, & Venselaar, 2010). These countries operate large informal e-waste processing sectors practicing highly hazardous methods to recycle components from electronic wastes (Sthiannopkao & Wong, 2013). The informal process includes peddlers collecting abandoned electronic devices in return for payments which are then transferred to recycling workshops. E-wastes are also collected from dumped wastes via sorting, which comes as a result of a lack of awareness among users. E-waste after collection is sorted and processed for reuse or recovery of materials. Collected wastes are sent for reuse via background channels by simple upgradation, repairing, or refurbishing the devices. The second-hand market is more popular in villages and rural areas and the government must allocate extra funds to channelize these markets in a planned and organized manner (Z. Wang, Zhang, Yin, & Zhang, 2011). Policies like "traded-home appliances" in China, along with on-site collection methods, are a step toward organizing the sectors. Consumers must be educated on the harmfulness of discarded e-wastes and be encouraged to follow proper e-waste management guidelines. Incentives must be provided to encourage consumers and conduct awareness programs to follow such guidelines. Moreover, technologies and their know-how must be shared among countries across borders. For example, Robot Daisy, designed by Apple, can dismantle an iPhone in a matter of minutes. Such technologies can help the e-waste management sectors to organize the act in a more time-saving and environmentally friendly manner. While EPR is mostly used by the producer for the recovery of more valuable components, the act must be (more importantly) extended for collecting cheaper and toxic components of recycled devices. Also further initiatives such as tech giant were all in the recent news for not shipping charger bricks with their new phone lineup in the name of "saving the planet by reducing e-waste." Similar approaches might be encouraged for all manufacturers in the smartphone domain to decide mutually for introducing unified charging technologies. In this way, any charger can be used to charge any phone and eventually will reduce the volume of e-waste (Figure 2.1).

2.2.1 INTERNET OF THINGS TO AID E-WASTE MANAGEMENT

In recent research works the IoT has been projected as a major tool to tackle e-waste problems. IoT networks have been reported to be used for data management, e-waste collection, and e-waste tracking. Here in this section we will analyze few such techniques to utilize IoT for e-waste management.

Major problems associated with e-waste management are caused due to inefficient documentation and legislation, low and irregulated collection of waste, and illegal movement. There is also difficulty in enforcement of eco-friendly design and lack of cooperation of the people. The massive amount of data generated during the production, consumption, and disposal of electrical and electronic equipment (EEE)

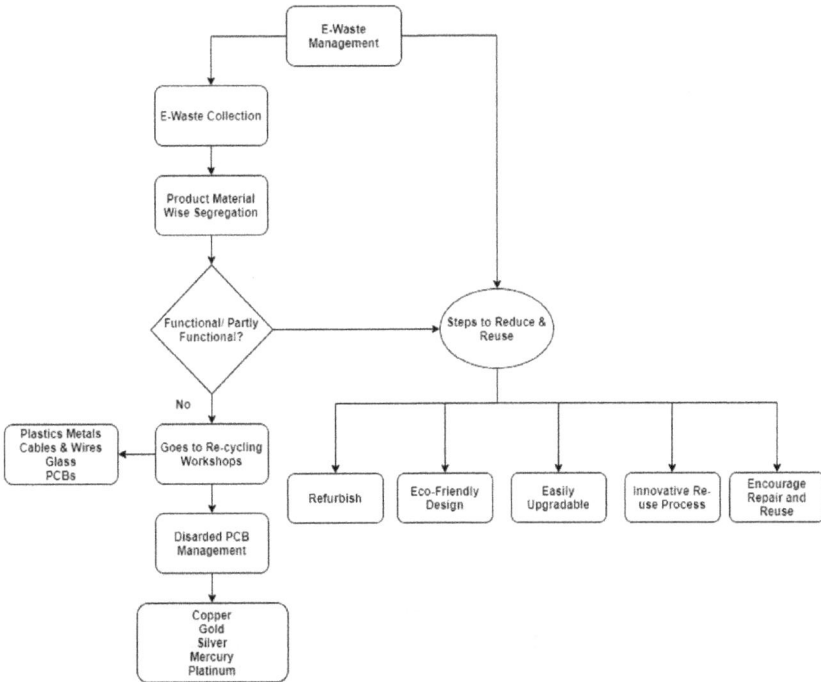

FIGURE 2.1 A flowchart describing the general modes of e-waste management.

fits the characteristics of big data. The IoT and big data can act as solutions to the WEEE Management problems.

> *Big Data in WEEE Management:* Recycling and reverse logistics of e-waste generates a huge database and requires adequate efficiency in storage and analysis algorithm for valid implementation. Following are key aspects of big data application in waste management.
> *Data Volume and Velocity:* Because of rapid advancement in the technological field, socio-psychological aspects and service upgradations and the consumption of EEE have accelerated at a huge rate. Surveys have reported much more than 7 billion mobile phones users around the world and the useful life of a mobile phone in general convention of usage is at around two years.
> *Data Complexity:* The diversity and volume of data generated from various types of sources creates a database of high complexity. The database may contain data values of all types such as characters, natural language, numbers, and Booleans.

The information related to WEEE management, generated from various devices located and diverse regions, is accumulated and rendered by the administrators through IoT. Wang and Wang (L. Wang, Wang, Gao, & Váncza, 2014; X. V. Wang & Wang, 2019) and works by other authors (Gu, Ma, Guo, Summers, & Hall, 2017; X. V. Wang & Wang, 2017; X. V. Wang, Wang, Mohammed, & Givehchi, 2017) have proposed cloud-based solutions where WEEE recycle/recovery process are installed as adaptable cloud services. The techniques are applicable in both the production and disposal of electronic equipment. This enables responsible authorities to monitor the real-time situation and keep track of movements in processes. The data accumulated can be utilized by all sections of people working at different levels of e-waste management. RFID and other tagging devices are used to maintain track of the various electronic equipment deployed across a network. Wireless sensor networks (WSNs), which can be defined as networks of spatially and technically diversified set of autonomous sensors implemented with low-power data acquisition arrangement and wireless communication capability, play a major role in managing data generated in IoT. The use of the WSN in the working field of the informal sector can generate proper information on informal WEEE recycling. The overall task of implementing big data and IoT for this purpose can be divided into three major layers: acquisition, processing, and analysis. Proper utilization of the IoT and the big data technologies in waste management have enough potential to improve the current system, optimizing business, and reducing waste hazards. Major challenges facing its implementation include the cost of implementation or operation, data collection, and data processing.

The application of IoT devices for easy collection of waste has been a popular topic of research and development in modern times. In a Smart City, the collection of waste is a crucial point for the environment and its quality should be considered seriously. The main countermeasure to environmental pollution in terms of a Smart City is the IoT-enabled waste collection (Medvedev, Fedchenkov, Zaslavsky,

Anagnostopoulos, & Khoruzhnikov, 2015). Issues connected to dynamic waste collection could be divided into two main problems:

i. Scheduling of waste collection.
ii. Routing the trucks for optimum transportation.

E-waste collection systems enhanced with IoT services can be enabled for dynamic scheduling and routing in a Smart City. Also, a design of a cloud system for organization of waste collection process and application for waste truck drivers and managers and an onboard surveillance system that raises the process of problem reporting and evidence collection to a higher level. Similarly, waste truck-owning companies may use a platform for organizing and optimization of their business process in general without serious investments in developing, deploying, and supporting their own system. Such a system must include effective dynamic routing based on IoT data for the truck fleet. These kinds of systems aim to provide services for different kinds of stakeholders involved in this area – from city administrations to citizens. Still, the design focuses mostly on providing software as a service (SaaS) to commercial waste management companies. Development of applications for city administrations, municipal staff, recycling factories, and other stakeholders is planned to be done in the future.

The above approaches are inevitable for efficient management of the rapidly increasing electronic waste. Keeping an eye on the harmful effect of electronic wastes, government organizations, industries, and citizens must comply with norms to emaciate the harmful effects of e-waste. While the above methods are to be implemented on a large scale, the reuse of IoT devices must be encouraged at the consumer end. In the next section we present some easy-to-implement applications for the reuse of abandoned or partially damaged mobile phones.

2.3 DIY TECHNIQUES FOR REUSING MOBILE PHONES

Mobile phones, being the most commonly used IoT devices in households, contribute to a major chunk of e-waste. Here in this chapter, different but easy-to-implement techniques to reuse old and obsolete mobile phones are described. Mobile phones are often discarded to adopt new technology or partial dysfunctionality to the device. These devices can be reused for different purposes, as described here. The first application uses mobile phones as a webserver for storing useful data or for hosting own website.

2.3.1 SECURITY SURVEILLANCE SYSTEM USING ABANDONED MOBILE PHONES

Surveillance systems are one of the most important/integral parts of today's homes, especially with the rapid increases of crimes and robberies. Originally CCTV systems were used for surveillance systems, but they had their own disadvantages, such as they were difficult to upgrade and not easily scalable (if needed). They also require additional money and time for installation and calls for regular maintenance. It can be a much better option for old and discarded mobile phones to be reused as a security camera. In the surveillance systems these old mobile phones are set to act as a camera and provide a point of access for internet.

The hardware requirement for designing such a system would be:

- An old phone which acts as a camera.
- A router.
- A phone with an enabled internet connection to view the live footage.

The main task in such a system is to view the live/recorded footage of the camera on other devices such as phones, tablets, or PCs. So, in order to view the footage there are three basic and distinct things required to get the system working and up for the task.

1. The first and the most important thing to view the home camera on any mobile device is linking up for a dynamic DNS (Dynamic Name System) service. This is a necessary step as the IP address provided by our ISP, or Internet Service Provider, is a dynamic IP address, which means it is not a fixed one and keeps on changing either with time or changes every time the router is switched ON and OFF. With a changing IP address, it is not possible to be connected to the camera and thus not possible to view the footage on mobile devices. So, a dynamic DNS service eliminates the issue of the home IP address constantly changing. In a DNS database, a domain name is mapped or joined with an IP address and the dynamic DNS will update the new IP address (as it changes), which is mapped to the domain name, which is why no matter how many times the IP changes it would be able to access the custom hostname. There are many dynamic DNS services available nowadays, but not all of them are supported by a router. This is because a router requires embedded coding in order to support a particular DNS service. One can check or find out all the supported dynamic DNS services by simply logging into the router and then clicking on the menu for the dynamic DNS, where a dropdown list will show all the supported dynamic DNS services by that specific router.

A host name can be created on one of the supported dynamic DNS services and that host name needs to be added onto the dynamic DNS menu by logging onto the router. This finally enables the dynamic DNS service on the router.

On the other hand, one can also purchase a static IP (an IP that does not change) directly from the ISP to eliminate the issue of a constantly changing IP address, but this can be a rather expensive option.

2. The second step would be to open a port on the router for the camera so that the video from the camera can reach the route. This can easily be done by noting the IPv4 address (which is a 32-bit number that uniquely identifies a network interface) of the camera, and also the port number needs to be noted (which is generally port 81).

These details can easily be found out on the phone by simply going to Settings and then About Phone, and this is where the option for IP address and the required information, i.e. the IPv4 address, can be found.

Now, a new virtual webserver on the router is created, and this can easily be done by logging onto the router and going to the access tab, where the IPv4 address of the camera (that was noted earlier) is entered along with the port number. This means that a portal where we will be able to see the live footage of the security camera is created. This system can easily be scaled by repeating the above procedures multiple times and adding a different IP address and port number to the router for each and every camera.

Now, the camera is streaming security video in real time to the port that has been specified on the router and the router is connected to the dynamic DNS provider. This makes the flowchart of the system something like this (Figure 2.2).

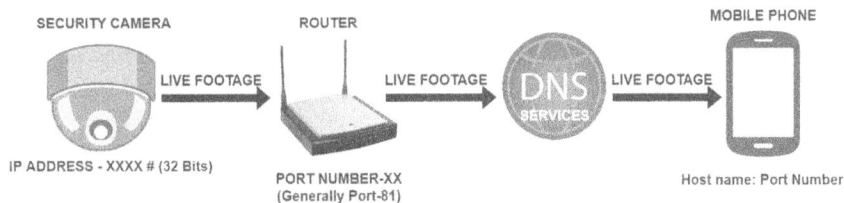

FIGURE 2.2 Block diagram for implementation of security camera using used mobile phones.

3. The third and last step to get the security system ready is to view the footage live on a mobile device. This could be done on any mobile device that has a web browser and supports an internet connection. For this, a web browser needs to be accessed and we need to type in the host name that was created while enabling the dynamic DNS service followed by a ":" and the port number at which the live footage is streaming.

For example, if the host name is viewlivecamera.com and the port number the camera is streaming the video to is 81, viewlivecamera.com:81 should be used to direct the browser to the open port that has been created and then the live footage from the camera can be viewed.

Additionally, this footage could also be viewed on an app instead of using the web browser. The same steps must be followed in order to get the system working, i.e. the IPv4 address and the port number of the camera must be configured in the app and then the system is ready.

2.3.2 HOME AUTOMATION

Mobile devices have been an important/integral part of our day-to-day life for the last few years, providing various facilities, and over the years mobile phone manufacturers have increasingly added prominent features on mobile devices. With the help of these smartphones, one can do many works with or without the internet, like for instance, creating a home automation system that uses the Android mobile devices to communicate with the system over long ranges with the help of Wi-Fi and give commands to the system to switch ON/OFF electrical equipment such as lights, fans, air conditioners, etc., at home or at any given place. All of these commands can easily be sent from mobile devices through an Android application, which can easily be used/operated by any user. Then, the system responds to the commands by the user by taking actions as specified by the user through the application, i.e. switching ON a bulb that has been instructed by the user. Lastly, the user will also be able to track the status of the tasks that are being performed by the system. For instance, as the system turns ON the bulb as specified by the user, it will send a response back to the user, notifying him that the task has been performed. Thus, such a home automation system would make our home more organized and smarter and improve our lifestyle.

To achieve this, the home appliances must be controlled with a microcontroller equipped with Wi-Fi modules. The Wi-Fi module must connect to the same Wi-Fi network as the mobile phone. This allows the data packets to be sent to the microcontroller, which reads the instruction, processes it as per programming, and then performs the specified task using relays or other different types of switches. Lastly, as the task is completed a signal is sent back from the appliances to the mobile application over the WI-FI network, to notify the user that the assigned task has been completed and the microcontroller is ready to perform the next task.

The requirements for this home automation system are:

- *Mobile device with working Wi-Fi module, in order to give commands:*
  A wireless local area network is established to connect multiple devices wirelessly. This technology features a long range and allows multiple devices to connect simultaneously. The mobile phone should tether its Wi-Fi signal and create an access point (hotspot if internet is required) for other appliances to connect.
- *Android application installed on the user's mobile device:*
  Android is the operating system that is used by mobile phones and this operating system provides applications to wirelessly connect over Wi-Fi or Bluetooth. Additionally, the app must meet the requirements of Android to be able to execute without any difficulties. Many such applications could be found over the internet that provides us with widgets that can be used as switches. Moreover, the interface of these apps could be modified according to the preference of the user.

Various web applications are available to build such a home automation UI as they already include readymade widgets incorporated within them, which can be directly used for a project by simply customizing them according to one's needs and preferences. This way a lot of precious time is saved, as no new application needs to be built from scratch, yet similar results can be achieved.

One such application is the *MIT App Inventor*, which is an open-source web application for Android. The app uses a graphical interface where users can simply create an application by dragging and dropping visual objects to their project. After the app has been designed, one can simply download it on his/her Android phone using the QR code displayed on the website or by downloading its APK and later installing it on his/her smartphone.

To build such an end-to-end application using the MIT App Inventor, the simple steps that one needs to perform are given as follows:

i. The first step would be to open up the website https://appinventor.mit.edu/ and click on "Create Apps."
ii. Secondly, click on "Start a new Project"; if the steps have been correctly followed the user would see the UI on the screen with all the visual objects displayed to the left of the screen.
iii. Lastly, once the user has customized and included all the necessary objects in the application, the user needs to click on "Build" and choose either of the two options: the QR code or the APK.

Once all the steps have been completed, the application can simply be installed on the Android device and can be used to control the home appliances.

- *Microcontroller with Wi-Fi module:*
  This is a circuit designed to govern a specific operation in any embedded application. The microcontroller has multiple digital and analog input/ output pins that can be used to connect various devices. These are easily powered by a DC battery and need to be programmed in order to perform the tasks specified by the user. The Wi-Fi module can be a built-in feature of the microcontroller board (such as Raspberry Pi) or may be an added module (such as ESP modules for Arduino). In the case of additional Wi-Fi modules, they communicate with the microcontroller using serial commu- nication protocols (such as UART). This would enable the embedded plat- form to act as a slave module and connect to the mobile phone's WLAN. In this kind of a system the authentication code can be generated from the mobile application and needs to be incorporated within the code, in order to connect the microcontroller with the safety.
- *Relays or other kinds of switches:*
  This is a switch that is electrically operated and they use electromagnets for their functioning. These are used wherever the system has to be controlled using low power and they perform the switching application using a semi- conductor device. These switches are used to match the power requirements of the appliances. Moreover, the relays or drivers can also be used to control the intensity of the appliances.

Some of the applications that this system can be used for are:

- *Opening and closing doors:* This system can be used to unlock doors with- out the keys with the help of the Android application. Moreover, this is a more secure way as anyone with the keys cannot open the door.
- *Control water sprinklers in gardens:* Using this system we can control when and how much water will be sprinkled in the garden. This way the water sprin- klers can be controlled remotely and also the health of the garden increases as we have control over the water supply that is harmful to the garden (Figure 2.3).

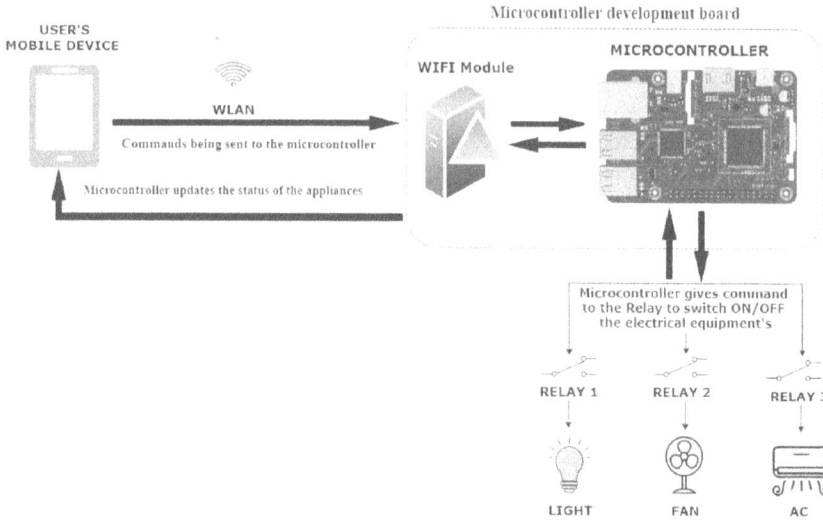

FIGURE 2.3 Block diagram to explain the working of mobile phones for home automation.

2.3.3 MOBILE PHONES USED AS SERVER

Websites are hosted and stored on special computers called servers. A static webserver, which is also called a "stack," consists of hardware with storage such as a computer running an HTTP (Hypertext Transfer Protocol) server, which is the software component. It is called static because the server will transmit its hosted files as they are to your browser. The data stored in the server can be accessed over the internet following a proper communication protocol. An old smartphone can also be used as a webserver to host websites and webpages. When we talk about a complete and an efficient IoT system, a webserver plays a very important/integral role in creating one. The main function of this webserver is to store web pages and files and transmit or broadcast them over the internet for anyone who requests to visit that webpage, or in other words we can call it a powerful computer that is capable of storing and transmitting information via the internet to anyone who wishes to see it. On a basic level, when a browser requests a file that is hosted on a webserver, the browser requests the file via HTTP, which is an internet protocol that is used for transmitting documents. When the request reaches the correct and specific (hardware) webserver, the (software) HTTP server accepts it, finds the requested document or file, and sends it back to the browser, through HTTP only. A simplified flowchart of the same can be seen in Figure 2.4.

Thanks to the technological advancements over time, it is now possible to develop and deploy new web services on smartphones and all of this has been possible due to the powerful processing and increased memory addressing capacity that is present in the smartphone today. The requirements of creating an HTTP server using a mobile phone are as follows:

1. A smartphone with an active internet connection.
2. An application on the smartphone that creates a webserver using the phone's IP address.

DOWNLOAD THE REVERSE PROXY TOOL AND SET UP A PUBLIC URL

MOBILE PHONE WHICH
ACTS AS THE SERVER HOST

DOWNLOAD AND SET UP THE APP THAT
CREATES A WEB SERVER USING PHONE'S
IP ADDRESS

APP DISPLAYS THE URL
https://192.168.0.160:8080,
8080 REPRESENTS THE
PORT NUMBER AT WHICH
THE SERVER IS RUNNING

1. CLICK ON LOCAL SERVER
SETTINGS

2. CLICK ON HTTP PORT

3. SET PORT AT WHICH THE
SERVER IS CURRENTLY RUNNING

localhost:portnumber TYPED ON THE WEB BROWSER OF THE
SMARTPHONE TO VIEW THE CONTENT OF THE SERVER ON
THE SERVER HOST

ENTER THE PUBLIC
URL ON THE WEB
BROWSER TO VIEW
THE CONTENTS ON
THE WEB BROWSER

COMPUTER CONNECTED TO THE PUBLIC INTERNET
TO VIEW THE CONTENTS OF THE WEB SERVER

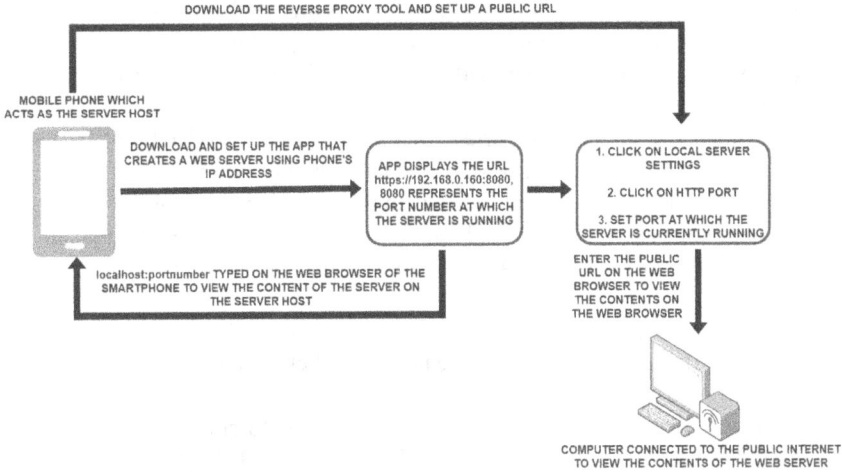

FIGURE 2.4   Block diagram to describe the use of old mobile phones as a server.

3. A reverse proxy tool that connects local servers to the public internet so that the contents of the server are visible to the world.

The basic steps to convert a smartphone into an HTTP server:

1. The first and foremost step is downloading an application on the smartphone (the one we intend to create a server on) that creates a webserver using the phone's IP address. There are many such applications available and one can easily find them on any mobile app distribution platform.
2. Each and every app will have a different user interface, but the only thing that needs to be done now is to start the server by clicking on the button that can be found on the main page of the app. As soon as the server starts the status of the server will be displayed as "RUNNING."
3. Additionally, the app will be displaying a URL through which the contents of the server can be viewed on the host smartphone.
   An example of the URL is https://192.168.0.150:8080. In the URL, 8080 represents the port number at which the server is running.
4. To view the contents of the server on the host smartphone, there are just two simple steps that need to be followed:
   • Open the browser on the host smartphone
   • Just type "localhost:portnumber"; as the port number is 8080 in most cases, we type "localhost:8080"
   • Now all the files that are stored on the internal storage of the phone can be viewed on the web browser on the smartphone.

To connect the local servers to the public internet, the few steps that need to be followed are:

1. Firstly, a reverse proxy tool needs to be downloaded on the smartphone. The main function of this application is to connect the local servers to the

public internet. Again, many such applications are available and can easily be downloaded from any mobile app distribution platform.

2. As soon as the app has been downloaded it asks for a public URL that needs to be filled by the user.
3. The second and the most important step is to change the HTTP port in the reverse proxy tool to the port at which the server is currently running. For this:
   * Click on the Local Server Settings in the reverse proxy tool
   * Then click on HTTP port and set it to the port at which the server is currently running. In this case it is set to 8080 as that is the port on which the server is currently running.
4. Now, it is all set and the webserver is ready to be deployed to the world. The only thing that needs to be done now is to click "Enable" on the reverse proxy tool.
5. To view the contents of the webserver, one needs to simply open a web browser on any other device than the host device and simply type in the public URL that was set up in step 2.

2.4   CONCLUSION

The chapter summarizes the ill effects of electronic waste, which is increasing at an uncontrollable rate with the rapid advancement of technology. It has been projected that while IoT devices contribute to a major volume of electronic waste products, the same can be utilized to manage the problem. Various techniques at the organizational level have been described to manage electronic wastes. The use of big data analysis along with cloud architecture and IoT devices can create a sustainable infrastructure to manage electronic wastes. In addition to these, three DIY techniques have been elaborated on to encourage reuse among consumers.

REFERENCES

Agyei-Mensah, S., & Oteng-Ababio, M. (2012). Perceptions of health and environmental impacts of e-waste management in Ghana. *International Journal of Environmental Health Research.* https://doi.org/10.1080/09603123.2012.667795

Cairns, C. N. (2005). E-waste and the consumer: Improving options to reduce, reuse and recycle. In *IEEE International Symposium on Electronics and the Environment* (pp. 237–242). https://doi.org/10.1109/isee.2005.1437033

Gu, F., Ma, B., Guo, J., Summers, P. A., & Hall, P. (2017). Internet of things and Big Data as potential solutions to the problems in waste electrical and electronic equipment management: An exploratory study. *Waste Management, 68,* 434–448. https://doi.org/10.1016/j.wasman.2017.07.037

Kiddee, P., Naidu, R., & Wong, M. H. (2013). Electronic waste management approaches: An overview. *Waste Management, 33*(5), 1237–1250. https://doi.org/10.1016/j.wasman.2013.01.006

Medvedev, A., Fedchenkov, P., Zaslavsky, A., Anagnostopoulos, T., & Khoruzhnikov, S. (2015). Waste management as an IoT-enabled service in smart cities. In *Lecture Notes in Computer Science (including subseries Lecture Notes in Artificial Intelligence and Lecture Notes in Bioinformatics)* (Vol. 9247, pp. 104–115). Springer Verlag. https://doi.org/10.1007/978-3-319-23126-6_10

O'Donovan, P., Leahy, K., Bruton, K., & O'Sullivan, D. T. J. (2015). An industrial big data pipeline for data-driven analytics maintenance applications in large-scale smart manufacturing facilities. *Journal of Big Data, 2*(1), 25. https://doi.org/10.1186/s40537-015-0034-z

Pinto, V. N. (2008, May 1). E-waste hazard: The impending challenge. *Indian Journal of Occupational and Environmental Medicine.* https://doi.org/10.4103/0019-5278.43263

Qu, T., Lei, S. P., Wang, Z. Z., Nie, D. X., Chen, X., & Huang, G. Q. (2016). IoT-based real-time production logistics synchronization system under smart cloud manufacturing. *International Journal of Advanced Manufacturing Technology, 84*(1–4), 147–164. https://doi.org/10.1007/s00170-015-7220-1

Sisinni, E., Saifullah, A., Han, S., Jennehag, U., & Gidlund, M. (2018). Industrial internet of things: Challenges, opportunities, and directions. *IEEE Transactions on Industrial Informatics, 14*(11), 4724–4734. https://doi.org/10.1109/TII.2018.2852491

Song, Q., & Li, J. (2014, December 1). Environmental effects of heavy metals derived from the e-waste recycling activities in China: A systematic review. *Waste Management.* Elsevier Ltd. https://doi.org/10.1016/j.wasman.2014.08.012

Sthiannopkao, S., & Wong, M. H. (2013). Handling e-waste in developed and developing countries: Initiatives, practices, and consequences. *Science of the Total Environment, 463*, 1147–1153.

UNEP. (2009). UNEP. Recycling – from e-waste to resources. NEP Division of Technology, Industry and Economics, Sustainable Consumption and Production Branch. United Nations Environment Programme (UNEP), Nairobi, Kenya.

Union, C. (2004). Frequently Asked Questions About Cell Phone Portability. Bryan, Colleen. "Number Portability for Consumers: Taking Your Wireless Number With You." Loy. Consumer L. Rev. 16 (2003): 267.

Wang, L., Wang, X. V., Gao, L., & Váncza, J. (2014). A cloud-based approach for WEEE remanufacturing. *CIRP Annals - Manufacturing Technology, 63*(1), 409–412. https://doi.org/10.1016/j.cirp.2014.03.114

Wang, X. V., & Wang, L. (2017). A cloud-based production system for information and service integration: An internet of things case study on waste electronics. *Enterprise Information Systems, 11*(7), 952–968. https://doi.org/10.1080/17517575.2016.1215539

Wang, X. V., & Wang, L. (2019). Digital twin-based WEEE recycling, recovery and remanufacturing in the background of Industry 4.0. *International Journal of Production Research, 57*(12), 3892–3902. https://doi.org/10.1080/00207543.2018.1497819

Wang, X. V., Wang, L., Mohammed, A., & Givehchi, M. (2017). Ubiquitous manufacturing system based on Cloud: A robotics application. *Robotics and Computer-Integrated Manufacturing, 45*, 116–125. https://doi.org/10.1016/j.rcim.2016.01.007

Wang, Z., Zhang, B., Yin, J., & Zhang, X. (2011). Willingness and behavior towards e-waste recycling for residents in Beijing city, China. *Journal of Cleaner Production, 19*(9–10), 977–984.

Widmer, R., Oswald-Krapf, H., Sinha-Khetriwal, D., Schnellmann, M., & Böni, H. (2005). Global perspectives on e-waste. *Environmental Impact Assessment Review, 25*(5), 436–458.

Wong, M. H., Wu, S. C., Deng, W. J., Yu, X. Z., Luo, Q., Leung, A., … Wong, A. H. (2007). Export of toxic chemicals – A review of the case of uncontrolled electronic-waste recycling. *Environmental Pollution, 149*(2), 131–140. https://doi.org/10.1016/j.envpol.2007.01.044

Zoeteman, B. C. J., Krikke, H. R., & Venselaar, J. (2010). Handling WEEE waste flows: On the effectiveness of producer responsibility in a globalizing world. *International Journal of Advanced Manufacturing Technology, 47*(5–8), 415–436. https://doi.org/10.1007/s00170-009-2358-3

3 Creating a Reliable IIoT Framework to Prioritize Workplace Safety in Industries Involving Hazardous Processes

Nishant Sharma and Parveen Sultana H.
Vellore Institute of Technology
Vellore, India

CONTENTS

DOI: 10.1201/9781003145004-3

3.1   INTRODUCTION

Internet of Things (IoT) is a fairly popular term in the modern-day world. In simple words, IoT aims at building up a network of everyday objects. These interconnected objects will perceive their environment, analyze sensed data, communicate relevant information with other such objects, and create a stable and reliable ecosystem for any application use case. There are many domains that will benefit from their integration with the paradigm of IoT. One such domain is that of industries. As sensors of the Industrial IoT (IIoT) generate data about various variables of interest in an industry, some of the benefits may be (1) gaining a better understanding of the industrial workflow from the generated data, (2) using the insights gained from the data to optimize the said workflow, (3) monitoring the state of inventory, and (4) creating a better and safer working environment for the people at work, among others.

The chapter considers as a central theme, namely the process of manufacturing, safe-handling, and storage of toxic industrial substances of economic value, and the safe disposal of toxic wastes from a large-scale industry. DGFASLI (2020) lists many such industries that involve hazardous processes. Since the onset of large-scale industrial manufacturing, we have witnessed many industrial disasters in India and in the world. The disasters become even more catastrophic when there are toxic and life-threatening compounds involved. Two of the worst industrial disasters in the world are the Bhopal gas tragedy of 1984 and the Chernobyl nuclear power plant disaster of 1986. Bhopal tragedy claimed a lot of lives and injured many more. Broughton (2005) puts the number of deaths at a minimum of 3,800. Chernobyl disaster, even though it claimed fewer direct lives, is a cause of latent mortality among young people who were exposed to the radiation. Even 20 years later, a significant correlation was found between the exposure of a young child or adolescent to the radiation of the Chernobyl plant and thyroid cancer (Cardis, 2006). A recent similar mishappening occurred in Vishakhapatnam (Wikipedia, 2020), where the mismanagement of the temperature for the safe storage of styrene monomer led to the vaporization of the compound, which, when inhaled by the people in its vicinity, led to symptoms such as breathing problems, irritation of eyes, and unconsciousness. In some cases, the symptoms proved to be fatal.

The safety of workers in such hazardous environments and the safety of the natives of the areas surrounding the industrial plant are of paramount importance. The ambition in optimizing industrial output to improve profits must be closely complemented with a serious caution regarding the prevention of industrial disasters. Chandra (2019) reviews the laws and regulations that are encompassed in the Indian Constitution that aim at providing post-accidental compensations for the workers working in such hazardous environments. However, there exists technology today that can help build a framework that would prioritize a proactive preventive approach to disaster management and a quick response in case a failure occurs. There is a positive discussion in the industry regarding these ideas. Srinivasan (2020) has reviewed these ideas for the Oil and Gas Operational Technology and expressed optimism for these state-of-the-art technologies. The industry and the government can come together in motivating and assisting each other as we progress into the future.

Consider the three approaches to hazard prevention and mitigation. The first approach is that of human vigilance. The second is to have a semi-automated system for monitoring and management of industrial workflow that is supplemented by human intervention whenever such a need arises. The third approach is to have a fully autonomous hazard management and mitigation unit that needs no human intervention. It can diagnose faults and failures and take corrective actions to mitigate those problems. Although the third approach may be seen as the long-term and desirable goal, it would take a myriad of innovations that are built upon each other to reach a level of reliability where the society can be comfortable with these systems being assigned such an important task all on their own. Up until such a reliable framework is realized, a collaboration of computerized and human effort will have to do. Eventually, however, a reliable automated system has a lot to offer. This chapter aims at discussing how such a reliable automated framework can be achieved.

Let us consider a basic example to get us started. Consider a carrier pipe carrying some toxic liquid in an industrial plant, as shown in Figure 3.1. The carrier pipe has barriers at regular length intervals to contain the flow of the toxic liquid if such a need arises. A watchful eye monitors the physical state of the carrier pipe. Consider that this watchful eye has determined a leakage in the carrier pipe that could lead to an untimely evacuation of the toxic liquid into the environment. The watchful eye tells the brain about this information. The brain, which can think, analyze, and make decisions, concludes that this is a cause for grave concern. The brain understands that if this concern is not quickly addressed, it could lead to a catastrophe. It then activates a barrier that is at the farthest distance short of the leakage point in

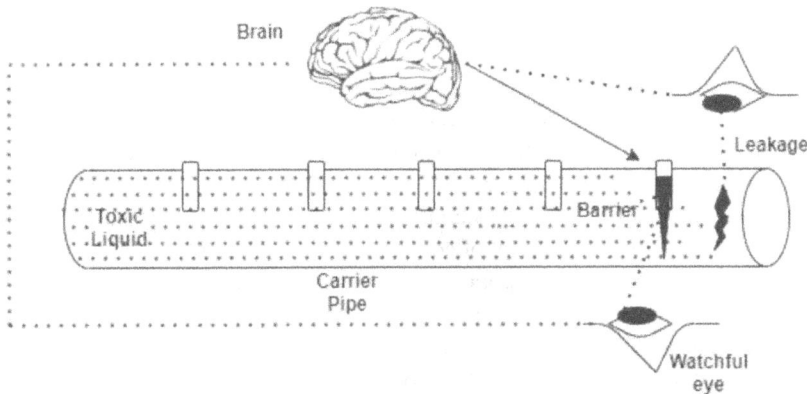

FIGURE 3.1   Smart actuation system.

an attempt to bring the toxic liquid to a steady state. It is at this point in time that another watchful eye is woken up. This eye is responsible for ensuring that the barrier has been successfully established. If the barrier has not been successfully established, then the brain is made aware and further steps may be taken. An example of such a step may be the dissemination of caution notification to a nearby public office or generation of alarms to do loss minimization.

Clearly, the watchful eyes in our example are the sensors that perceive their environment. The brain is the one responsible for providing the control decisions. The barrier is the actuator in our example. Fortunately, the system does not rest until it has determined that the actuation has been successfully established. We may employ in the range of three to seven sensors to monitor the state of each actuator (Kopetz, 2011). The number of sensors may increase if an actuator has to be monitored for more variables or for better sensor fusion based on the sensitivity of its designated task.

Unfortunately, both the physical components, namely the sensors and actuators, and the software systems are prone to failures. The urgent *safety* mechanism, therefore, in the monitoring and management systems for the hazardous industries can be achieved by solving the problem that the system should either reliably remain operational for a predetermined amount of time or that it may survive a certain subcomponent failure and still remain functional in a minimalistic sense in order to prevent a catastrophe from happening. One of the key parameters that can guide us in the right direction is the idea of minimizing risk. Even though failures are inevitable both in physical components and in software systems, the quicker the system switches the process flow to a safe state in the event of a failure, the more successful it will be in minimizing gross damages to both life and property. An important aspect to consider in the implementation of the *safety* mechanism should be to properly justify at what point can the *safety* algorithm rest comfortably after a failure has occurred, has been successfully diagnosed, and is being remedied. Another important characteristic of the *safety* mechanism is that it should be able to detect failures with a probability close to 1 if a failure has occurred, and no sub-component failure should be alien to the mechanism. Research in the domains of Machine Learning, Data Science, and Big Data Analytics can help to create better predictive and insightful models that can be employed to make the *safety* mechanism in such systems more reliable.

When considering monitoring physical wear and tear, we could take two approaches. As a first approach, we could monitor the wear and tear of the physical components by using some other physical devices designated with that task. These monitoring devices will again be at risk of failure. We could, as a second approach, use mathematical and natural science models to predict when a physical component may fail and shift the replacement of components on failure policy to a periodic and predictive system maintenance policy. We could also use a collaboration of both of these approaches depending on the sensitivity of the task. If we want the replacement of the components to be automated and quick, more modularity may be introduced into the system design. However, innovative engineering would be needed to design modular systems in such a way so that they provide the same reliability as tightly coupled systems design. If the occurrence of the physical failure may be masked using mechanisms built-in software or using redundancy, then such transparency is welcome.

In this chapter, we survey the existing literature for architectures related to IIoT and propose an architectural view that emphasizes the computational needs of a safety-critical system. We later discuss various ways to ensure reliability in such a system and to do fault management. We propose a conditional probability-based knowledge database for fault management in IIoT.

3.2 RESOURCES FOR FURTHER READING

General Electric (GE) coined the term industrial internet from Evans (2012) as the third wave of innovation and change in the evolution of the industrial sector. They have illustrated graphical representations for the key elements, namely costs, benefits, and data loop, etc., while describing the core concepts of the industrial internet. The resource also lays out the various opportunities and challenges relating to industrial internet.

Another great resource from GE (Bohm, 2018) discusses the implementation strategies for the industrial internet. This resource is aimed specifically at the developers. The resource discusses the world of IIoT applications, the key ideas in Operational Technology, the concepts and benefits of industrial edge technology, the requirements for a reliable IIoT architecture, and the usage of various tools and techniques to create such an architecture. It also emphasizes using simulation-based analysis to refine the IIoT system and framework.

Kopetz (2011) is a definitive guide as the introduction to the world of distributed and embedded design for applications requiring real-time features. The book is comprehensive about the principles behind real-time systems; however, it has a special focus on the applications that are safety critical. Design principles explained in this book can be carried over to the industry. For the industrial applications that require hard real-time and safety-critical guarantee, this book provides guidelines.

A subsection of this chapter discusses the parameter of safety in automated industrial monitoring and management systems as the conditional that if there is a failure that has occurred in the automated system, then the probability that this fault is successfully identified is close to 100%. This concept of failure identification is reviewed and discussed in terms of conditional probability in a later subsection of this chapter. An excellent introductory course on probability theory and on conditional probability in particular is found in (Blitzstein, 2019). The fault diagnosis must, however, be made within hard deadline constraints.

Taking an informed and educated control decision based on analytics, strategic planning with a proper contextual understanding of the historical data, and with newer information continuously being perceived in real-time requires an innovative usage of big data. Not only is the storage aspect of data relevant, but how must that data be queried is relevant as well. White (2012) provides an excellent resource on big data management and querying. White (2012) also provides insights into two of the distributed data management paradigms: Hadoop Distributed File System (HDFS) and Apache HBase. HDFS systems are more appropriate for batch processing and for processing a large amount of data to generate insights similar to a background process. HBase, however, is more apt for real-time, data-driven, and quick response applications. For multi-criticality systems like industry, these data management and processing systems may provide suitable mechanisms to write and access big data. On time-critical systems, a balance must be found for the querying of data. Models that prioritize timely response as well as optimize decision-making should be preferred.

3.3   BUILDING A SUITABLE SAFETY-CRITICAL ARCHITECTURE FOR IIoT

A process flow for futuristic industrial internet has been illustrated in Figure 3.2. Modular functions that do specific tasks may be identified for such a system. The base layer of any such system would consist of sensors that collect data about various measurands. This raw data needs to be processed such that it is converted into *good* data. By good data we mean that such data be clean and complete and may then be used to extract useful information. The process of extracting useful information from good data is the third essential step in the stated process flow. This extracted information is then verified against a knowledge base for its historical significance. By historical significance, we mean if such information amounted to

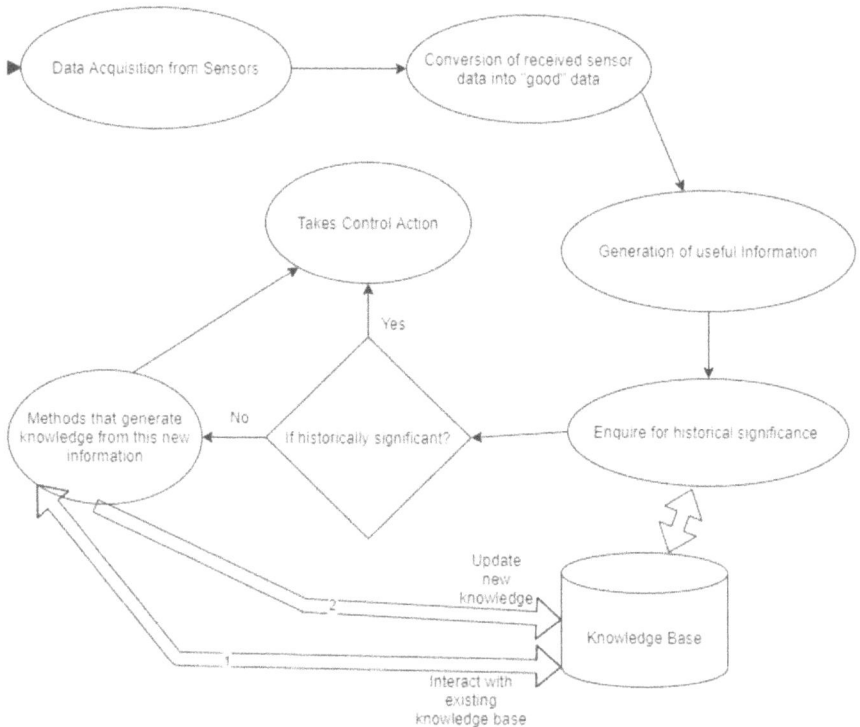

FIGURE 3.2   Futuristic industrial internet process flow.

an important control action in the past. If so, then we may initiate such a control action. However, if that is not the case, then we need methods that would generate new knowledge with respect to this newer information. This must be done taking the current knowledge base into account. A function for new knowledge generation can be stated as (3.1).

$$K_{new} = F\left(K_{old}, I_{new}\right) \tag{3.1}$$

where K_{new} is the new knowledge being created, K_{old} is the past knowledge, and I_{new} is the newer information that has been extracted in the previous steps. As this new knowledge becomes available, it must later be updated to the knowledge base. If this new knowledge demands a control action to be taken, then such an action must be taken.

3.3.1 AMBIENT INTELLIGENCE

The concept of ambient intelligence is such that given the information gathered in real time by using sensors and a knowledge base with historical context, a system with such intelligence may be able to take decisions to help the actors of a particular environment. The actor may be a human or it may be another system that is designated to perform a complementary task. One important characteristic of a system with ambient intelligence is that such a system should be transparent to the other actors of the environment. This ambient intelligence paradigm has a *disappearing* nature. The technologies that are disappearing in nature seamlessly integrate themselves into everyday life so much so that they become immanent. Ambient intelligence though ubiquitous and pervasive also has a key aspect of decisiveness integrated into it, i.e. the systems that are ambient intelligent demonstrate the ability of cognition, analysis, and decision making.

3.3.2 SELF-AWARE SMART SENSORS AND ACTUATORS

It is here that we introduce the abstract concept of self-awareness in smart sensors and actuators. Sensors and actuators are electronic components and, like all physical things, are heavily prone to failure. A self-aware smart sensor/actuator, although designated for a specific purpose, may also monitor its own well-being and request preventive maintenance action in case a failure is imminent as a result of wear and tear in its sub-system. An analogy of a patient and a doctor may be presented here to drive home the concept of self-awareness. A human being, even though he/she may not know the cause or the corrective measures for an ailment that they are experiencing, must, however, be self-aware enough to know when something does not feel right with their bodily function. They must also have the cognition to seek out a suitable doctor who, through their vast experience and knowledge base, may be able to offer an appropriate corrective measure. They (the doctor) may also be able to tell them if the ailment is terminal. Unlike humans, if the overall sensor/actuator unit is understood to be terminally failing, it would be completely ethical and heavily preferred to replace it with a newer component.

3.3.3 COMPUTING PARADIGMS FOR INTERNET OF THINGS

The three major computing paradigms for IoT are cloud, fog, and edge. In the cloud paradigm, the data is stored on multiple servers, and although cloud has the property that the data may be accessed by the end users from anywhere, it still has issues of latency. Fog and edge computing paradigms bring the data computation and analysis closer to the components of the system where the control actions based on data analysis will be acted upon. The two paradigms of fog and edge computing are closely interrelated. Edge computing means that the data processing is done at the edge of the network, where most data is generated to increase efficiency in applications requiring faster responses. Fog computing, on the other hand, extends the cloud by placing small servers strategically between the edge and the cloud so that some data can be processed at these small servers if needed. Edge computing requires better edge resources as compared to fog computing. Both fog and edge computing paradigms bring the data processing and analysis closer to the *data acquisition* layer that hosts the many sensors of a system. Fog and edge computing paradigms are essential because they facilitate real-time data-driven decision-making. Most of the complex systems such as industry are inevitably multi-criticality systems, where some state changes will need immediate attention, whereas other analysis-based control actions are going to be the best effort. Therefore, hybrid solutions combining cloud and fog/edge will be needed for such multi-criticality systems.

3.4 PROPOSED SAFETY-CRITICAL ARCHITECTURAL VIEW FOR IIoT

Bohm (2018) recommends the developers to understand the industrial edge. Younan (2020) puts forward an architecture for futuristic IoT. Lee (2015) presents a five-layered architecture for cyber-physical systems. Based on such recommendations, an IIoT architecture may be visualized as in Figure 3.3.

3.4.1 DATA ACQUISITION LAYER

The data acquisition layer is composed of various sensors: S_1 ... S_n. These sensors measure their respective measurands and pass the measured data to the data acquisition layer. It would be desirable to develop drivers for these sensors that capture both the aspects of self-awareness and smartness. In that way, the overall reliability of the acquired data would increase since a self-aware system will be able to qualify the quality of its measurement.

3.4.2 DATA PREPROCESSING LAYER

The data preprocessing layer is the layer that verifies whether the data received by the acquisition layer is *good* or not. By good data we mean that the data is clean and complete and can be used by the immediate upper layer to generate information. It would be desired to have data preprocessing as close to the acquisition as possible. Therefore, Figure 3.3 presents both the data acquisition layer and the data preprocessing layer as one collective. This responsibility of data preprocessing would

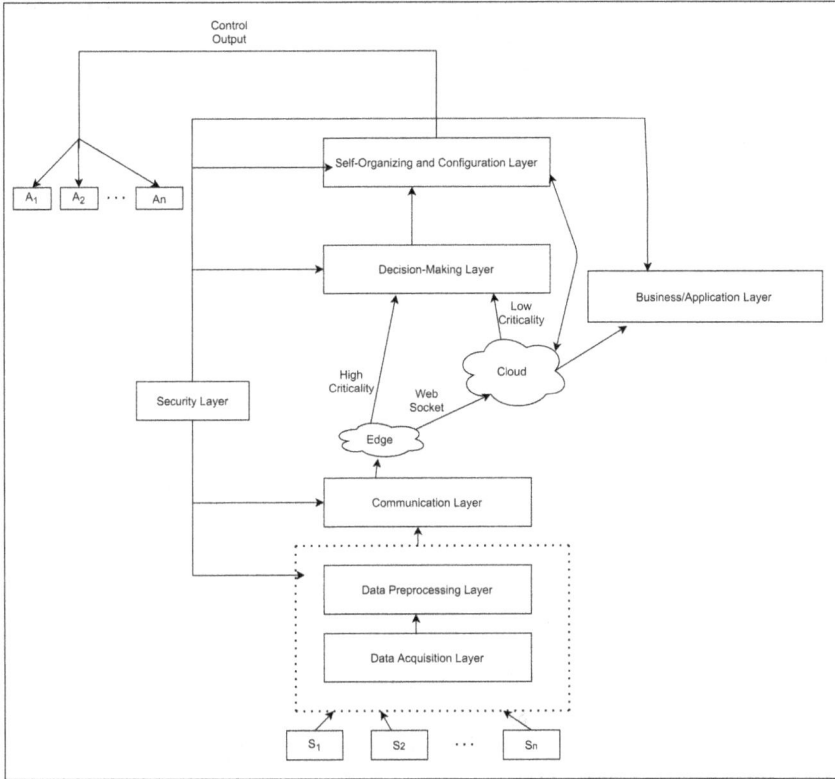

FIGURE 3.3   Safety-critical architecture for IIoT.

ideally lie with the autonomous sensor unit that would consist of the sensing technology as well as the preprocessing capability. However, the smarter and self-aware sensors in this design with added preprocessing capabilities are still expected to be small form factor machines.

There are two purposes that are being addressed using this architecture in industry. The first purpose being addressed is the acquisition and processing of data with respect to the controlled object, which, apart from the diagnosis of failure in the physical world, can also provide insights for optimizing business-related processes. The other aspect is the monitoring of self in these systems to prevent misinterpretation of an actual disaster condition. This layer may also provide separation for these two types of purposes.

Data that is obtained from various sensors from different vendors may also be heterogeneous in structure. The data preprocessing layer must also provide a standard system-readable structure to this heterogeneous data.

3.4.3   Communication Layer

The communication layer provides the paradigms necessary to transfer the preprocessed good data to the upper decision-making layer. Bohm (2018) recommends the

Modbus protocol for the communication between the data acquisition and preprocessing layer and the edge. Modbus is the de-facto standard in edge communication facilitating serial communication between Modbus server and Modbus clients. There are many versions of Modbus protocol facilitating TCP, UDP, and RTU (Remote Terminal Unit) communication, among others.

3.4.4   DECISION-MAKING LAYER

This layer sees a bifurcation. One aspect of the bifurcation aims at self-monitoring and failure prevention. The other (business/application layer discussed next) aims at creating applications to get better insights into the process being monitored and using those insights to optimize workflow. In the proposed design, the decision-making layer is the layer that takes in the data and processes it to generate useful information that can then be used to take an appropriate decision in case a system or component failure is going to occur with a higher probability. The purpose of this layer is to train, test, and validate models so that they can be applied to the received data and some useful predictions may be made. Determining what prediction models are better suited for what kind of an environment is something that would need careful attention. Whether those models are created by experienced data scientists for generalized cases or a new model must be created to better suit the needs of a particular environment should be subject to careful domain analysis. The decision-making aspect of an industrial multi-criticality system necessitates that the critical decisions made for failure management in the more safety-critical aspects of such a system be done as close to the smart components as possible and as quickly as possible. This would require data processing and analysis at the edge to assist in making quick decisions that mitigate failures for the controlled object. Other decisions of decreasing criticality or the training of the models using Big Data Analytics may be disposed to the cloud. Therefore, the decision-making layer in Figure 3.3 interacts with both the edge and the cloud.

3.4.5   BUSINESS/APPLICATION LAYER

There is another reason for data collection in the industry and that is its usage for increased revenue generation. The data collected is used to gain insights from it using various statistical methods and to use those insights to optimize the industrial workflow and create better and far-reaching opportunities. This aspect of the generation of insightful knowledge from the data collected for the purposes of increasing efficiency and providing useful applications that improve the industrial workflow processes is taken care of by this layer. The industry may interact with the trusted third-party business ecosystem using this layer.

3.4.6   SELF-ORGANIZING AND CONFIGURATION LAYER

This is the layer that maintains the rules and regulations against which the decision made by the decision-making layer must be analyzed and, if necessary, be acted upon. This layer sees the whole picture and has a complete understanding of the entire system. The layer proactively interacts with the knowledge base on the cloud. The mechanisms and paradigms that are built into this layer make informed choices on when an automated action must be taken, who must be informed at what time for assistive help, and so on. The newer knowledge and insights gained through the whole fault management process flow may be stored back in the knowledge base at the cloud for future reference. It is this layer that should also have the mechanisms in place for when a sub-component failure occurs and revert to a fail-operational sub-routine (discussed later).

3.4.7 SECURITY LAYER

Although security would be one of the most important aspects of the IIoT Architecture, and some or another form of it must be implemented in each of the layers to maintain the confidentiality, integrity, and authenticity (CIA) triad, we have chosen to keep this layer as abstract. The central idea is that the security layer must safeguard all communication that occurs between the layers and the outside internet.

3.5 IMPROVING FAULT TOLERANCE IN THE SAFETY-CRITICAL IIoT ARCHITECTURE

Recall the example from the Introduction where a carrier pipe was carrying some toxic liquid in some industrial plant. Such a system will definitely have state-of-the-art sensors and sophisticated computer programs that carefully analyze the received data. Although the main purpose of the real-time monitoring system set up here is to prevent a fatal accident from happening, this problem quickly reduces to the problem of the reliability of the monitoring system. Obviously, there are many benefits to letting a *reliable* computerized system do the necessary vigilance. As automated computerized monitoring systems come into action, the value of human vigilance will decrease. This will happen because the chances that a program running in an infinite loop, day and night, and 24/7, being able to detect an aberration in the controlled object is much higher than those with a human being. Also, the response time to act on such an aberration is also magnitudes of values faster than the response time for such a measure by a human being. Even though the initial system design may demand a collaboration of both the human and the computer, as the technology evolves, it would be cost intensive to maintain the mechanisms needed for human interference and the need for human interference may actually prove to be potentially dangerous. Rather, it would make more economical sense to invest in the evolution of this computerized system, maintain appropriate redundancy, and develop better failure detection and management mechanisms.

An important distinction when considering the monitoring of the control object in the design of the real-time systems and applications for industrial disaster prevention is to classify the class of industrial disaster prevention systems as either fail-safe or fail-operational systems.

3.5.1 FAIL-SAFE VS. FAIL OPERATIONAL SYSTEMS

If we were content in reaching a safe state for the controlled object when the system monitoring the controlled object detects a failure in any of its sub-components or a software glitch, then the class of systems that our monitoring and management system belongs to are fail safe. Upon complete system failure an external program called watchdog (discussed later) may bring the controlled object into the safe state. A safe state (in an identified system failure), for instance, for a high-temperature nontoxic liquid flowing through a pipe, would be to close the barrier to contain most liquid. Clearly, a fail-safe condition in such a system is not a bad design choice if the rate of system failures is low. However, there are systems where it is not possible to identify a safe state for the controlled object. If the system monitoring or managing such a controlled object fails, it will lead to a disaster. Such a class of monitoring and management systems is referred to as fail operational. A classic example of a fail-operational system is the fly-by-wire system introduced into passenger aircraft. In fly-by-wire systems, even the inputs initiated by the flying pilots must comply with the "environmental awareness" of the aircraft for them to be executed. A decision made by a pilot may be vetoed against if the software program running on the fly-by-wire system determines that such a control action may lead to harm. Clearly, a complete system failure in such a system will be catastrophic. A minimum functionality is absolutely necessary for a fail-operational system. Considering the central theme of this chapter, the systems monitoring the flow and release of toxic substances in an industrial plant must fall under the category of "somewhat" fail-operational systems. We may not find ourselves comfortable in just putting the controlled object in a safe state and then allow the system to fail. Such a failure would demand a quick and necessary intervention. There must be processes that occur in the after to ensure a safe recourse in such a scenario.

A fail-operational systems design is presented in Figure 3.4. A larger and complex real-time monitoring system (consisting of many sub-systems and complex protocols) monitors the controlled object through sensors and actuators. Let us call this system A. If a sub-component failure is observed in A, then A relinquishes its control to B (discussed later). A should have the self-awareness to notify the external systems that are responsible for the management of the failures in A so that those

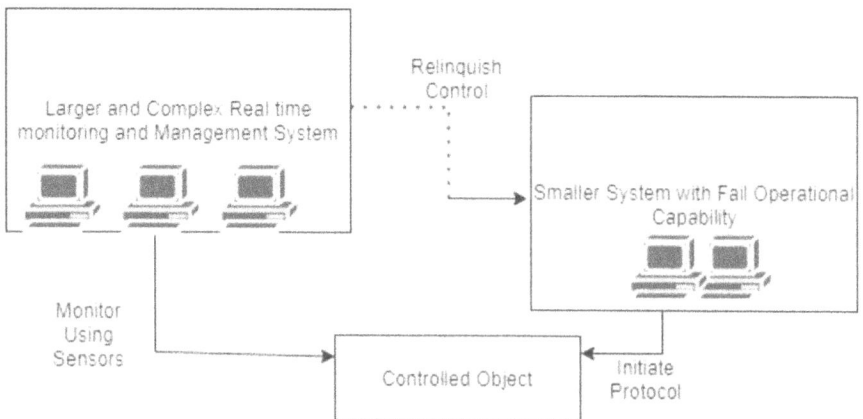

FIGURE 3.4   Fail-operational systems design.

external systems can initiate the maintenance protocols. If a complete system failure occurs and A goes down, an external mechanism (Watchdog) forces the relinquishment of control from A to B. In both cases, a smaller and basic system B with minimal functionality takes over. B must at least keep the basic functions relating to the application available so that the system does not fail completely.

3.5.2 WATCHDOG

A real-time system may employ a software process that is external to the monitoring system designated solely to monitor the lifeline of the real-time system. This external process is called the *watchdog*. A periodic ping may serve the purpose of communicating to the watchdog and detecting that the watched system is operating properly and has not succumbed to a total system failure. If the ping is not received within a pre-determined time interval, a failure may be presumed and the controlled object may be brought into the safe state. In the context of a fail-operational system, for the example discussed throughout the chapter, the watchdog program should initiate a post-routine that, after having forced relinquishment of control from the failed system to the minimalist system, also initiates maintenance protocols for the failed system. An example of a minimalistic functionality for an industry dealing with hazardous substances may just be to send a critical notification to an appropriate authority seeking loss minimization. An external watchdog program is presented in Figure 3.5.

In Figure 3.5, an external watchdog program monitors the operational status of a real-time monitoring and management system designated with the task of monitoring the controlled object. If, after unsuccessful pings to the monitoring and management system, the watchdog program has decided that the system has failed, then the watchdog program signals immediately for bringing the controlled object into safe state.

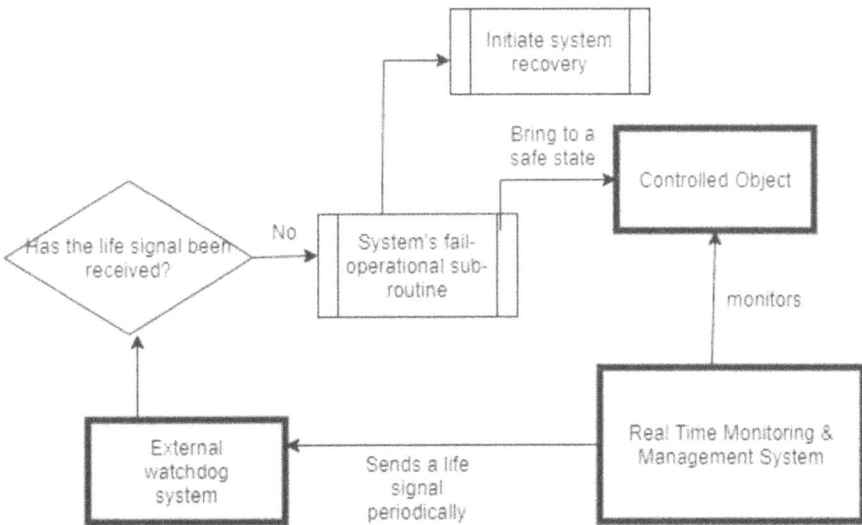

FIGURE 3.5 Watchdog program.

However, considering that we want "somewhat" fail-operational capability for our example, an additional responsibility has been placed on the watchdog. The watchdog program must initiate a post-safe-state sub-routine that initiates investigation and maintenance protocols for the failure in the main system and ensures the successful recovery of the main system.

3.5.3 CASCADING FAILURES

There is one key point that is absolutely essential to the success of fail-operational systems. There should absolutely be no cascading failures. If the larger and more complex real-time monitoring and management system in our discussion experiences a sub-component or system failure, the error should not cascade to the basic minimalist sub-system. Utmost care must be taken to ensure that such an abstraction exists between the two systems.

3.6 COGNITIVE ANALYTICS AND KNOWLEDGE GENERATION

Since the monitoring and management system comprises self-aware and smart sensors and actuators it is expected to work in a harsh industrial environment and be capable of or assist in self-organization, configuration and healing, etc. This system should be able to constructively contribute to the overall safety outlook of the industrial workflow and provide the confidence of reliability by being self-aware about failures and by seeking recovery whenever possible. As these systems get more and more advanced, the autonomic management for these self-aware and smart sensors and actuators may be cognitive enough to partake in advanced reasoning powered by their historical experiences as stored in the knowledge base (local or global based on need and implementation) and their current situational awareness. The quickness and the merit of the recourse that is sought in case of a failure in the system determine the success of the monitoring and management system. One of the central concepts in this cognition is that the probability that the system will diagnose self-failure given that a failure has occurred must be close to 1. This central concept is best explained in terms of conditional probability.

3.6.1 CONDITIONAL PROBABILITY

Let us briefly review the concept of conditional probability. Consider the following definitions:

3.6.1.1 Definition 1

A **sample space (S)** is a set of all possible outcomes of an experiment.

3.6.1.2 Definition 2

An **event (A)** is a subset of the sample space S: $A \subseteq S$. If P represents probability, then $P(A)$ is the probability of the occurrence of event A.

In this section, we would like to introduce the concept of conditional sample space. Consider a sample space (S_1) of tossing a fair coin three times. The sample space S_1 therefore is {HHH, HHT, HTH, HTT, THH, THT, TTH, TTT}. However, this is not a sample space that exhibits conditional behavior. In this sample space S_1, the tossing of coins is mutually exclusive and the outcomes are independent of each other. Let us introduce a conditional attribute to our sample space. Consider the following statement: "Throw a fair coin three times. If the first coin toss outcome is a head, then and only then is the coin tossed a second time. Similarly, if the second coin toss outcome is a head, then and only then is the coin tossed a third time." The resulting sample space S_2 is {T, HT, HHH, HHT}. The sample space S_2, as opposed to S_1, shows conditional behavior.

Now, consider a query on the conditional sample space S_2: "What is the probability that a total of three coin tosses are observed given that the first coin toss outcome was a head?" The relevant equation is given as (3.2).

$$P(x|y = y*) = P(x, y = y*)/P(y = y*) \qquad (3.2)$$

| is the symbol that is read as "given that." The formula read as a statement would be: "The probability of the occurrence of event x given that the event y as $y*$ has already occurred is the probability of the occurrence of the event x in conjunction with the event y as $y*$ divided by the probability of the occurrence of the event y as $y*$."

A cursory look at the sample space S_2 reveals that there are three events where the first coin toss outcome is a head, of which there are two events where a total of three coin tosses happened. Therefore, the probability of the occurrence of our conditional query is 2/3. Now, let us verify this pre-analysis by using the formula given by (3.2). Clearly, from the sample space, the probability of three outcomes being observed and the first outcome being a head is 2/4 = 1/2. This is the $P(x, y = y*)$ (numerator) in the right-hand side (RHS) of (3.2), where x denotes the event of three outcomes being observed and $y = y*$ denotes the event that the first outcome has been observed and it is a head. Also, the probability of the first outcome being a head is 3/4. This translates to $P(y = y*)$ (denominator) in the RHS of (3.2). Therefore, the required conditional probability is (1/2)/(3/4) = (2/3) as expected.

There may be types of failures that would form the starting point for a series of cascading failures in fail-safe and fail-operational systems. Following such types of failures would make more sense as opposed to following all possible combinations of failures.

3.6.2 LAW OF TOTAL PROBABILITY

The most important law when considering conditional probabilities is the law of total probability. A brief overview of this law is provided next. Consider a finite sample space S. We partition the sample space S into events $A_1...A_n$ as shown in Figure 3.6 such that:

1. The events do not overlap. Mathematically, $(A_i \cap A_j = \emptyset)$, where $1 \le i < j \le n$, $A_i \in S$, $A_j \in S$.
2. All the events add up to the sample space S, i.e. $U (A_i) = S$, where $1 \le i \le n$, $A_i \in S$.

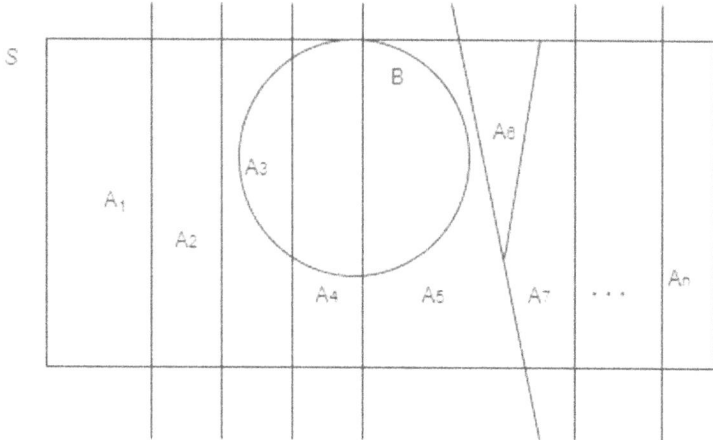

FIGURE 3.6   Partition of sample space S and the event B.

Suppose we want to find the probability that an event B has occurred. Here, $P(B) = \sum P(B \cap A_i)$, where $1 \le i \le n$,

$$\Rightarrow P(B) = \sum P(B|A_i) \times P(A_i), \text{ where } 1 \le i \le n \qquad (3.3)$$

Equation (3.3) is the law of total probability. Stated in words, the law of total probability tells us that the probability of an event B occurring within a sample space partitioned into n, "A_i," parts is the sum total of the products of the probability that B could occur given that some event A_i has occurred times the probability that the event A_i could occur.

3.6.3 EXTENDED CONDITIONAL PROBABILITY

There may be two or more events that may lead to a singleton event occurring. We may be interested in such a probability too. Let A be the event that we want to calculate the probability of occurrence given that events B and C have already occurred. This is given by (3.4).

$$P(A|B \cap C) = P(A \cap B \cap C)/P(B \cap C) \qquad (3.4)$$

B and C may have occurred independently of each other. However, there may be more complicated cases to consider. B and C may not be conditionally independent. If B and C are conditionally dependent, then we must consider the case where A occurs given (B occurs given C has occurred) has occurred, or A occurs given (C occurs given B has occurred) has occurred.

Therefore, either B can occur independently or it can occur as a consequence of being conditionally dependent on C and vice versa. There may be many variables that exhibit conditional dependencies among themselves and if so, they must carefully be taken into account.

3.6.4 FAULT MANAGEMENT

Now that we have reviewed the concept of conditional probability, we may be able to better appreciate the analysis that could assist in the understanding and prevention of failures. There may be two ways to look at it: top to bottom or bottom to top.

3.6.4.1 Top to Bottom Failure Identification

A failure event A has occurred. In the top to bottom failure identification, we must look at this event A and try to figure out the combination of events $B_1, B_2, ..., B_n, n \geq 0$ that could have led to the occurrence of A. We go down the path and stop only when we discover an atomic failure localized to one sub-component. We can use logical operators to express the combination of sub-component faults that lead to the said failure. Consider a fault expression given by (3.5).

$$\left(\left(D \wedge E \to B\right) \wedge \left(F \vee G \to C\right)\right) \to A \qquad (3.5)$$

Equation (3.5) specifies the if-then relationship. It also maintains independence of the occurrence of events on the RHS of the conditional. For example, if the events D and E happen, then B must happen; however, B may happen even though D and E have not happened. If that case has arisen where B has occurred independently and without the occurrence of D and E, then we may be lacking in the full understanding of the failure sub-structure and further analysis may need to be done, and the knowledge base may be updated appropriately.

3.6.4.2 Bottom to Top Failure Identification

This is the technique in which the weaker sub-components of a system design are recognized and an analysis is done to figure out when these weaker components would fail and how this failure will cascade upwards. This technique would require experience and intuition as to what can go wrong. A log can also be maintained to standardize the process of fault diagnosis and management.

3.6.4.3 Simulation-Based Fault Analysis

One of the ways to ensure the reliability of a safety-critical system is to create computerized models and to run simulation-based analysis. Software replicas of the various sub-components of a system that try to capture all real-world interactions are created: the interactions between sub-system components and the interactions of the system with the outer environment. Insights gained from simulating these interactions can be extremely useful. For example, consider a model that uses the concepts of physics and material science to create the idea of wear and tear in a component. Simulations that are run for the models of each sub-component of a system can provide us insights into which sub-component may be more vulnerable and may need maintenance and repair and at what point in time. For instance, if we were to carry a corrosive liquid through a carrier pipe, it would be good to know that the carrier pipe was made of a compound that the liquid could not easily corrode. Similarly, physical damages can happen to the sensors that are deployed in harsh industrial environments because the materials that the sensors are physically made of are not resistant to the corrosive exhausts from the industry. For extremely toxic environments, sensors may be engineered with more resistive elements to increase their lifetime. Otherwise, they may be replaced in a regular preventive and maintenance fashion using the insights gained from the simulation-based analysis. Evidently, such simulations are cost effective as no components are physically damaged, and as we have determined, they are heavily insightful. Multiple "what-if" scenarios can be analyzed using such simulations. The data received in real time and the knowledge generated over time from the real-time monitoring and management systems could be fed as simulation parameters to reflect the nature of the physical objects at play accurately. As symbiosis, the insights gained from the simulation model can then be used to engineer better systems for the task at hand.

3.6.4.4   A Conditional Probability Knowledge Database for Fault Management in IIoT

Using simulation-based analysis and data gained over time in the real world, a conditional probability-based knowledge database can be generated and maintained that gives quantitative values signifying what condition or a combination of conditions could lead to an occurrence of failure. An instance of such a probabilistic query for a system S could be to find the probability of failure of S at some time t in the nearest future given that some n numbers of conditions are true for S. An alert may be raised at the time when such a query evaluates to true. Some of the systems may display tendencies for spontaneous failure. In other systems, however, a careful conditional probability-based approach can assist in changing the policy related to fault management from post-failure to preventive failure management. The top to bottom and bottom to top failure identification processes can both benefit from such a knowledge database. Of course, with such a vast database, care must be taken to develop knowledge generation and querying models that do not get completely bogged down by unnecessary processing and that they have a reasonable response time.

3.7   SUMMARY

IIoT brings with it a host of opportunities and challenges. On the one hand, it offers solutions to problems that plague industries, while on the other hand, it requires a radical shift in how the industry operates. There are situations in the industrial workflow where a well-engineered automated system will be a better disaster manager than the best of humans. One such situation, as discussed in this chapter, is that of monitoring and managing an industrial ecosystem that deals with toxic substances and minimizes risks by developing systems that are reliable and robust. There are efforts going on both in academia and industry to realize such infrastructure. As we move forward in the direction of real-time monitoring of operational technology in the industries, it would be reasonably expected from the industrial world to embrace and encourage the development of such systems. These systems, when integrated into industries, will help safeguard and value human life in the industrial ecosystem above all.

REFERENCES

Blitzstein, J. K., & Hwang, J. (2019). *Introduction to Probability*. Boca Raton, FL: CRC Press.

Bohm, R. (2018). *Industrial Internet of Things for Developers*. New Jersey, USA: Wiley.

Broughton, E. (2005). The Bhopal disaster and its aftermath: A review. Environmental Health, 4(1), 1–6.

Cardis, E., Howe, G., Ron, E., Bebeshko, V., Bogdanova, T., Bouville, A., … & Drozdovitch, V. (2006). Cancer consequences of the Chernobyl accident: 20 years on. Journal of Radiological Protection, 26(2), 127.

Chandra, A. (2019). I, Robot: The impact of automation on heavy industry worker safety in India. IJSSER, 4(3), 2302–2310.

DGFASLI. (2020). List of Industries involving hazardous processes. Retrieved from https://dgfasli.gov.in/book-page/list-industries-involving-hazardous-processes

Evans, P. C., & Annunziata, M. (2012). Industrial internet: Pushing the boundaries. General Electric Reports, 488–508.

Lee, J., Bagheri, B., & Kao, H. A. (2015). A cyber-physical systems architecture for industry 4.0-based manufacturing systems. Manufacturing Letters, 3, 18–23.

Kopetz, H. (2011). *Real-Time Systems: Design Principles for Distributed Embedded Applications*. New York: Springer Science & Business Media.

Srinivasan, R., & Somasundaram, S. (2020). O&G's Operational Technology to Enter Digital Domain with Automated Systems, Robots. Wipro. Retrieved from https://www.wipro.com/en-IN/oil-and-gas/o-g_s-operational-technology-to-enter-digital-domain-with-automa/

White, T. (2012). *Hadoop: The Definitive Guide*. Sebastopol, CA: O'Reilly Media, Inc.

Wikipedia. (2020). Vishakhapatnam Gas Leak. Retrieved from https://en.wikipedia.org/wiki/Visakhapatnam_gas_leak

Younan, M., Houssein, E. H., Elhoseny, M., & Ali, A. A. (2020). Challenges and recommended technologies for the industrial internet of things: A comprehensive review. Measurement, 151, 107198.

4 Parkinson Disease Prediction and Drug Personalization Using Machine Learning Techniques

M. S. Hema
Anurag University
Hyderabad, India

K. Meena
Vel Tech Rangarajan Dr Sagunthala R&D
Institute of Science and Technology
Morai, India

R. Maheshprabhu
Aurora's Scientific and Technological Institute
Hyderabad, India

M. Nageswara Guptha
Sri Venkateshwara College of Engineering
Bengaluru, India

G. Prema Arokia Mary
Kumaraguru College of Technology
Coimbatore, India

CONTENTS

DOI: 10.1201/9781003145004-4

4.1   INTRODUCTION

4.1.1   INTRODUCTION TO PARKINSON DISEASE

Parkinson disease (PD) is a neurodegenerative disease that occurs due to insufficient level of dopamine in the human brain. The human brain induces a chemical called dopamine, which is responsible for the motor activities of the human body. If the dopamine level is low, it will lead to PD. The common symptoms of PD are categorized as motor symptoms, secondary motor symptoms, and non-motor symptoms [1]. The motor symptoms are tremor, bradykinesia, rigidity, and postural inability. The secondary motor symptoms are micrographia, freezing, mask-like expression, and unwanted acceleration. The non-motor symptoms are mood disorder, sleep disorder, loss of sense of smell, constipation, and Lewy body. The cause of PD has not yet been identified. It may be due to genetic factors or environmental factors such as rural living, well water drinking, chemical exposure, and pesticides exposure. Then, there exists no definite procedure to diagnose PD. PD is diagnosed based on the symptoms, clinical trials, and number of laboratory tests. In spite of all these procedures, sometimes there is a chance of misdiagnosis that leads to inappropriate treatment. To mitigate the above issue, machine learning techniques are used to predict PD. This will aid the doctor in taking the decision regarding treatment. Also it helps for the personalization of medication.

4.1.2   OVERVIEW OF MACHINE LEARNING TECHNIQUES

Machine learning is a division of artificial intelligence that educates the computer system to self-learn and make decisions automatically. It builds the model based on the

Raw dataset input

Data Preprocessing

Feature Selection

Classification and Prediction

Prediction output

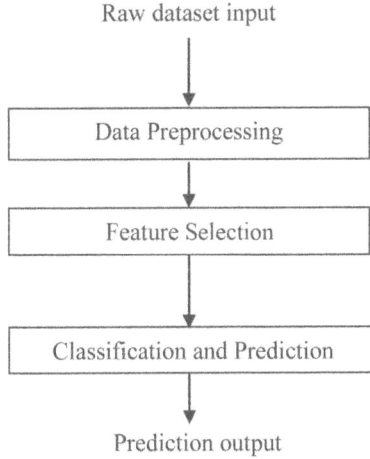

FIGURE 4.1   Steps in machine learning techniques.

data and expected output. Then, the decision of the future input will be obtained based on this model. Some of the applications of the machine learning techniques are bioinformatics, web search, finance, robotics, social media networks, and information extraction. Figure 4.1 shows the processing steps of machine learning techniques.

Machine learning techniques comprise the following processing steps:

1. Data preprocessing
2. Feature selection
3. Classification and prediction

4.1.2.1   Data Preprocessing

It is the first step of machine learning techniques. Raw data is collected from different data sources through the Internet of Things (IoT). The raw data has null values, incomplete values, incorrect values, and duplicate values. The preprocessing process converts raw data into ready-to-use data and enhances the data quality.

4.1.2.2   Feature Selection

All features available in the dataset are not required for classification and prediction. The extraction of required features from the original data is called feature selection. Some of the feature selection methods are the filter method, wrapper method, and embedded method.

4.1.2.3   Classification and Prediction

There are three types of machine learning techniques available for classification and prediction: (1) supervised learning, (2) unsupervised learning, and (3) reinforcement learning.

Supervised Learning
In a supervised learning algorithm, the input data includes desired outputs called label data. Supervised learning is classified into classification and

regression algorithms. In classification, the output or labeled data is in non-continuous form. Some of the classification algorithms are Naïve Bayes, support vector machine (SVM), K-nearest neighbor (K-NN), decision tree classification, and random forest classification. The labeled data is in continuous form in regression. The regression algorithms are further categorized into simple linear regression, multiple linear regression, polynomial linear regression, and ridge regression.

Unsupervised Learning

The input data does not include the desired output. The input data are grouped based on some rules. The commonly used clustering algorithms are k-means clustering, k-medoids clustering, hierarchical clustering, Gaussian mixture, neural networks, and hidden Markov model.

Reinforcement Learning

A reinforcement learning agent is activated to learn in an interactive environment. The agent will get rewards based on its own action. The feedback is collected from its agent's experiences and given to the next step for execution. The commonly used reinforcement algorithms are state–action–reward–state–action and Q-learning.

Nowadays, machine learning techniques are widely used in bioinformatics. They are used to classify the disease and classify the stages of the disease. The main contribution of the chapter is to predict PD and cluster Parkinson patients for drug recommendation.

4.2 LITERATURE SURVEY

Data preprocessing and feature selection methods are reviewed in this chapter. This chapter addressed the data preprocessing issues such as unknown feature values, discretization, and data normalization techniques. It also consolidated the feature selection methods such as filter methods and wrapper. The performances of these approaches are compared [1]. Data preprocessing techniques are proposed to improve the data quality. The authors mentioned that preprocessing is done to understand the problems in data and to enhance the quality of data. They have considered two applications for experimentation: semiconductor manufacturing and aerospace application [2]. A simple compression approach is proposed to detect the noise in training data. C4.5, k-NN, and CN2 algorithms are used to eliminate the noise from the training data. The experimental results showed that the accuracy of machine learning techniques is improved when using training data with noise detection [3]. A methodology was proposed to eliminate data inconsistencies. A Markov model was used to resolve the data inconsistencies [4]. Feature selection approaches were consolidated. Parameters such as simplicity, stability, number of reduced features, classification accuracy, storage, and computational requirements are considered to choose the feature selection algorithm [5]. The maximal information compression index clustering algorithm was proposed for feature selection, which is suitable for large datasets. The feature similarity was found using pairwise similarity instead of searching. The searching time and performance were improved. The entropy measure was used for handling redundancy reduction and information loss [6]. A correlation-based feature selection approach was proposed to select the features for machine learning techniques. Further, two extended versions of the correlation-based feature selection approach were used to select the features. The first extension was pairwise feature selection, and the second version was adding the weight for each feature. The weight was calculated using the RELIEF algorithm. The features with the weight approach gave better results when compared to the other two methods [7]. An information gain-based feature selection approach was proposed. This method eliminates similar and redundant features. The authors observed that ranking-based feature selection methods reduced the computational complexity when compared to subset-based feature selection. The information gain approach improved the classification accuracy [8]. Comparisons among various features selection approaches were presented. The four classification algorithms, namely IB1, Naïve Bayes, C4.5 decision tree, and RBF network, were used for classification. The rank-based feature selection algorithms gave better accuracy [9]. A novel clustering-based unsupervised feature selection approach was proposed for feature selection. The features were grouped under cluster labels and similar features were eliminated. The representation of samples, clustering, and feature selection was proposed as single framework [10]. A novel framework was proposed for feature selection and classification. The framework combined deep learning, feature selection, inference, and data analytics for Alzheimer's disease. The experimental results showed that the proposed approach improved the prediction accuracy [11]. The proposed approach compared the traditional feature selection methods with apriori feature selection. The LAASO, information gain, principal component analysis (PCA), ridge regression, and apriori feature selection methods were taken for comparison [12]. The rough set method was proposed for feature selection in pattern recognition. The feature projection and dimensionality reduction were made by using PCA [13]. An information entropy–based incremental rough set feature selection approach was proposed. With more features added to the dataset, this approach will select features with very less time [14]. A rough set approach was proposed to select the features in the mixed data. This approach is based on information entropy. Features were selected based on the entropy and wrapper method [15]. Fuzzy rough set combined with steady-state genetic algorithms was used for feature selection. The proposed methodology was used for both instance selection and data reduction.

The results showed that the feature selection approach improved the performance of the classifier [16]. A novel online streaming approach, called OFS-A3M feature selection, was proposed. The advantage of this approach is that it does not require any domain knowledge. It selects the feature with low redundancy, high correlation, and high dependence among the features [17]. CEBARKCC and quick reduct algorithm were proposed to select the feature set along with the ant lion optimizer. The proposed approach generates a pool of random solutions to select the features. Using this information, the feature set is selected from the original set [18]. The chi-square with PCA method was proposed for dimensionality reduction. The heart disease dataset was taken for experimentation. The proposed approach performed better than other methods [19]. An approach involving fuzzy PCA (FPCA), along with SVM, was proposed for feature selection. FPCA was used to select the first set of features and then the features were classified using SVM. After feature selection, the missing data was handled using the fuzzy c-means approach [20]. Filter-based univariate feature selection and PCA were proposed for feature selection and extraction of features, respectively [21]. The author presented two methods, namely PCA and t-statistics. PCA was used for dimensionality reduction, whereas t-statistics was used to extract the relevant features from the reduced dimensional features [22]. In the proposed approach, PCA and recursive feature elimination techniques were used for feature reduction. The random forest classifier was used for classification and the synthetic minority oversampling technique was used for balancing the dataset classes. The cervical cancer dataset was taken for experimentation [23]. The authors used the recursive feature elimination method to select the feature subset from the original dataset. It is a nonlinear feature selection method. A nonlinear classifier was used for classification [24]. Intrinsic mode function-based features were proposed to detect PD. The speech signal dataset was taken for classification. The features of the speech signal were illustrated using empirical mode decomposition. Intrinsic mode function cepstral coefficient was used to represent the speech characteristics [25]. A study was presented on how environmental factors cause PD. The environmental factors are well water drinking, pesticides exposure, chemical exposure, and rural living [26]. A multi-variant stacked autoencoder approach was proposed for PD detection. It has three layers, namely the input layer, output layer, and hidden layer. Here, the autoencoder was proposed to perform both encoding and decoding [27]. Automatic classification of tremor severity for Parkinson patients was proposed. The data were recorded from PD patients using wearable devices such as wristwatches. This wearable device has an accelerometer and a gyroscope for data recording. Decision tree, SVM, discriminant analysis, random forest, and K-nearest neighbor algorithms were used for classification [28]. PD prediction was proposed using machine learning techniques. SVM was used for PD prediction. The L1-norm SVM technique was used for feature selection [29]. A 3D deep convolution neural network was proposed for PD analysis [30]. A deep learning algorithm was proposed for PD diagnosis. The author explored various possibilities of deep learning used for PD analysis [31].

A deep learning approach was proposed to perform the multi-label classification of chronic disease. The proposed approach was used to identify the peculiarity of each feature and the impact of the feature on disease. Hypertension, diabetes, fatty liver, and healthy individuals' dataset was taken for experimentation. The performance of the proposed approach is compared with a traditional SVM algorithm's performance [32]. A comparative analysis of different classification algorithms was implemented using the PD dataset. Logistic regression with multi-class classification was found to give better performance when compared to other algorithms [33]. A novel approach was proposed to diagnose PD. The k-means clustering was used for feature selection. The k-means clustering selected the features based on the weights assigned to each feature. The artificial neural network approach was proposed for classification [34]. The authors developed a professional, community-based network to train PD. A number of physiotherapists are available in the network. The PD patients are clustered based on the severity of PD and other factors. They recommend therapizing for PD patients based on the clusters [35]. A method was proposed for future extraction and clustering of PD. Acceleration signal and surface electromyograms of Parkinson's patients were taken for experimentation. PCA was used for dimensionality reduction. The iterative k-means algorithm was used for clustering [36]. A clinical decision support system was developed to support individual patients' stimulation and medication of PD patients. Three machine learning algorithms, namely random forest, SVMs and Naïve Bayes, were used for implementation [37]. A methodology was proposed to personalize the medication for disease subgroups. The patients are divided into subgroups for precise medication [38]. A study was presented to find the patient similarity for personalization of medication for the patients. A total of 279 articles were taken for analysis. The analysis is done using data types used, findings, data analytics methods, and applications considered [39].

4.3 PROPOSED METHODOLOGY

The main objective of the proposed methodology is to classify and group PD patients. The outcome of the classification helps the doctor diagnose the PD patients accurately. The grouping or clustering of PD patients is used for drug personalization. The proposed methodology is shown in Figure 4.2.

The proposed methodology has five steps: (1) data collection, (2) data preparation, (3) dimensionality reduction, (4) classification, and (5) clustering. The data is taken from the PPMI dataset. During data preprocessing the null data is filled with the mean value of the attribute. After preprocessing the data undergoes dimensionality reduction. If the dataset has more features, then the following difficulties may occur:

• It becomes very difficult to differentiate between relevant and irrelevant features
• The processing time will be more
• It may degrade the performance of the classification algorithm
• It may lead to algorithm complexity.

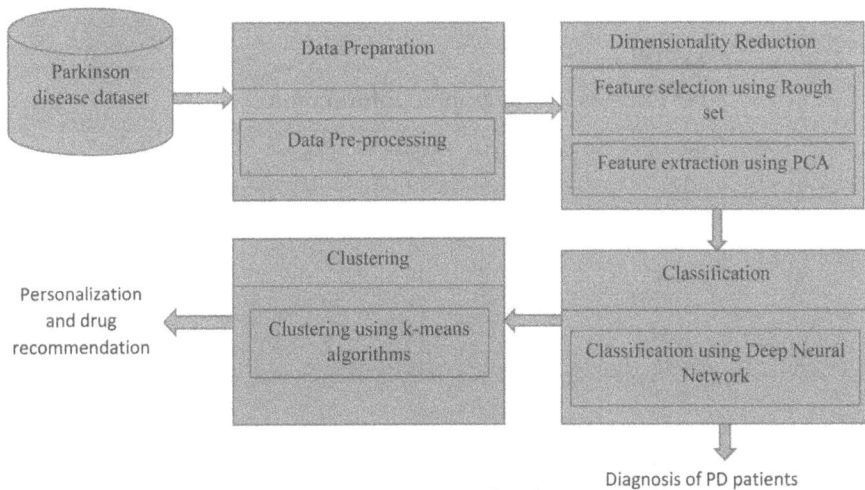

FIGURE 4.2 Structure of proposed methodology.

Dimensionality reduction is the solution for the above problems. The rough set algorithm is used for feature selection. PCA is used for feature extraction. After dimensionality reduction, the required features subset selected is given to the classification algorithm. The deep neural network is used for classification. After classification, the PD patients' output is given as input for the clustering algorithm to get groups. The k-means clustering algorithm is used for clustering. Finally, the k-means clustering algorithm gives the cluster.

4.3.1   FEATURE SELECTION USING ROUGH SET

4.3.1.1   Rough Set Introduction

Rough set is an approximation of the convention set (crisp set). It provides the lower and upper approximation of the original set. The rough set deals with uncertain data or information. The rough set has a close association with the knowledge discovery process, and it can be used in several ways for knowledge discovery, such as data reduction, feature selection, feature extraction, rule generation, and knowledge discovery.

Basic notations of the rough set are the following.

4.3.1.1.1   Decision Table

In an information system, the data is represented in the form of tables. It is represented by (O, A), where O is the set objects and A is the set of attributes. Both (O, A) are non-empty finite sets. An information system table that includes the class label is called a decision table. The decision table is a pair of (O, A U {C}) where C is the decision attribute. The PD patients' demographic information is shown in Table 4.1 as an example.

Note: The above example has 12 objects and 5 attributes. Here, "Class" is the decision attribute. In column Gender, 0 denotes male and 2 denotes female.

TABLE 4.1
PD Patients' Demographic Information

Pat. No.	Birth Year	Age	Gender	Class
3403	1941	79	2	1
3400	1971	49	0	2
3402	1964	56	2	1
3406	1975	45	2	1
3407	1945	75	2	2
3409	1947	73	2	1
3408	1972	48	2	1
3051	1939	81	2	1
3451	1956	64	2	1
3050	1960	60	0	2
3502	1941	79	2	1
3101	1962	58	2	1

4.3.1.1.2 Indiscernibility

A decision table has many objects that represent the same features. It makes the decision size big and complex. The response time may be increased. The reduction of duplicate objects in a decision table is called indiscernibility. The representation of indiscernibility is shown in equation 4.1.

$$IND(P) = \{(a,b) \in O^2 / \forall_{x \in A}, x(a) = x(b)\} \tag{4.1}$$

where IND(P) is indiscernibility of decision table, a and b are objects that are duplicates of each other by attributes from A. The indiscernibility is calculated for the data shown in Table 4.2.

Here the above example objects {A1, A5} have same values and objects {A2, A4, A7} also have same values. So, only representative objects will be taken. It will reduce the table size.

4.3.1.1.3 Approximation

The rough set has two types of approximations to find out the crisp set: lower approximation and upper approximation.

4.3.1.1.3.1 Lower Approximation

It includes the set of objects which are definitely classified as a member of target set T. It is denoted in equation 4.2.

$$\underline{R}(T) = \{B \in O/R : B \leq T\} \tag{4.2}$$

where $\underline{R}(T)$ is the lower approximation, B is a subset of objects from original set O.

TABLE 4.2
PD Patient's Data

Label	Birth Year	Age	Gender	Class
A1	1941	79	2	1
A2	1972	48	2	2
A3	1960	60	2	1
A4	1972	48	2	2
A5	1941	79	2	1
A6	1947	73	2	1
A7	1972	48	2	2
A8	1939	81	2	1
A9	1956	64	2	1
A10	1960	60	2	1

Note: IND(A) = ({A1, A5}, {A2, A4, A7}, {A3, A10}, {A6}, {A8}, {A9}).
In column Gender, 2 denotes female.

4.3.1.1.3.2 Upper Approximation Upper approximation includes a set of objects that are possibly classified as a member of target set T. It is shown in equation 4.3.

$$\bar{R}(T) = \cup\{B \in O/R : B \cap T \neq \varnothing\} \qquad (4.3)$$

where $\bar{R}(T)$ is the upper approximation.

The target set of the Table 4.1 is

$$T = (\{A1, A4, A5, A8, A9\}, \{A5, A3, A7, A10\})$$

$$R = \{r1, r2, r3, r4, r5\}$$

$$IND(A) = \{\{A1, A5\}, \{A2, A4, A7\}, \{A3, A10\}, \{A6\}, \{A8\}, \{A9\}\}$$

$$\underline{R}(T) = (\{A1, A5, A8, A9\}, \{A3, A5, A10\})$$

$$\bar{R}(T) = (\{A1, A2, A4, A5, A7, A8, A9\}, \{(A3, A2, A4, A5, A7, A10\})$$

4.3.1.1.4 Positive Region

The positive region comprises the definite elements in the target set. The positive region is calculated using the formula shown in equation 4.4.

$$Pos(T) = \cup\underline{R}(T) \qquad (4.4)$$

where Pos(T) is the positive region of target set T. The positive region of given example is

$$Pos(T) = \{A1, A3, A5, A8, A9, A10\}$$

4.3.1.1.5 Negative Region

The negative region comprises elements that are not in the target set. The positive region is calculated using the formula shown in equation 4.5.

$$Neg(T) = \cup - \cup\bar{R}(T) \qquad (4.5)$$

where Neg(T) is the negative region of the target set. The negative region of the given example is

$$Neg(T) = \{A6\}$$

4.3.1.1.6 Boundary Region

The boundary region comprises elements that may or may not be present in the target set. The formula to calculate the boundary region is shown in equation 4.6.

$$BR(T) = \bar{R}(T) - \underline{R}(T) \qquad (4.6)$$

where BR(T) is the boundary region of the target set. The boundary region of the given example is

$$BR(T) = \{A2, A4, A7\}$$

The target set T is a non-crisp set with respect to R because the boundary region is non-empty. So, target set T is called a rough set.

4.3.1.2 Feature Selection

In an information system, not all features are required for classification. The selection of required subset features from the original features is called feature selection. In rough set the required feature subset is called reduct. The attribute subset $\alpha \leq \beta$ then preserves the indiscernibility of β if the attribute α-β is dispensable. The rough set is able to generate a number of subsets from the original set. The minimal subset is called the reduct of the information system. The reduct should not contain any dispensable attributes.

Let α is a subset of β and x belongs to α.

x is dispensable in α for the condition shown in equation 4.7.

$$IND(\alpha) = IND(\alpha - \{x\}) \tag{4.7}$$

Otherwise x is indispensable in α.

If all the features of α are indispensable, then set α is independent.

Subset α of β is a reduct of α if IND(α)=IND(β) and α is independent.

4.3.2 FEATURE EXTRACTION USING PRINCIPAL COMPONENT ANALYSIS

PCA is a famous technique used for dimensionality reduction. It extracts the K number of relevant features from the original dataset and creates new independent features for analysis.

The steps involved in PCA are (1) standardization, (2) covariance matrix calculation, (3) computing eigenvector, (4) choosing top K features, and (5) reorientation of data.

4.3.2.1 Standardization of Data

In PCA, data standardization is a mandatory step. The data is standardized using the mean, variance, and standardized deviation of the data. Data scaling is performed. The formula used for data scaling is shown in equation 4.8.

$$S = \frac{y - \mu}{\sigma} \tag{4.8}$$

where S is the scaled value, y is the initial value, μ is the mean of the data, and σ is the standard deviation of the data. The data scaling makes the analysis equal.

4.3.2.2 Covariance Matrix Computation

Covariance is one of the measures and indicates the dependency among two variables. The covariance is computed using the formula shown in equation 4.9.

$$M = Cov(A,B) = \frac{1}{n-1}\sum_{i-1}^{n}\left(A_i - \bar{A}\right)\left(B_i - \bar{B}\right)$$ (4.9)

where A and B are two variables or features, n is the number of instances, and M is the covariance matrix.

4.3.2.3 Computing Eigenvector

Eigenvector is a vector that is used for linear transformation. It is a non-zero vector. The vector of matrix M is represented in equation 4.10.

$$M\vec{v} = \lambda\vec{v}$$ (4.10)

where λ is the eigenvalue. λ is used for linear transformation. The rewritten of the eigenvector is shown in equation 4.11.

$$\vec{v}\left(M - \lambda I\right) = 0$$ (4.11)

where I is the identity matrix.

The eigenvector will be calculated based on the eigenvalues.

4.3.2.4 Choosing Top-k Features

Sort the eigenvector values in decreasing order and top-K features are selected.

4.3.2.5 Reorientation of Data

Finally, the data are transformed from the original space into a new principal component subspace. It is represented in equation 4.12. Let T is the scaled data matrix.

$$Z = T' \times Y$$ (4.12)

where Z is the data orientation in new subspace, T' is transpose matrix of T, and Y is the feature vector.

Finally, the data points are projected into a new subspace using computed principal components.

4.3.3 Deep Learning

Deep learning is a subfield of machine learning techniques and is based on artificial neural networks. The artificial neural network is based on the neurons in the human brain.

Why deep learning?

- When the dataset is very large then deep learning outperforms when compared to other techniques.

- Deep learning performs well with very less domain understanding about the features.
- Deep learning outperforms in image classification, speech recognition, and other complex problems.

4.3.3.1 Introduction to Deep Learning

Deep learning has three layers, namely the input layer, hidden layer, and output layer. The neural network has only one hidden layer. But a deep neural network has more than one hidden layer. The deep neural network is shown in Figure 4.3.

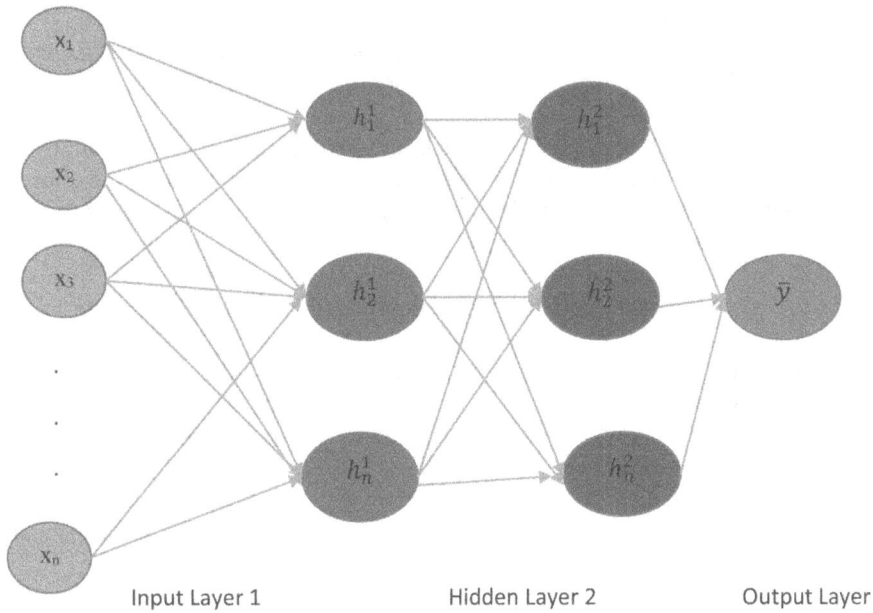

FIGURE 4.3 Deep neural network.

4.3.3.1.1 Learning Process of Deep Learning Network

The learning of deep neural networks is an iterative process. It has two steps: forward propagation and backward propagation.

4.3.3.1.1.1 Forward Propagation

Step 1: Set of inputs $\{x_1, x_2, x_3, ..., x_n\}$ and corresponding weights (w_1, w_2, $w_3...w_n$) are given as input to the input layer. It is represented in equation 4.13.

$$O = \sum_{i=1}^{n} w_i \times x_i \tag{4.13}$$

Step 2: The weights and inputs are multiplied and it is added to bias. It is represented in equation 4.14.

$$z_1 = O + bias \qquad (4.14)$$

Step 3: The above result is given the activation function of the first hidden layer. It is represented in equation 4.15.

$$h_1^1 = Activation\_function(z1) \qquad (4.15)$$

Step 4: The output of the first hidden layer is given to the second hidden layer. The input is given to the function of the second layer.
Step 5: The process is repeated until the output layer is reached. The output layer will give the final output of the learning of that iteration.

4.3.3.1.2 Backward Propagation Backward propagation is learning from the output layer to the input layer. It calculates the loss function to estimate the error of the iteration. The error is the deviation between the actual output gotten from the forward propagation and desired output.

Step 1: The error of the output layer is calculated, then the weight and bias are adjusted accordingly.
Step 2: The last hidden layer is calculated. Then, the weight and bias are adjusted accordingly.
Step 3: This is process is repeated layer by layer backward until it reaches the input layer.

The forward and backward processes are executed iteratively until the actual output is very close desired output.

4.3.3.1.3 Activation Functions
The activation function is used in every layer of the deep neural network. If the activation function is not used in the deep neural network, then it is simply a linear regression model. Types of activation Types of activation of functions are described below.

4.3.3.1.3.1 Binary Step Activation Function It is used in a threshold-based classifier. If the input is greater than a specified threshold, then the neurons are activated; otherwise, neurons are deactivated. The mathematical representation of the binary step function is shown in equation 4.16.

$$A(z) = \begin{cases} 1, & z \geq 0 \\ 0, & z < 0 \end{cases} \qquad (4.16)$$

where $A(z)$ is the activation function and z is the input. It is used for binary classification. The gradient value of this activation function is zero.

4.3.3.1.3.2 Linear Activation Function It is proportional to the input. Mathematically, it is as represented in equation 4.17.

$$A(z) = \partial z \tag{4.17}$$

where ∂ is constant. This activation function will improve the error because the gradient value is the same for every iteration.

4.3.3.1.3.3 Sigmoid Function It is a widely used function for binary classification. It is a nonlinear function. It is represented in equation 4.18.

$$A(z) = \frac{1}{1 + e^{-z}} \tag{4.18}$$

4.3.3.1.3.4 Tanh Function It is similar to the sigmoid function except that the Tanh function is symmetric in origin. It is represented in equation 4.19.

$$A(z) = \left(\frac{2}{1 - e^{-2z}} \right) - 1 \tag{4.19}$$

4.3.3.1.3.5 Rectified Linear Unit It is a nonlinear activation function. It is represented in equation 4.20. The main advantage of Rectified Linear Unit (ReLU) is that all the neurons are not activated at a time. The neurons are deactivated if the output of the linear transformation is less than zero.

$$A(z) = \max(0, z) \tag{4.20}$$

4.3.3.1.3.6 Leaky ReLU Leaky ReLU is an advanced version of ReLU. It defines an extremely small linear component of z if the linear transformation output is less than zero. It is represented in equation 4.21.

$$A(z) = \begin{cases} 0.01z, \ z < 0 \\ z, \ z > 0 \end{cases} \tag{4.21}$$

The sigmoid activation function is used for binary classification. Mostly, ReLU and leaky ReLU activation functions are used for hidden layers.

4.3.3.2 Derivatives in Deep Learning

Let

al[0] is input parameter of the input layer
al[1], al[2], al[3] ...., al[l] is activation functions of each layer
wl[1], wl[2], wl[3] ...., wl[l] is weights of each layer
bl[1], bl[2], bl[3], ... Bl[l] is bias of each layer
zl[1], zl[2], zl[3], ..., zl[l] is output of each layer

$\overline{y}$ is output

da1[1], da1[2], da1[3], ..., da1[l] is error of each layer
dw1[1], dw1[2], dw1[3], ————, dw1[l] is change in weights of each layer
da1[1], da1[2], da1[3], ..., da1[l] is change in bias of each layer
dz1[1], dz1[2], dz1[3], ..., dz1[l] is change in output of each layer

The forward and backward propagation processing methods are presented in Figure 4.4.

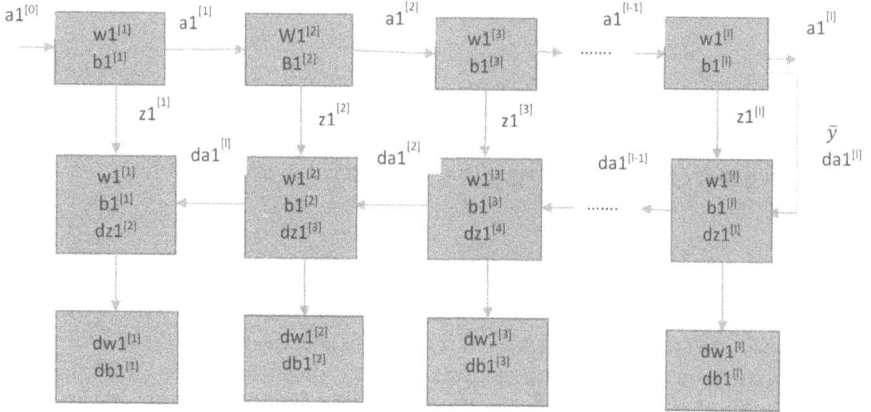

FIGURE 4.4 Processing of deep neural network.

4.3.3.2.1 Forward Propagation

The layer 1 output is calculated using the formula shown in equation 4.22.

$$z1^{[1]} = w1^{[1]} \times a1^{[0]} + b1^{[1]} \tag{4.22}$$

where $a1[0]$ is the actual input of the deep neural network.

The first hidden layer output is calculated using the formula shown in equation 4.23.

$$z1^{[2]} = w1^{[2]} \times a1^{[1]} + b1^{[2]} \tag{4.23}$$

$a1[1]$ is calculated using the formula shown in equation 4.24.

$$a1^{[1]} = g^{[1]}\left(z^{[1]}\right) \tag{4.24}$$

where $g[1]$ is the activation function of the layer.

In general, the layer 1 output is calculated using the formula shown in equation 4.25.

$$z1^{[l]} = w1^{[l]} \times a1^{[l-1]} + b1^{[l]} \tag{4.25}$$

$a1[l]$ is calculated using the formula shown in equation 4.26.

$$a1^{[l]} = g^{[l]}\left(z^{[l]}\right) \tag{4.26}$$

where $g[l]$ is the activation function of the output layer.

4.3.3.2.2 Backward Propagation

Backpropagation is used to calculate the error of each layer's output. It starts from the output layer.

The error of layer 1, the output layer, is calculated using the formula presented in equation 4.27.

$$dz1^{[l]} = da1^{[l]} \times g^{[l]\prime} \left(z^{[l]} \right) \tag{4.27}$$

The output layer difference in activation function calculated using the formula is shown in equation 4.28.

$$da1^{[l]} = w1^{[l+1]^T} \times dz1^{[l+1]} \tag{4.28}$$

The value of $da1[1]$ is taken directly. It is given in equation 4.29.

$$da1^{[l]} = \frac{-y}{a1} + \frac{(1-y)}{1-a1} \tag{4.29}$$

It is the first input of backpropagation, where y is the output of the layer and $a1$ is the activation function.

The difference is that the weight of layer 1 is calculated using the formula shown in equation 4.30.

$$dw1^{[l]} = dz1^{[l]} \times a^{[l-1]} \tag{4.30}$$

The difference in activation function in the first hidden layer is calculated using the formula shown in equation 4.31.

$$da1^{[l-1]} = w1^{[l]^T} \times dz1^{[l]} \tag{4.31}$$

The difference bias is shown in equation 4.32.

$$db1^{[l]} = dz1^{[l]} \tag{4.32}$$

The error is calculated for all the layers from the output layer to the input layer using the above formula.

After the completion of the first iteration the weight and bias modification is done using backpropagation values. The new weight calculation formula is shown in equation 4.33.

$$w[i] = w[i-1] - \partial dw[i-1] \tag{4.33}$$

where i represents the iteration number and ∂ learning rate.

The new bias is calculated using the formula shown in equation 4.34.

$$b[i] = b[i] - \partial db[i-1] \tag{4.34}$$

where i represents the iteration number and ∂ learning rate. This is called gradient descent. After the calculation of gradient descent, iteration is performed continuously until the actual output is close to the desired output.

4.3.3.3   k-Means Clustering

k-Means clustering is an unsupervised learning approach. It groups the unlabeled numeric data into groups. It is widely used for clustering because of its simplicity and efficiency. It is an iterative algorithm.

Steps in k-means clustering algorithm:

Choosing k-values: k is the number cluster or groups. The user needs to choose the number cluster or groups, i.e. their k value. If the user does not have an idea about the correct k value, then the k value will be selected based on the trial and error method.

Assigning data points to cluster: Each cluster has a random centroid. The nearest data points are assigned to each cluster. To identify the closeness of the data point and cluster centroid, the distance between data points and each cluster centroid is calculated. Then, the data points are assigned to a cluster that has a minimum distance between all the clusters. The distance is calculated using Euclidean distance. The formula of the Euclidean is shown in equation 4.35.

$$d(y_i, c_i) = \sqrt{\sum_{i=1}^{n} (y_i - c_i)^2} \tag{4.35}$$

where d is the distance between data points and centroid of the ith cluster, y_i is ith data point, and c_i is the centroid of the ith cluster.

The new centroid is calculated using the formula shown in equation 4.36.

$$c_i = \frac{1}{n_i} \sum_{j=1}^{n_i} y_j \tag{4.36}$$

where c_i is the centroid of the ith cluster, n_i is the number of data points of the ith cluster, and y_j is the jth data point of the ith cluster.

The first and second steps are repeated until no data point needs reassignment.

4.4   EXPERIMENTATION SETUP

The PPMI dataset is taken for experimentation. The demographics data, family history, motor symptoms, and non-motor symptoms data are taken for experimentation. The dataset has 72 attributes and 10,000 instances. There are two class labels in the data set, namely class 0 and class 1. Class 1 represents PD patients and class 0 represents healthy individuals. The dataset is divided into a training dataset and a testing dataset. Totally, 80% of the data is taken for the training set and 20% of the data is taken for the testing set. The hardware used for this experimentation is a personal laptop with 16 GB RAM and Windows 10 operating system. The performance matrices used to assess the performance of the proposed methodology are accuracy, specificity, and sensitivity.

Accuracy
The formula to calculate accuracy is shown in equation 4.37.

$$Accuracy = \frac{(TP + TN)}{(TP + TN + FP + FN)} \tag{4.37}$$

where TP is true positive, i.e. the PD patients are correctly predicted as PD patients. TN is true negative, i.e. the healthy individuals are correctly predicted as healthy individuals. FP is false positive, i.e. the healthy individuals are wrongly predicted as PD patients. FN is false negative, i.e. the PD patients are wrongly predicted as healthy individuals.

Sensitivity
The formula to calculate sensitivity is shown in equation 4.38.

$$Sensitivity = TP/(TP+TN) \qquad (4.38)$$

Specificity
The formula to calculate sensitivity is shown in equation 4.39.

$$Specificity = TN/(TP+TN) \qquad (4.39)$$

Precision
The formula to calculate precision is shown in equation 4.40.

$$Precision = TP/(TP+FP) \qquad (4.40)$$

4.4.1 PRINCIPAL COMPONENT ANALYSIS

The output of the PCA is shown in Figure 4.5. The output shows that component one is more important than other components. Add any one component from component two to component nine in addition to the component one.

4.4.2 DEEP LEARNING IMPLEMENTATION

The original dataset and reduced features dataset are taken for experimentation. R language is used for implementation. The Deepnet H2O package is used for the

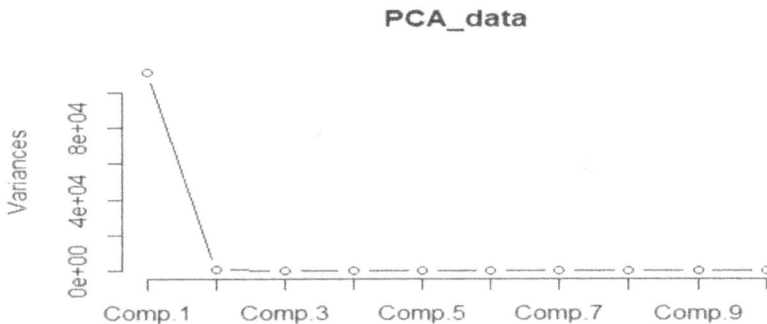

FIGURE 4.5 PCA for PD dataset.

implementation of a deep neural network in R. The sigmoid activation function is used in the output layer because the objective of the proposed methodology is to predict the presence of disease, i.e., binary classification. The ReLU activation function is used in hidden layers. In the first iteration, the weight is initialized as a non-zero random value and bias is initialized as zero. The learning rate is defined as a small positive number to minimize the cost. If the sigmoid function value is greater than 0.5, then they are classified as PD positive, otherwise healthy individuals. Figure 4.6 illustrates the deep neural network with the input, input layer, hidden layers, and output layer.

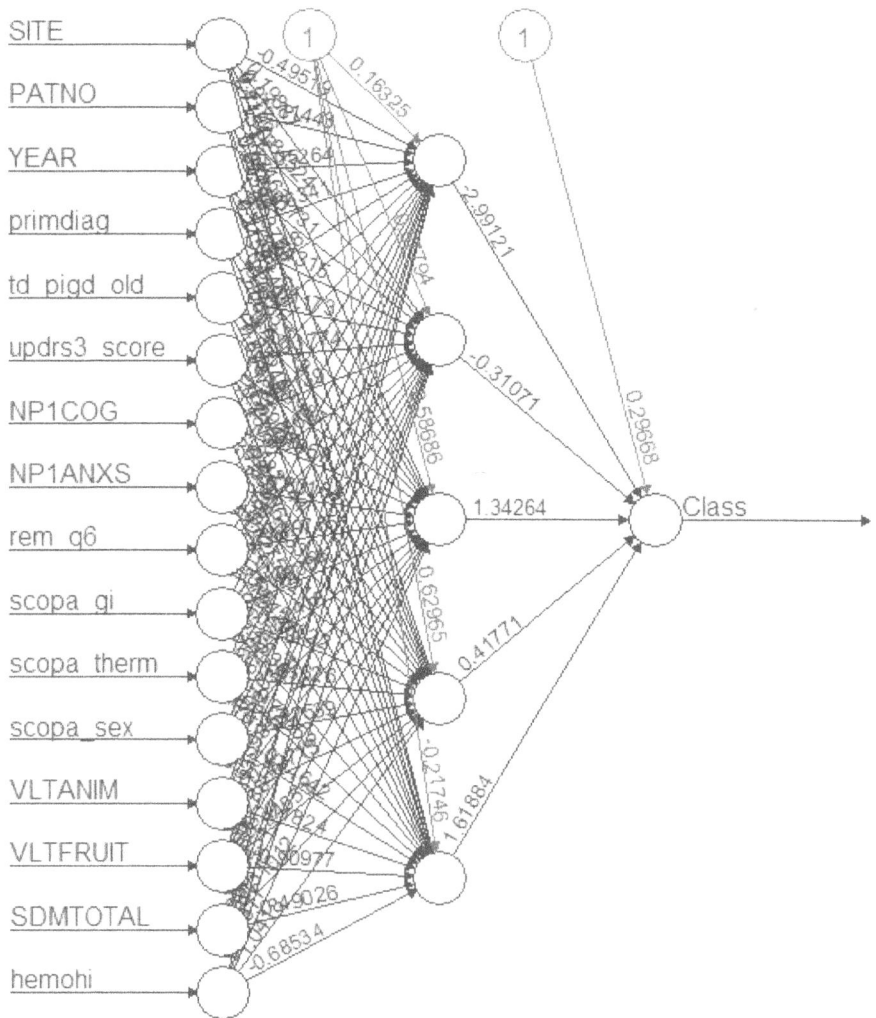

FIGURE 4.6   PD dataset prediction using deep neural network.

TABLE 4.3

Performance of the Classifiers

Classifier	Accuracy	Sensitivity	Specificity	Precision
Deep neural network with all features	0.994	0.9954	0.9914	0.9954
Deep neural network with dimensionality reduction	0.997	1.0000	0.9914	0.9954
Random forest with all features	0.96	0.97	0.87	0.94
Random forest with dimensionality reduction	0.95	0.98	0.88	0.95
SVM with all features	0.90	0.91	0.87	0.88
SVM with dimensionality reduction	0.88	0.90	0.87	0.88

4.4.3 CLASSIFICATION AND PREDICTION

The results of the deep neural network are compared with those of random forest and SVM classifiers. It is shown in Table 4.3.

Table 4.3 shows that the deep neural network with dimensionality reduction obtained 99.7% accuracy. It performs better when compared to the other two methodologies. The dimensionality reduction method improves the response time of the classifier. Doctors can use this deep neural work for decision making in addition to their clinical diagnosis.

4.4.4 CLUSTERING USING K-MEANS CLUSTERING ALGORITHM

The optimal number k value is determined by using elbow and silhouette methods. The output of the elbow and silhouette methods is shown in Figures 4.7 and 4.8, respectively.

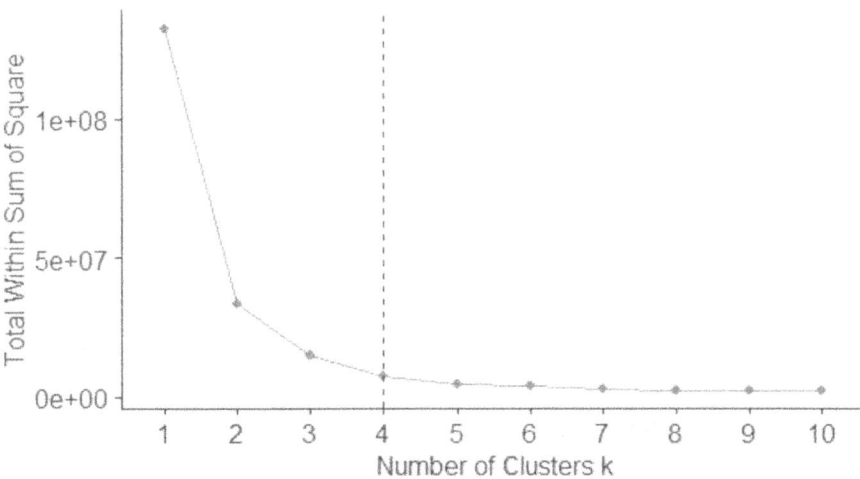

FIGURE 4.7 Number cluster selection using the elbow method.

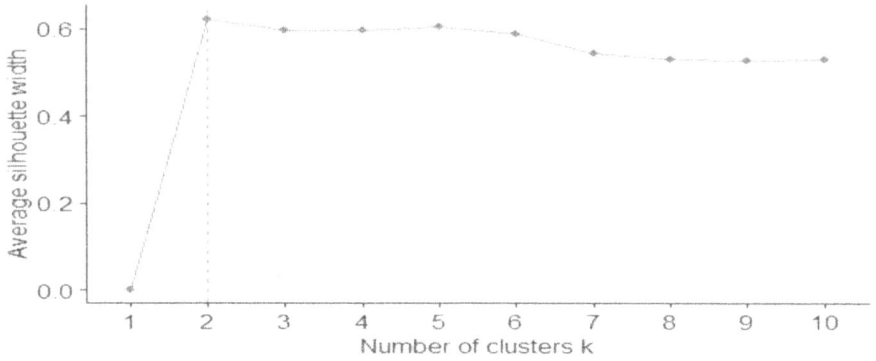

FIGURE 4.8   Number cluster selection using the silhouette method.

The results of the elbow and silhouette methods show that the optimal number of clusters is 4 and 2, respectively. The result of the clustering is presented in Figure 4.9.

The results show that either two clusterings or four clusterings give a clear output. Based on the clustering of the patients, the personalized drug can be recommended for each cluster of patients.

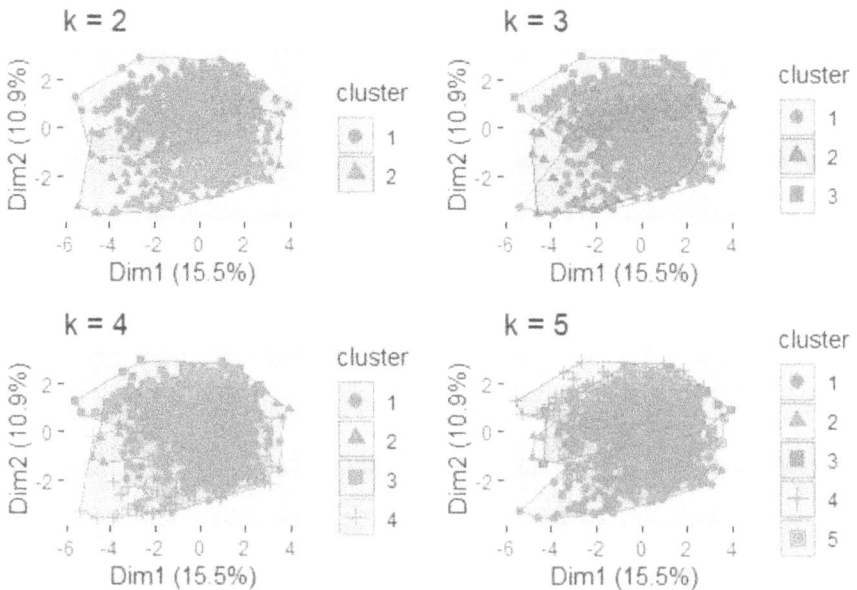

FIGURE 4.9   PD patients' clustering using a k-means clustering algorithm.

4.5 CONCLUSION

PD patients' prediction and clustering methodology are proposed and implemented. The prediction outcome helps the doctor to predict the onset of PD in patients accurately. The outcome of the clustering is used for personalized drug recommendations to PD patients. The rough set approach is used for feature selection and PCA is used for feature extraction. The deep neural network approach is implemented for the prediction of PD patients. The k-means clustering is used for the clustering of PD patients. Patients' demographic data, family history, motor symptoms, and non-motor symptoms data are taken for experimentation. The deep neural network outperforms random forest and SVM classifiers. In future, the causes of PD may be identified and real data from PD patients may be collected and experimented.

REFERENCES

1. Kotsiantis, S. B., Dimitris Kanellopoulos, and P. E. Pintelas. "Data preprocessing for supervised leaning." International Journal of Computer Science 1.2 (2006): 111–117.
2. Famili, A., et al. "Data preprocessing and intelligent data analysis." Intelligent Data Analysis 1.1 (1997): 3–23.
3. Gamberger, Dragan, Nada Lavrac, and Saso Dzeroski. "Noise detection and elimination in data preprocessing: experiments in medical domains." Applied Artificial Intelligence 14.2 (2000): 205–223.
4. Saranya, K., M. S. Hema, and S. Chandramathi. "Data fusion in ontology based data integration." International Conference on Information Communication and Embedded Systems (ICICES2014). IEEE, 2014.
5. Chandrashekar, Girish, and Ferat Sahin. "A survey on feature selection methods." Computers & Electrical Engineering 40.1 (2014): 16–28.
6. Mitra, Pabitra, C. A. Murthy, and Sankar K. Pal. "Unsupervised feature selection using feature similarity." IEEE Transactions on Pattern Analysis and Machine Intelligence 24.3 (2002): 301–312.
7. Hall, Mark Andrew. "Correlation-based feature selection for machine learning." (1999).
8. Win, Thee Zin, and Nang Saing Moon Kham. "Information gain measured feature selection to reduce high dimensional data." Seventeenth International Conference on Computer Applications. ICCA, 2019.
9. Novaković, Jasmina. "Toward optimal feature selection using ranking methods and classification algorithms." Yugoslav Journal of Operations Research 21.1 (2016).
10. Zhu, Pengfei, et al. "Subspace clustering guided unsupervised feature selection." Pattern Recognition 66 (2017): 364–374.
11. Liu, Yuanyuan, et al. "Deep feature selection and causal analysis of Alzheimer's disease." Frontiers in Neuroscience 13 (2019): 1198.
12. Kaushik, Shruti, et al. "Comparative analysis of features selection techniques for classification in healthcare." MLDM 2 (2019).
13. Swiniarski, Roman W., and Andrzej Skowron. "Rough set methods in feature selection and recognition." Pattern Recognition Letters 24.6 (2003): 833–849.
14. Liang, Jiye, et al. "A group incremental approach to feature selection applying rough set technique." IEEE Transactions on Knowledge and Data Engineering 26.2 (2012): 294–308.
15. Zhang, Xiao, et al. "Feature selection in mixed data: a method using a novel fuzzy rough set-based information entropy." Pattern Recognition 56 (2016): 1–15.

16. Derrac, Joaquín, et al. "Enhancing evolutionary instance selection algorithms by means of fuzzy rough set based feature selection." Information Sciences 186.1 (2012): 73–92.
17. Zhou, Peng, et al. "Online streaming feature selection using adapted neighborhood rough set." Information Sciences 481 (2019): 258–279.
18. Mafarja, Majdi M., and Seyedali Mirjalili. "Hybrid binary ant lion optimizer with rough set and approximate entropy reducts for feature selection." Soft Computing 23.15 (2019): 6249–6265.
19. Garate-Escamilla, Anna Karen, Amir Hajjam E. L. Hassani, and Emmanuel Andres. "Classification models for heart disease prediction using feature selection and PCA." Informatics in Medicine Unlocked 19 (2020): 100330.
20. Dzulkalnine, Mohamad Faiz, and Roselina Sallehuddin. "Missing data imputation with fuzzy feature selection for diabetes dataset." SN Applied Sciences 1.4 (2019): 362.
21. Raihan-Al-Masud, Md, and M. Rubaiyat Hossain Mondal. "Data-driven diagnosis of spinal abnormalities using feature selection and machine learning algorithms." PLos One 15.2 (2020): e0228422.
22. Rahman, Md Asadur, et al. "Employing PCA and t-statistical approach for feature extraction and classification of emotion from multichannel EEG signal." Egyptian Informatics Journal 21.1 (2020): 23–35.
23. Geetha, R., et al. "Cervical cancer identification with synthetic minority oversampling technique and PCA analysis using random forest classifier." Journal of Medical Systems 43.9 (2019): 286.
24. Aich, Satyabrata, et al. "Prediction of Parkinson disease using nonlinear classifiers with decision tree using gait dynamics." Proceedings of the 2017 4th International Conference on Biomedical and Bioinformatics Engineering. 2017.
25. Karan, Biswajit, Sitanshu Sekhar Sahu, and Kartik Mahto. "Parkinson disease prediction using intrinsic mode function based features from speech signal." Biocybernetics and Biomedical Engineering 40.1 (2020): 249–264.
26. Kanagaraj, S., M. S. Hema, and M. Nageswara Gupta. "Environmental risk factors and Parkinson's disease–A study report." International Journal of Recent Technology and Engineering (IJRTE) 7 (2018) 412–415.
27. Nagasubramanian, Gayathri, et al. "Parkinson data analysis and prediction system using multi-variant stacked auto encoder." IEEE Access 8 (2020): 127004–127013.
28. Jeon, Hyoseon, et al. "Automatic classification of tremor severity in Parkinson's disease using a wearable device." Sensors 17.9 (2017): 2067.
29. Haq, Amin Ul, et al. "Feature selection based on L1-norm support vector machine and effective recognition system for Parkinson's disease using voice recordings." IEEE Access 7 (2019): 37718–37734.
30. Pianpanit, Theerasarn, et al. "Neural network interpretation of the Parkinson's disease diagnosis from SPECT imaging." arXiv Preprint arXiv:1908.11199 (2019). Doi: 10.1109/JSEN.2021.3077949.
31. Gottapu, Ram Deepak, and Cihan H. Dagli. "Analysis of Parkinson's disease data." Procedia Computer Science 140 (2018): 334–341.
32. Maxwell, Andrew, et al. "Deep learning architectures for multi-label classification of intelligent health risk prediction." BMC Bioinformatics 18.14 (2017): 523.
33. Kanagaraj, S., M. S. Hema, and M. Nageswara Gupta. "Machine learning techniques for prediction of Parkinson's disease using big data." International Journal of Innovative Technology and Exploring Engineering (IJITEE) 8.10 (2019): 3788–3791.
34. Gürüler, Hüseyin. "A novel diagnosis system for Parkinson's disease using complex-valued artificial neural network with k-means clustering feature weighting method." Neural Computing and Applications 28.7 (2017): 1657–1666.

35. Munneke, Marten, et al. "Efficacy of community-based physiotherapy networks for patients with Parkinson's disease: a cluster-randomised trial." The Lancet Neurology 9.1 (2010): 46–54.
36. Rissanen, Saara M., et al. "Surface EMG and acceleration signals in Parkinson's disease: feature extraction and cluster analysis." Medical & Biological Engineering & Computing 46.9 (2008): 849–858.
37. Shamir, Reuben R., et al. "Machine learning approach to optimizing combined stimulation and medication therapies for Parkinson's disease." Brain Stimulation 8.6 (2015): 1025–1032.
38. Espay, Alberto J., Patrik Brundin, and Anthony E. Lang. "Precision medicine for disease modification in Parkinson disease." Nature Reviews Neurology 13.2 (2017): 119–126.
39. Parimbelli, Enea, et al. "Patient similarity for precision medicine: a systematic review." Journal of Biomedical Informatics 83 (2018): 87–96.

5 IoT and Deep Learning-Based Prophecy of COVID-19

K. Meena
Vel Tech Rangarajan Dr Sagunthala R&D
Institute of Science and Technology
Morai, India

R. Raja Sekar
Kalasalingam Academy of Research and Education
Srivilliputhur, India

CONTENTS

DOI: 10.1201/9781003145004-5

5.1   INTRODUCTION

Motivation

The main motivation of this chapter is to detect the COVID-positive cases accurately in a cost-effective manner within a short time span using long short-term memory (LSTM). In order to predict the severity of disease we used deep learning (DL) architectures and tested the result against the five convolutional neural network (CNN) architectures. Due to time constraints we have conducted the experiment of the dataset with these five CNN architectures, and in the future we have planned to extend it to classify stages of severity.

COVID-19

The coronavirus entitled SARS-CoV-2 emerged in December 2019, headed for the initiation of the deadly disease of respiratory infection called COVID-19. It proved to be a complicated disease and it came into sight in diverse forms and levels of cruelty ranging from mild to brutal, which leads to multi-organ failure, rigorous pneumonia, and bereavement. The amount of the populace infected with the virus is rising hastily. As per a report by the World Health Organization (WHO), the number of positive cases and deaths reported were 34,804,348 and 1,030,738, respectively, till October 04, 2020 across 235 countries [1–4]. The major confirmed positive cases of COVID-19 were found among the patients with brutal respiratory, hypertension, and cardiovascular syndrome and senior citizens. This recommends the universe to propose a new model with minimum cost and time to detect the infected cases much earlier using modern technologies. This deadly virus can easily spread from one person to another; keeping this in mind we provide a solution to reduce human intervention in predicting the stages of severity in infected patients using artificial intelligence (AI) techniques.

5.1.1   INTRODUCTION TO AI

AI is a promising tool in the meadow of medical imaging and has contributed forcefully to the battle against COVID-19 [5]. The conventional imaging workflow completely depends on human labor, whereas AI facilitates secure, precise, and proficient imaging resolution without human participation. AI plays a vital role in cracking complex problems in various fields, such as engineering [6–8], medicine [9–12], economy [13], and psychology [14].

5.1.2   JUSTIFICATION OF AI TECHNOLOGIES IN THE PREDICTION OF COVID-19

Modern COVID-19 applications powered by AI are used to take X-ray and computed tomography (CT) scan images of lungs and segment the affected region from the provided lung images. Apart from prediction, the proposed model also identifies the stages of severity. This demonstrates how extreme and to what level AI plays a crucial role in the advancement of health care organizations worldwide [7]. Hence keeping the value of human life in mind during this pandemic situation, AI tools have been utilized to help physicians in predicting the severity of the virus in less time and with high accuracy.

On the other hand, an additional dilemma that physicians and researchers need to deal with is large volumes of data, termed big data. Big data in the healthcare division plays a major role in clinical management, particularly in the field of radiology, by using DL techniques [15]. DL is the division of machine learning (ML); it consists of several layers of artificial neural networks (ANNs) so as to provide a diverse analysis of the data being fed into input layers. In this chapter, we propose a deep convolutional neural network (DCNN), an advanced tool to assist the physician in identifying and predicting the stages of the COVID-19 virus via chest X-ray and CT scan images of lungs. The proposed study is based on the Recurrent Neural Network (RNN) and it is assembled by using multiple parallel layers with varying kernel sizes to sense global and local features and by linking the residuals to further layers to pass information. The model is trained with 3000 chest X-ray images to predict COVID-19 cases. In addition, the results are compared with those of other popular DL algorithms. DCNN identifies COVID-19-positive cases accurately in an extremely short time and helps in early diagnosis. Also, chest X-ray is cost effective when compared to other radiographic images.

5.1.3 ORGANIZATION OF THE CHAPTER

The rest of the chapter is organized as follows. Section 5.2 summarizes the literature survey of COVID-19 prediction. Section 5.3 includes the proposed architecture. Section 5.4 elaborates on the dataset, experimental setup, and performance measures of the proposed model. Section 5.5 includes the conclusion and future enhancement.

5.2   LITERATURE SURVEY

The study and prediction of COVID-19 were seriously considered due to its cruel effect. It is hard to identify exposed personnel since symptoms cannot be determined instantly. AI is an alternative tool to predict COVID-19 compared to the usual time-consuming methods. Even though there are plentiful studies of COVID-19, this chapter pays attention to ML and DL in predicting and diagnosing patients infected by COVID-19 through chest X-ray and CT scan images of the lung. ML is the subset of AI that involves training machines with statistical models in order to learn and detect patterns from training data. The test data were then compared against the generated pattern obtained from trained data to measure the accuracy.

5.2.1   COVID-19 DIAGNOSIS USING DEEP LEARNING

The utilization of ML methodologies was rapidly increased in the health care field [16–18]. DL is the subset of ML that produces higher accuracy when compared to other ML techniques. The CNN has been found to be the most promising technique to produce good results compared to other DL models [19, 20].

5.2.2 X-Ray Diagnosis Using Deep Learning

X-ray apparatus are used to scan parts of the body when people are affected by lung diseases, injuries, and bone dislocations, whereas CT scans are superior X-ray devices used by doctors for identifying internal organ damages [21]. X-rays are preferred because they are cost effective and less harmful to radiation when compared with CT scans [22, 23].

Hassanien et al. [23] tested 40 (25 COVID-confirmed cases and 15 negative cases) chest X-ray images using multi-level thresholding and support vector machine (SVM) classifier. They achieved the following performance: 95.76% sensitivity, 99.7% specificity, and 97.48% accuracy in detecting COVID-19.

To identify COVID-19, Narin et al. [22] trained the system using a CNN model with 50% chest X-rays of COVID patients and 50% without COVID patients. They compared the result with three CNN models: ResNet-50, Inception-ResNet-v2, and Inception-v3.They concluded that among the three, ResNet-50 produces the best accuracy of 98%.

In addition, Hemdan et al. [24] developed a model named COVIDX-Net, which used 50 X-ray images, out of which 25 were COVID-positive cases and 25 were COVID-negative cases. The model was evaluated using seven DL models: DenseNet, Inception-ResNet-v2, Inception-v3, modified VGG19, MobileNet, ResNet-v2, and Xception. They obtained the best performance using DenseNet and VGG19 with an F1-score of 91%.

Sethy and Behera [25] proposed a new model with 11 deep-learning models: AlexNet, VGG16, VGG19, GoogLeNet, ResNet-18, ResNet-50, ResNet-101, Inception-v3, Inception-ResNet-v2, DenseNet-201, and XceptionNet. The features were extracted from chest X-ray images by using the above 11 deep-learning algorithms and were classified as positive and negative cases using an SVM classifier. The obtained result proved to be good with 95.38% accuracy using ResNet-50.

Huang et al. [26] compared several DL models like CNN, multi-layer perceptron (MLP), LSTM, and gated recurrent units (GRUs). They compared the result of the above models using mean absolute error (MAE) and root mean square error (RMSE). Finally they concluded that CNN produces very good accuracy among the three.

5.2.3 CT Scan Diagnosis Using Deep Learning

In the year 1972, Godfrey Hounsfield and Allan Cormack developed computerized axial tomography (CAT), also called a CT scan. It is a computerized x-ray method that provides high-clarity 3D images, whereas X-rays provide 2D images [22]. It was especially used for scanning soft tissues and blood vessels. CT scans are very expensive when compared to X-rays.

Gozes et al. [27] developed a DL model to robotically predict COVID-19 using 157 CT scans of foreign patients from China and the United States. The proposed system used Resnet-50-2 and produced an accuracy of 99.6% with 98.2% sensitivity and 92.2% specificity.

Wang et al. [28] modeled a DL approach with 90 CT scan images, among which 44 were COVID positive and 50 were COVID negative. They used a modified network inception model and obtained an accuracy of 82.9% with specificity and sensitivity values of 80.5% and 84%, respectively.

Fu et al. [29] implemented a classification system to find OID'19 using ResNet-50 to detect COVID-19. They tested their model with 60,427 CT scans of 918 patients. Among these 60,427 images, 14,944 were COVID-19 patients and 15,133 were non-COVID-19 patients with pneumonia. They performed several tests for several lung diseases. They obtained specificity, sensitivity, and accuracy values of 98.9%, 98.2%, and 98.8%, respectively.

Xu et al. [30] analyzed whether testing COVID-19 at the initial stage using real-time reverse transcription–polymerase chain reaction (RT-PCR) was inaccurate in predicting positive cases. In order to predict COVID-19 at its early stage DL techniques were used. The experiments were conducted with 618 CT samples and analysis was done with ResNet-18 and ResNet DL techniques. The accuracy obtained was 86.7%.

5.3 ARCHITECTURE OF PROPOSED MODEL

5.3.1 SEVERITY PREDICTION MODULE

The X-ray images of the lung were given as input to the model. The images were pre-processed using the pre-processing module. Then the pre-processed X-ray images were trained using the recurrent neural networks (RNNs). We trained the model using two architectures of RNN: LSTM and GRU. The features obtained from both architectures were independently used to predict COVID-19. If the result of the prediction was positive, then we detect the severity of the patient using their CT scan since CT scans provide more details when compared to X-ray images. Also if the prediction of COVID-19 was negative, then there was no need to take the CT scan. The determination of the severity of COVID-19 was done with CNNs with several architectures like LeNet-5, AlexNet, VGG-16, GoogLeNet, and ResNeXt-50. The results obtained from individual architecture were tabulated and compared to detect the severity of the disease. It is shown in Figure 5.1.

5.3.2 COVID-19 PREDICTION MODULE

In this module we predict whether the patient is affected by COVID or not using chest X-ray images with the help of RNN algorithms.

5.3.2.1 Pre-Processing

To remove the unnecessary details from chest X-rays, we perform resizing, shuffling, and normalization on the images in order to extract the needed information for further processing. Data augmentation, like rotation and scaling, was also carried out. The images were resized to 224×224 pixels.

5.3.2.2 RNN Feature Extraction Module

In CNNs, the outputs and inputs are independent of each other. To solve this problem, we use RNN [31, 32], in which the outputs from k1 stride are fed as input to the

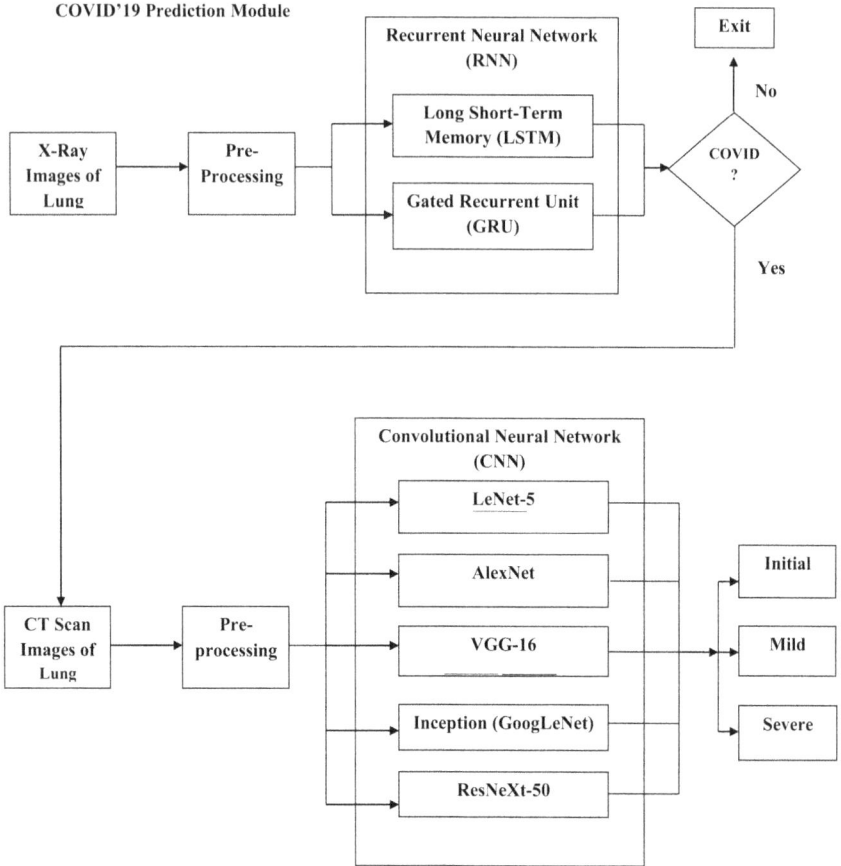

FIGURE 5.1   Architecture of the proposed model.

k^{th} stride. RNNs are widely used because of their hidden layer, which plays a vital role in remembering the sequence of information. It remembers the sequence of calculated information using a "memory." The same weight and bias are applied to all the hidden layers, which diminishes the complication of rising parameters. Here x, h, and y indicate input, activation function, and output, respectively, and W_{hx}, W_{hh}, and W_{yh} indicate weight applied to the hidden layer, input layer, and output layers, respectively. It is shown in Figure 5.2. The obtained output from the last stride is compared with the actual output and the error is calculated, and then error (gradient) is back propagated to the network to adjust the weight accordingly to obtain the expected result. RNN was not able to remember the long sequence of information. Thus the flow of gradient in RNN leads to two major problems: vanishing gradient and exploding gradient [33]. The gradient is computed by recurrent multiplication of derivatives. So, if the generated gradient is too small it causes a vanishing gradient, and if it is greater than the threshold, it leads to an exploding gradient.

To overcome the above problems, we were moving to LSTM and GRU

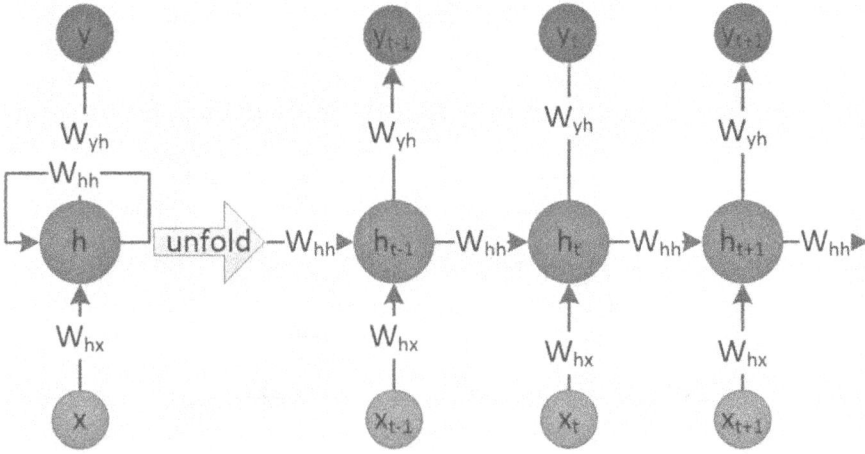

FIGURE 5.2 Architecture of RNN.

5.3.2.3 Long Short-Term Memory

LSTM is one type of architecture used in RNN, which was introduced by Hochreiter and Schmidhuber in 1990 to overcome the problem of short-term memory. LSTM can remember long sequences of information with the help of gates [34]. These gates control the flow of information within the net by eliminating unnecessary information from preceding steps and feeding the needed information to the succeeding steps [35–38]. The architecture of LSTM is shown in Figure 5.3.

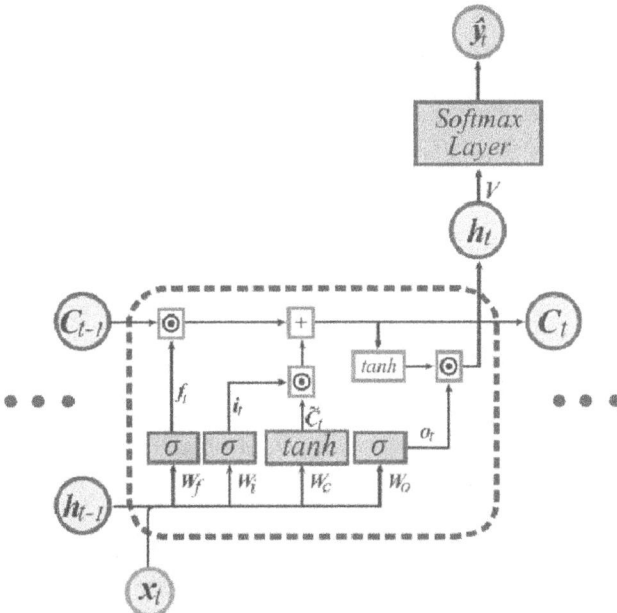

FIGURE 5.3 LSTM architecture.

LSTM consists of three gates: forget gate, input gate, and output gate. These three gates determine the cell state (act as memory), which is the central part of LSTM. The information is added or removed from the cell based on the gates. x_t denotes current input. C_t indicates the content of the latest cell state and C_{t-1} denotes the cell state of the previous LSTM unit. h_t denotes the current output and h_{t-1} represents the previous LSTM unit's output. W_f, W_i, and W_o are the weights applied to forget gate, input gate, and output gate, respectively, and b_i, b_f, and b_o are the biases applied to forget gate, input gate, and output gate, respectively [35]. It includes two functions: (1) sigmoid function (σ) and (2) Tanh function. The sigmoid activation function is used in all three gates, which converts the output value to stay in the range of 0 to 1. Similarly, the Tanh activation function converts the output to fall in the range of −1 to 1, which is used in the input and output gates. Using these functions the network can learn about the data itself, i.e. which part is important and which is not important. Therefore gates play a vital role in deciding which information to be kept and which one to be discarded.

5.3.2.3.1 Forget Gate

The inputs to forget gate are h_{t-1} and x_t. Depending on the value of inputs, the sigmoid function decides the value of C_{t-1}. If the output of σ was closer to 0, it indicates that it was useless and hence it could be discarded from C_{t-1}, and if it was closer to 1, it indicates that it was useful information, and hence it is kept in C_{t-1}. [35]. The equation for forget gate is given in equation (5.1).

$$f_t = \left(W_f \cdot \left[h_{t-1}, \ x_t \right] + b_f \right) \tag{5.1}$$

5.3.2.3.2 Input Gate

This gate is used to update the cell state. Inputs to the sigmoid function of the input gate are h_{t-1} and x_t. Similarly the same inputs are applied to the Tanh function. The outputs obtained from both the functions are multiplied and the sigmoid function finalizes what should be kept from Tanh output. The equations of input gate are given in equations (5.2) and (5.3).

$$I_t = \left(W_i \cdot \left[h_{t-1}, \right] + b_i \right) \tag{5.2}$$

$$\tilde{C}_t = \tanh\left(W_i \cdot \left[h_{t-1}, \ x_t \right] + b_i \right) \tag{5.3}$$

5.3.2.3.3 Output Gate

The sigmoid function gets the two inputs current input (x_t) and output of the preceding hidden layer (h_{t-1}). The updated cell state is given as input to the Tanh function. The outputs from sigmoid and Tanh functions are multiplied to obtain the next hidden layer's input. The equation is given below in equations (5.4) and (5.5).

$$O_t = \left(W_O \cdot \left[h_{t-1}, \ x_t \right] + b_O \right) \tag{5.4}$$

$$h_t = O_t \ \tanh \left(C_t \right) \tag{5.5}$$

5.3.2.3.4   Cell State

The cell state c_t was multiplied by the output obtained from the forget gate. The output from the forget gate decides whether to keep the cell state as it was earlier or to update it. Then the value of cell state c_t is added with the output obtained from the input gate in order to update the cell state to a new value. The equation of cell state was stated in equation (5.6).

$$C_t = f_t C_{t-1} + i_t \tilde{C}_t \qquad (5.6)$$

5.3.2.4   Gated Recurrent Unit

GRU is similar to LSTM. It uses only two gates, namely (1) reset gate and (2) update gate, whereas LSTM uses three gates. GRUs use only fewer gates; hence it was a bit faster than LSTM, but slight variation in performance [39, 40]. The update gate was similar to the input and forget gate of LSTM. The reset gate decides to what extent the past information had to be discarded. Unlike LSTM, GRU does not use the cell gate and output gate (Figure 5.4).

The equations of GRU cell are stated below from equations (5.7) to (5.10):

$$r_t = \text{sigm}\left(W_{xr} x_t + W_{hr} h_{t-1} + b_r\right) \qquad (5.7)$$

$$z_t = \text{sigm}\left(W_{xz} x_t + W_{hz} h_{t-1} + b_z\right) \qquad (5.8)$$

$$\tilde{h}_t = \tanh\left(W_{xh} x_t + W_{hh}\left(r_t h_{t-1}\right) + b_h\right) \qquad (5.9)$$

$$h_t = z_t h_{t-1} + \left(1 - z_t\right) \tilde{h}_t \qquad (5.10)$$

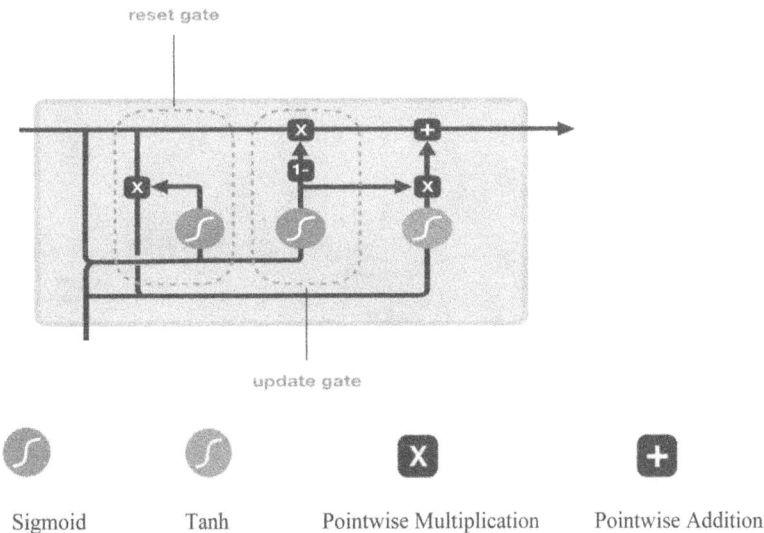

FIGURE 5.4   GRU architecture.

where r_t, z_t, x_t, and ht are the reset gate, update gate, input bias, respectively; sigm and tanh denote the sigmoid and tangent activation function, respectively.

5.3.3 SEVERITY PREDICTION MODULE

When COVID-19 is confirmed for a patient, we recommend the concern to take the CT scan and it is pre-processed. Then the processed image is analyzed using CNNs [40, 41]. The features parameters of CNN models are determined without human intervention by Grid Search. It added the advantage of using CNN models.

The CNN architecture is a brilliant architecture as it mimics the neural pattern of the human brain. It consists of several layers. Each layer has a set of neurons that analyze every portion of the image. CNN compares the image portion by portion. The portion it looks for is called filters or features. It extracts the image features and converts them to pictures of lesser dimensions with no loss of image characteristics. To do so, it uses the following layers.

5.3.3.1 Input Layer

The CT scan image is given as input to the input layer. Before feeding it to the input layer, the image is converted to a single dimension as a column matrix. For example, if the image dimension is 32×32, we have to reshape it into a single column with size 1024×1, i.e. if the training sample was "n" then the input dimension would be (1024, n).

5.3.3.2 Convolution Layer

The image features are extracted using this layer. Hence it is also called a feature extractor layer. It is mainly used to extract important features from the input image. This layer contains numerous filters that do convolution operations shown in Figure 5.5. Performing dot product operation between the image pixel (portion of the input image with the same size as the filter) and the filter is termed convolution operation. The output of this operation is a single integer of the expected output dimensions. For example, if the dimension of input image is 6×6 and the size of the filter is 3×3. Then the expected output dimension is 4×4. Then we glide the

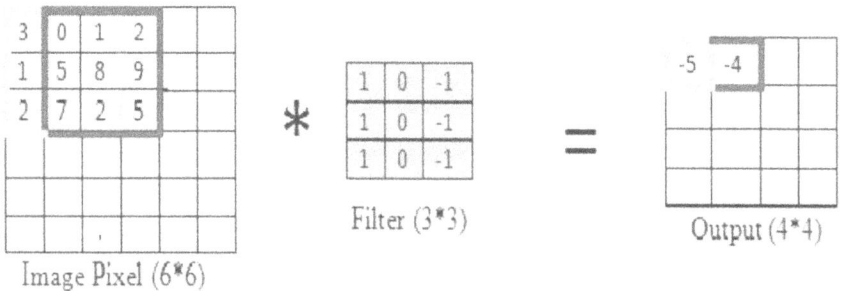

FIGURE 5.5 Convolution operation.

filter above the succeeding portion of the input image until we complete the entire image. Thus the output obtained from this layer becomes the input to the succeeding layer. The dot product operation is shown below. This layer also includes the ReLU (Rectified Linear Unit) activation function, which converts negative values to zero. It is applied to the output obtained from the convolution operation.

$$3 \times 1 + 0 \times 0 + 1 \times -1 + 1 \times 1 + 5 \times 0 + 8 \times -1 + 2 \times 1 + 7 \times 0 + 2 \times -1$$
$$= (3 - 1 + 1 - 8 + 2 - 2) = -5$$

$$0 \times 1 + 1 \times 0 + 2 \times -1 + 5 \times 1 + 8 \times 0 + 9 \times -1 + 7 \times 1 + 2 \times 0 + 5 \times -1$$
$$= (-2 + 5 - 9 + 7 - 5) = -4$$

5.3.3.3   Pooling Layer

This layer reduces the dimension of the image obtained from the convolution layer. This layer is also called downsampling. It is mainly used to reduce the computational overhead. Either we can perform average pooling (takes the average of all pixels) or max-pooling (takes the maximum of all pixels). For example, if the dimensions of the image to the input are 4×4, it converts them to 2×2 after maximum pooling. This is shown in Figure 5.6.

5.3.3.4   Fully Connected Layer (FC)

FC has three layers: FC input layer, FC layer, and FC output layer (softmax Layer). The input layer converts the output obtained from the pooling layer to a single vector. The FC layer applies to weight and bias to the features generated from the input layer. The softmax layer generates the probability, which is used for multi-class prediction.

5.3.4   ARCHITECTURES OF CNN

To predict the severity of COVID we have used the most popular five architectures of CNN and they are given as follows.

5.3.4.1   LeNet-5

This was introduced in 1998. It has two convolution layers, sub-sampling layers (pooling layer), and three FC layers. It contains nearly 60,000 parameters

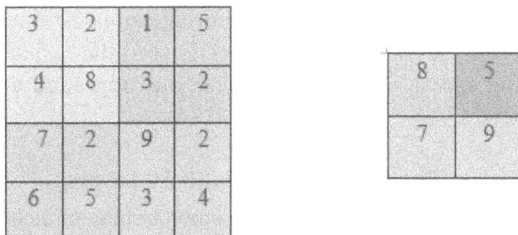

FIGURE 5.6   MaxPooling operation.

[42, 43]. The size of the convolution Filter is 5×5, and there are $6 \times (5 \times 5 + 1) =$ 156 weights in total, where +1 indicates the bias. So each pixel in Convolution Layer (C1) is connected to 5×5 pixels and 1 bias, resulting in a total connection of $156 \times 28 \times 28 = 122{,}304$.

5.3.4.2 AlexNet

This was introduced in 2012. AlexNet consists of five convolution layers and three fully connected layers, with nearly 60 million parameters, and it was the winner of ILSVCR (ImageNet Large Scale Visual Recognition Competition) in 2012 [44, 45]. The input size of the image is $224 \times 224 \times 3$. It is passed to the first convolution layer with an 11×11 filter size, 5×5 in layer two, and 3×3 in layers three to five. The window size of the max-pooling layer is 3×3 and it is used with stride 2, which is followed by three fully connected layers. It contains ten times more convolution layers than LeNet. This resolved the over-fitting issue in the dropout layers and also the size of the max-pooling network is reduced.

5.3.4.3 VGG-16 (Visual Geometry Group)

This architecture was developed in 2014 and was the first runner-up of ILSVRC 2014.It has 13 convolution layers and3 FC layers. The number of parameters of VGG is about 138 million [46, 47]. VGGNet is used to get better training time by diminishing the number of parameters (training variables) in the convolution layers. Thereby the learning rate is faster than AlexNet. The filter size of the convolution layer and max-pool layers are 3×3 and 2×2 with two strides.

5.3.4.4 GoogLeNet (Inception)

This was the winner of ILSVRC 2014 and was named after Prof. Yann LeCun's LeNet. When compared to VGGNet it has a minimum error rate [48–50]. It was released in many versions such asV1 (2014), V2 & V3 (2015), and V4 (2016) [51]. V1 consists of 1×1, 3×3, and 5×5 convolution layers along with a max-pooling layer. Totally it contains 22 layers with 5 million parameters with an error rate of 6.67%. In V2 to reduce the number of parameters from V1, batch normalization was done. In this version they converted 5×5 convolution layers to 3×3 convolution layers. It ultimately reduces cost by increasing accuracy with an error rate of 4.8%. V3consists of 48 layers and the number of parameters is increased to 24 million. It produces a high accuracy, thereby reducing the error rate to 3.58%, which is half the error rate of V1.V4: It was introduced in 2016 with the combination of Inception ResNet (Residual Network). Due to the residual association, it has a high training speed. This stem module was upgraded and it consists of 43 million parameters.

5.3.4.5 ResNeXt-50: (Residual Network with 50 Layers)

It was introduced in 2017. When compared to other architectures it has a lower error rate of 3.03%. It has 50 layers with about 25.5 million parameters [52, 53]. This solves the vanishing gradient problem when CNN moves deeper and deeper.

(a)

(b)

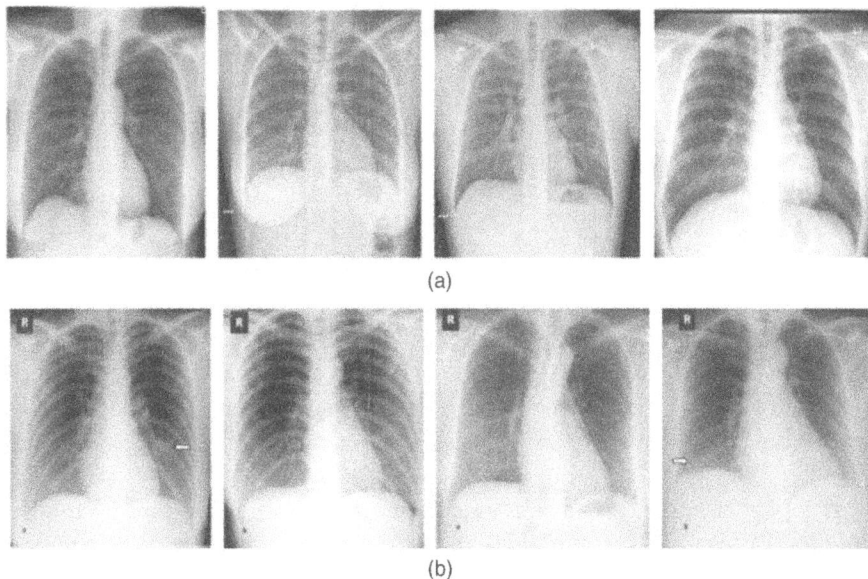

FIGURE 5.7   (a) Sample COVID-positive chest X-rays. (b) Sample COVID negative chest X-rays.

5.4   EXPERIMENTAL SETUP AND RESULT ANALYSIS

In this experimental setup, the datasets were split into training and testing sets at a ratio of 80:20. The results were obtained using the 5-fold cross-validation technique. The chest X-rays were pre-processed (Figure 5.7). Then the processed chest X-rays were given to LSTM and the GRU–RNN architecture in order to extract the features. Based on the features extracted we evaluated the performance using a confusion matrix. The obtained result is tabulated in Table 5.1. The confusion matrix presents the following.

TP indicates that COVID-positive cases were labeled correctly as positive. FP indicates that COVID-negative cases were labeled as positive. TN indicates that COVID-negative cases were labeled correctly as negative. FN indicates that COVID-positive cases were labeled as negative (Figure 5.8 and Table 5.2).

TABLE 5.1
Confusion Matrix

Confusion Matrix		Predicted	
		COVID +	COVID −
Actual	COVID +	TP	FN
	COVID −	FP	TN

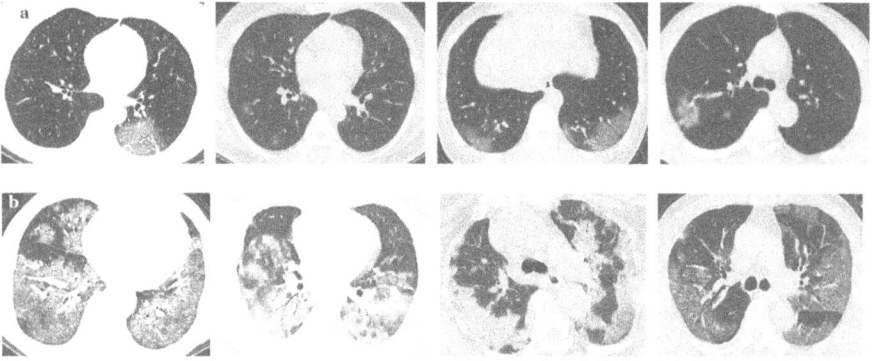

FIGURE 5.8   (a) Chest CT scan of mild COVID cases. (b) Chest CT scan of severe COVID cases.

A dataset of 1020 CT scans was used for result analysis. Among them 816 were used for training and 214 were used for testing. The datasets were pre-processed by reshaping the images to 224 × 224.The datasets were trained and tested with different CNN architectures such as LeNet, AlexNet, GoogLeNet, VGG-16, and ResNeXt-50. The following performance metrics were evaluated from the training and testing dataset.

$$\text{Sensitivity (Recall / True Positive Rate)}$$

$$= \text{True Positive} / (\text{True Positive} + \text{False Negative})$$

$$\text{Specificity (True Negative Rate)}$$

$$= \text{True Negative} / (\text{False Positive} + \text{True Negative})$$

$$\text{Accuracy} = \text{TP} + \text{TN} / (\text{TP} + \text{TN} + \text{FP} + \text{FN})$$

$$\text{where TP} - \text{True Positive, TN} - \text{True Negative,}$$

$$\text{FP} - \text{False Positive, and FN} - \text{False Negative}$$

TABLE 5.2
Performance Metrics of Various CNN Architectures

Architecture of RNN	Class	Accuracy (%)	Specificity (%)	Sensitivity (%)	F1-Score
LSTM	COVID-19	99.2	99.2	99.3	98.9
	Normal	99.8	99.7	100	99.7
	Time Taken to Extract Feature per Image: 0.369 (sec)				

Architecture of RNN	Class	Accuracy (%)	Specificity (%)	Sensitivity (%)	F1-Score
GRU	COVID	98.7	98.8	98.9	97.6
	Normal	99.3	99.3	99.2	99.2
	Time Taken to Extract Feature per Image: 0.284 (sec)				

TABLE 5.3

Performance Metrics of Various CNN Architectures

CNN Architecture	Group	Sensitivity	Specificity	Accuracy	AUC
LeNet	Training	80.63	86.76	83.70	0.928
	Validation	80.39	86.27	83.33	0.926
AlexNet	Training	93.38	71.81	82.60	0.913
	Validation	89.21	68.63	78.92	0.894
VGG-16	Training	94.85	79.41	87.13	0.955
	Validation	92.16	78.43	85.29	0.943
GoogLeNet	Training	97.30	89.95	93.62	0.990
	Validation	97.06	87.25	94.73	0.982
ResNeXt	Training	98.77	99.26	99.63	0.997
	Validation	98.04	99.02	99.03	0.994

Also we have calculated the Area Under Curve (AUC) using the ROC (Receiving Operating Characteristics) curve. The results are shown in detail in Table 5.3. The severity of the disease is indicated based on the level of infection of the chest scan and the clinical data obtained from the hospital.

The performance graphs for the RNN architecture for the GRU model and the LSTM model are shown in Figure 5.9 (a) and (b), respectively, whereas the performance graph for the CNN architecture model is shown in Figure 5.9 (c). The loss function for each model was calculated by cross-entropy loss or log loss function.

(a)

Prediction Using GRU

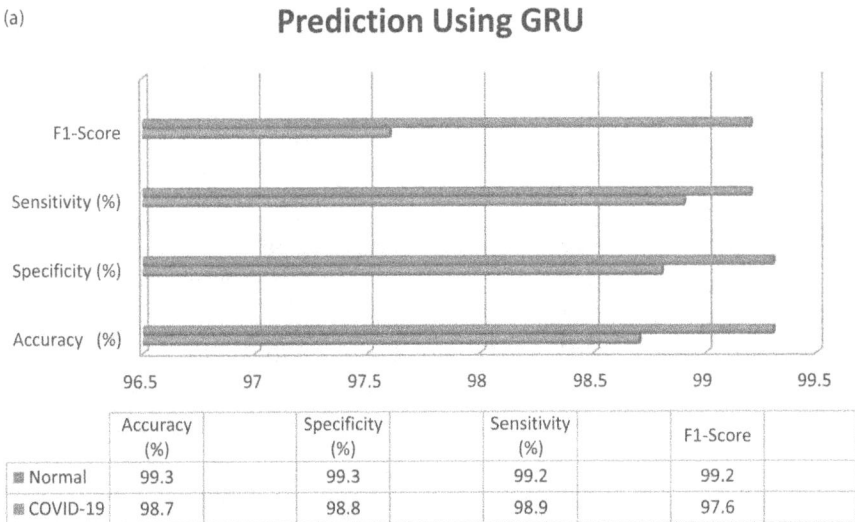

	Accuracy (%)	Specificity (%)	Sensitivity (%)	F1-Score
Normal	99.3	99.3	99.2	99.2
COVID-19	98.7	98.8	98.9	97.6

FIGURE 5.9   (a) Performance graph for GRU model. *(Continued)*

(b)

Prediction Using LSTM

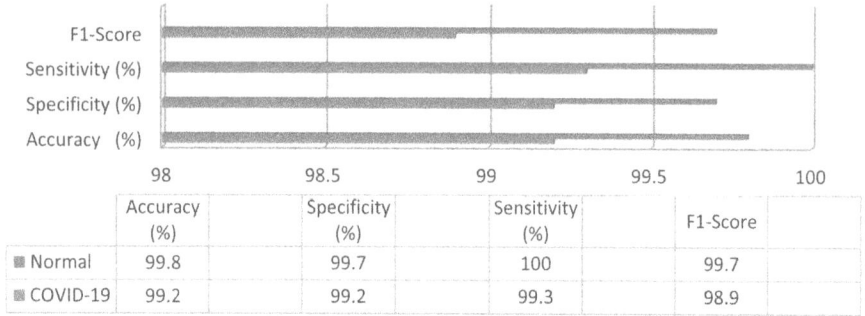

	Accuracy (%)	Specificity (%)	Sensitivity (%)	F1-Score
▦ Normal	99.8	99.7	100	99.7
▦ COVID-19	99.2	99.2	99.3	98.9

(c)

Prediction Using CNN Model

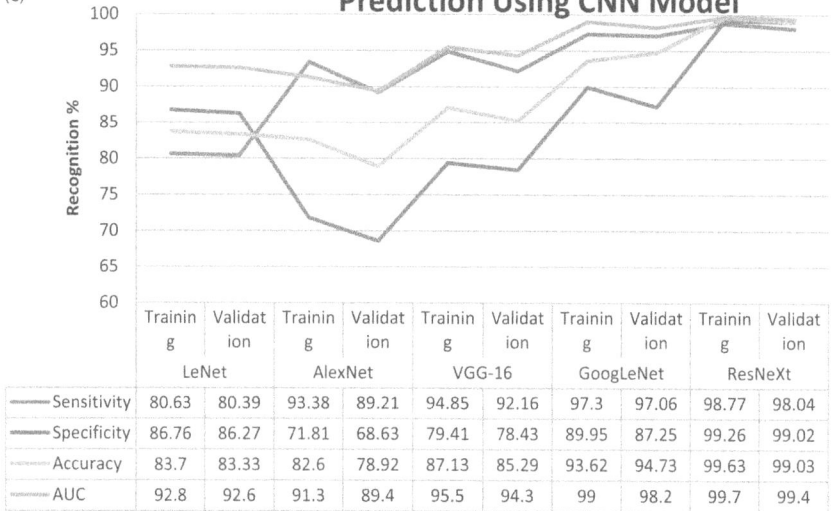

	Training	Validation	Training	Validation	Training	Validation	Training	Validation	Training	Validation
	LeNet		AlexNet		VGG-16		GoogLeNet		ResNeXt	
Sensitivity	80.63	80.39	93.38	89.21	94.85	92.16	97.3	97.06	98.77	98.04
Specificity	86.76	86.27	71.81	68.63	79.41	78.43	89.95	87.25	99.26	99.02
Accuracy	83.7	83.33	82.6	78.92	87.13	85.29	93.62	94.73	99.63	99.03
AUC	92.8	92.6	91.3	89.4	95.5	94.3	99	98.2	99.7	99.4

FIGURE 5.9 *(Continued)* (b) Performance graph for LSTM model. (c) Performance graph for CNN model.

5.5 CONCLUSION AND FUTURE ENHANCEMENTS

COVID-19 positive cases are rising day by day; being a deadly disease, it has killed a lot of people all over the world. Hence, we need to identify the infected individuals quickly and isolate them from others. To identify the individuals quickly, we used algorithms such as RNN and CNN. The chest X-rays of COVID-positive cases were obtained and they were trained using LSTM and GRU architectures of the RNN algorithm. It was found that LSTM produced very good results for the classification of COVID-positive and COVID-negative cases. The results obtained by LSTM are tabulated in Table 5.1. But it takes more time to extract features. GRU takes less time than LSTM for extracting features, but it has some performance degradation compared to LSTM. The performance metrics evaluated were specificity, sensitivity, accuracy, and F1-Score, which were 99.2%, 99.3%, 99.2%, and 98.9 for LSTM and 98.8%, 98.9%, 98.7%, and 97.6 for GRU, respectively.

Later the CT scans of the confirmed COVID cases were obtained and evaluated using various CNN architectures such as LeNet, AlexNet, VGG-16, GoogLeNet and ResNeXt-50. The following performance metrics were evaluated for the above CNN architectures sensitivity, specificity, accuracy and AUC. The experiment was conducted using 1020 CT scans of the chest. The obtained results are tabulated in Table 5.2. Among them ResNeXt-50 produced very excellent results with sensitivity: 98.77%, specificity: 99.26%, accuracy: 99.63%, and AUC: 99.7%. In this pandemic situation, we trust that the projected classification would help doctors predict the COVID cases quickly and safely by reducing their workload.

The future work that is in progress: we planned to segment the infected portion of lungs from CT scans and the data from clinical sources are incorporated into the model in order to predict the severity of the disease on the individual.

REFERENCES

1. C. Huang, et al., "Clinical features of patients infected with 2019 novel coronavirus in Wuhan, China," *Lancet*, vol. 395, no. 10223, pp. 497–506, 2020.
2. N. Chen, M. Zhou, X. Dong, J. Qu, F. Gong, Y. Han, Y. Qiu, J. Wang, Y. Liu, Y. Wei, J. Xia, T. Yu, X. Zhang, and L. Zhang, "Epidemiological and clinical characteristics of 99 cases of 2019 novel coronavirus pneumonia in Wuhan, China: a descriptive study," *Lancet*, vol. 395, no. 10223, pp. 507–513, Feb. 2020.
3. D. Wang, B. Hu, C. Hu, F. Zhu, X. Liu, J. Zhang, B. Wang, H. Xiang, Z. Cheng, Y. Xiong, and Y. Zhao, "Clinical characteristics of 138 hospitalized patients with 2019 novel coronavirus-infected pneumonia in Wuhan, China," *JAMA*, vol. 323, no. 11, pp. 1061–1069, 2020.
4. K. Liu, Y.-Y. Fang, Y. Deng, W. Liu, M.-F. Wang, J.-P. Ma, W. Xiao, Y.-N. Wang, M.-H. Zhong, C.-H. Li, G.-C. Li, and H.-G. Liu, "Clinical characteristics of novel coronavirus cases in tertiary hospitals in Hubei province," *Chin. Med. J.*, vol. 133, no. 9, pp. 1025–1031, May 2020.
5. L. A. Bullock Joseph, P. K. Hoffmann, L. Cynthia, and A. Luengo-Oroz Miguel, "Mapping the landscape of artificial intelligence applications against COVID-19," *arXiv:2003.11336*, 2020. DOI: 10.1613/jair.1.12162.
6. M. Jamshidi, A. Lalbakhsh, S. Lot, H. Siahkamari, B. Mohamadzade, and J. Jalilian, "A neuro-based approach to designing a Wilkinson power divider," *Int. J. RF Microw. Comput. Aided Eng.*, vol. 30, no. 3, Art. no. e22091, Mar. 2020.
7. M. Jamshidi, A. Lalbakhsh, B. Mohamadzade, H. Siahkamari, and S. M. H. Mousavi, "A novel neural-based approach for design of microstrip filters," *AEU-Int. J. Electron. Commun.*, vol. 110, Art. no. 152847, Oct. 2019.
8. M. B. Jamshidi, N. Alibeigi, A. Lalbakhsh, and S. Roshani, "An ANFIS approach to modeling a small satellite power source of NASA," in Proc. IEEE 16th Int. Conf. Netw., Sens. Control (ICNSC), May 2019,pp. 459–464.
9. Y. Mintz and R. Brodie, "Introduction to artificial intelligence in medicine," *Minim. Invasive Ther. Allied Technol.*, vol. 28, no. 2, pp. 73–81, 2019.
10. R. B. Parikh, Z. Obermeyer, and A. S. Navathe, "Regulation of predictive analytics in medicine," *Science*, vol. 363, no. 6429, pp. 810–812, Feb. 2019.
11. A. Becker, "Artificial intelligence in medicine: what is it doing for us today?" *Health Policy Technol.*, vol. 8, no. 2, pp. 198–205, Jun. 2019.
12. N. J. Schork, "Artificial intelligence and personalized medicine," in *Precision Medicine in Cancer Therapy*. Cham, Switzerland: Springer, 2019, pp. 265–283.

13. M. B. Jamshidi, M. Gorjiankhanzad, A. Lalbakhsh, and S. Roshani, "A novel multi-objective approach for detecting money laundering with a neuro-fuzzy technique," in Proc. IEEE 16th Int. Conf. Netw., Sens. Control (ICNSC), May 2019, pp. 454–458.

14. M. B. Jamshidi, N. Alibeigi, N. Rabbani, B. Oryani, and A. Lalbakhsh, "Artificial neural networks: a powerful tool for cognitive science," in Proc. IEEE 9th Annu. Inf. Technol., Electron. Mobile Commun. Conf. (IEMCON), Nov. 2018, pp. 674–679.

15. H. Zhao, G. Li, and W. Feng, "Research on application of artificial intelligence in medical education," in Proc. Int. Conf. Eng. Simulation Intell. Control (ESAIC), Changsha, China, vol. 8, Aug. 2018, pp. 340–342.

16. Y. Xu, Y. Wang, J. Yuan, Q. Cheng, X. Wang, and P. L. Carson, "Medical breast ultrasound image segmentation by machine learning," *Ultrasonics*, vol. 91, pp. 1–9, 2019.

17. A. Mesleh, "Lung cancer detection using multi-layer neural networks with independent component analysis: a comparative study of training algorithms," *Jordan Biol. Sci.*, vol. 10, Dec. 2017 pp. 239–249.

18. A. Mesleh, D. Skopin, S. Baglikov, and A. Quteishat, "Heart rate extraction from vowel speech signals," *J. Comput. Sci. Technol.*, vol. 27, pp. 1243–1251, 2012.

19. K. Suzuki, "Overview of deep learning in medical imaging," *Radiol. Phys. Technol.*, vol. 10, pp. 257273, 2017.

20. Y. LeCun, Y. Bengio, and G. Hinton, "Deep learning," *Nature*, vol. 521, pp. 436444, 2015.

21. *Difference between X-ray and CT Scan*. https://www.envrad.com/difference-between-x-ray-ct-scan-and-mri/

22. A. Narin, C. Kaya, and Z. Pamuk, "Automatic detection of coronavirus disease (COVID-19) using X-ray images and deep convolutional neural networks," arXiv Preprint *arXiv:200310849*, 2020.

23. A. E. Hassanien, L. N. Mahdy, K. A. Ezzat, H. H. Elmousalami, and H. A. Ella, "Automatic X-ray COVID-19 lung image classification system based on multi-level thresholding and support vector machine," *medRxiv*, 2020. https://doi.org/10.1101/2020.03.30.20047787

24. E. E.-D. Hemdan, M. A. Shouman, and M. E. Karar, "A framework of deep learning classifiers to diagnose COVID-19 in X-ray images," arXiv Preprint *arXiv:200311055*, 2020.

25. P. K. Sethy and S. K. Behera, "Detection of coronavirus disease (COVID-19) based on deep features," 2020.

26. C.-J. Huang, Y.-H. Chen, Y. Ma, and P.-H. Kuo, "Multiple-input deep convolutional neural network model for COVID-19 forecasting in China," *medRxiv*, 2020. https://doi.org/10.1101/2020.03.23.20041608.

27. O. Gozes, et al., "Rapid AI development cycle for the coronavirus (covid-19) pandemic: Initial results for automated detection & patient monitoring using deep learning CT image analysis," arXiv Preprint *arXiv:200305037*, 2020.

28. S. Wang, et al., "A deep learning algorithm using CT images to screen for Corona virus disease (COVID-19)," *medRxiv*, vol. 31(8): 6096–6104, 2020.

29. M. Fu, et al., "Deep learning-based recognizing COVID-19 and other common infectious diseases of the lung by chest CT scan images," *medRxiv*, 2020. https://doi.org/10.1101/2020.03.28.20046045

30. X. Xu, et al., "Deep learning system to screen coronavirus disease 2019 pneumonia," arXiv Preprint *arXiv:*200209334, 2020.

31. P.J. Werbos, "Backpropagation through time: what it does and how to do it," in Proc. IEEE, vol. 78, pp. 550–1560, 1990. https://doi.org/10.1109/5.58337.

32. Y. Bengio, P. Simard, and P. Frasconi, "Learning long-term dependencies with gradient descent is difficult," *IEEE Trans. Neural. Netw.*, vol. 5, p.157, 2014.

33. S. Hochreiter, "The vanishing gradient problem during learning recurrent neural nets and problem solutions," *Int. J. Uncertain. Fuzziness Knowledge-Based Syst.*, vol. 6, pp. 107–116, 1998. doi10.1142/S0218488598000094.

34. P. Liu, X. Qiu, X. Chen, S. Wu, and X. Huang, "Multi-timescale long short-term memory neural network for modelling sentences and documents," in Conf. Proc. – EMNLP 2015 Conf. Empir. Methods Nat. Lang. Process., pp. 2326–2335, 2015. https://doi.org/10.18653/v1/D15-1280.

35. X.-H. Le, H.V. Ho, G. Lee, S. Jung, "Application of long short-term memory (LSTM) neural network for flood forecasting," *Water*, vol. 11, p. 1387, 2019. [**Google Scholar**]

36. Chen, G. "A gentle tutorial of recurrent neural network with error backpropagation," *arXiv:1610.02583*, 2016. [**Google Scholar**]

37. K. M. Tsiouris, V. C. Pezoulas, M. Zervakis, S. Konitsiotis, D.D. Koutsouris,and D. I. Fotiadis, "A long short-term memory deep learning network for the prediction of epileptic seizures using EEG signals," *Comput. Biol. Med.*, vol.99, pp. 24–37, 2018. [**Google Scholar**]

38. H. Sak, A. Senior, and F. Beaufays, "Long short-term memory recurrent neural network architectures for large scale acoustic modeling," in Fifteenth Annual Conference of INTERSPEECH, Dresden, Germany, 6–10 September 2015. [**Google Scholar**]

39. Y. Yu, X. Si, C. Hu, and J. Zhang, "A review of recurrent neural networks: LSTM cells and network architectures," *Neural. Comput.*, vol. 31, pp.1235–1270, 2019. doi:10.1162/neco_a_01199.

40. D. Singh, V. Kumar, and M. Kaur, "Classification of COVID-19 patients from chest CT images using multi-objective differential evolution-based convolutional neural networks," *Eur. Clin. Microbiol. Infecti. Dis.*, vol. 39, pp. 1379–1389, 2020.

41. I. D. Apostolopoulos and T. A. Mpesiana, "Covid-19: automatic detection from X-ray images utilizing transfer learning with convolutional neural networks," *Phys. Eng. Sci. Med.*, vol. 43, pp. 635–640, 2020.

42. https://engmrk.com/lenet-5-a-classic-cnn-architecture/

43. https://www.kaggle.com/blurredmachine/lenet-architecture-a-complete-guide

44. https://iq.opengenus.org/architecture-and-use-of-alexnet/

45. Alex Krizhevsky, Ilya Sutskever, Geoffrey E. Hinton. ImageNet Classification with Deep Convolutional Neural Networks, 2012. Communications of the ACM Volume 60 Issue 6 June 2017 pp 84–90. https://doi.org/10.1145/3065386

46. D. Dansana, R. Kumar, A. Bhattacharjee, D. J. Hemanth, D. Gupta, A. Khanna, and O. Castillo, Early diagnosis of COVID-19-affected patients based on X-ray and computed tomography images using deep learning algorithm, pp. 1–9, Aug. 2020, doi: 10.1007/s00500-020-05275-y.

47. K. Simonyan and A. Zisserman, "Very Deep Convolutional Networks for Large-Scale Image Recognition," Published as a conference paper at ICLR, 2015.

48. https://medium.com/coinmonks/paper-review-of-googlenet-inception-v1-winner-of-ilsvlc-2014-image-classification-c2b3565a64e7

49. C. Szegedy, et al., *Going Deeper with Convolutions* Proceedings of the IEEE Conference on Computer Vision and Pattern Recognition (CVPR), pp. 1 -9, 2015.

50. https://towardsdatascience.com/review-inception-v4-evolved-from-googlenet-merged-with-resnet-idea-image-classification-5e8c339d18bc

51. https://towardsdatascience.com/a-simple-guide-to-the-versions-of-the-inception-network-7fc52b863202

52. https://blog.paperspace.com/popular-deep-learning-architectures-densenet-mnasnet-shufflenet/#:~:text=ResNeXt%20Architecture&text=As%20seen%20in%20the%20table,groups%20in%20the%20grouped%20convolution.

53. S. Xie, et al., "Aggregated residual transformations for deep neural networks," *arXiv:1611.05431v2* [cs.CV] 11 Apr.2017.

6 Machine Learning Applications and Challenges to Protect Privacy in the Internet of Things

Mahadev Gawas
Govt College of Arts, Science and Commerce
Sanquelim, India

Aishwarya Parab
Govt College of Arts, Science and Commerce
Quepem, India

Hemprasad Y. Patil
SENSE, VIT
Vellore, India

CONTENTS

6.1 INTRODUCTION

The Internet of things (IoT) has grown to be a vital part of our everyday life. The technology of IoT can facilitate modernizations to increase the value of life and possess the ability to identify and comprehend the current environment. IoT is considered to be the most promising field among others in the history of computing. Some of the rapidly growing trends in IoT are smart cities, smart homes, smart cars, smart health care, smart natural disaster management, and smart urban management, as shown in Figure 6.1.

The IoT is a world where all devices that we use in our daily lives are connected to a network. We can use them collectively to accomplish difficult tasks that involve a greater intellect level. The IoT devices are outfitted by embedded sensors, actuators, processors, and transceivers in order to achieve this level of intelligence and interconnection. IoT can be considered an amalgamation of various technologies that work together, rather than it being a single technology.

An IoT system is multifaceted and requires consolidative ordering. A greater challenge lies in keeping up the larger attack surface off the IoT structure. The fixes provided by the IoT are based on the incorporation of information technology to

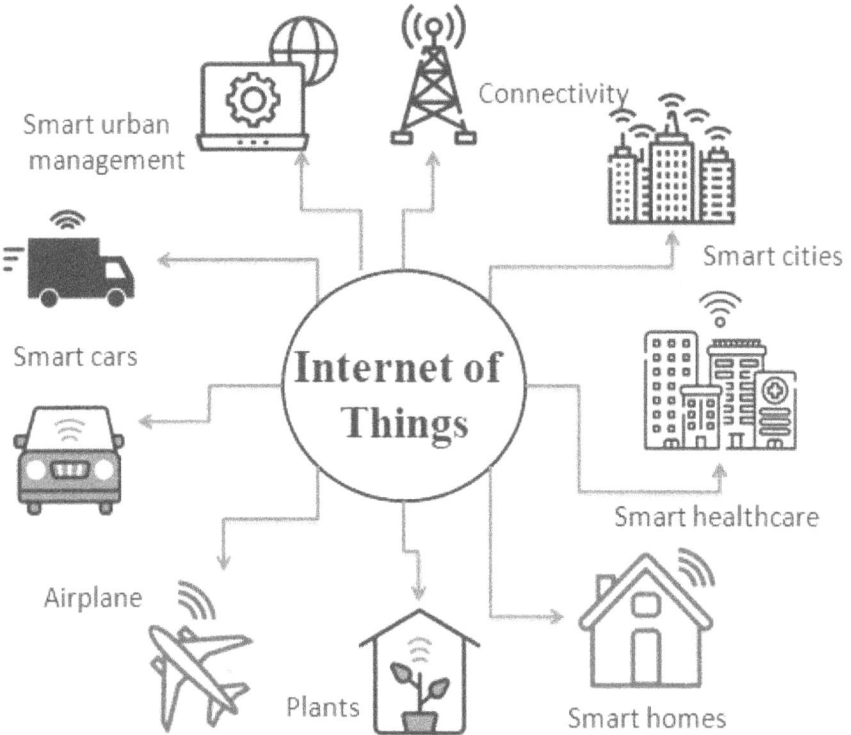

FIGURE 6.1 IoT applications.

store, retrieve, and process data and communications technology. For communication purposes between individuals or groups, it includes electronic systems.

The ease of access and probable information exchange collectively open tremendous additional opportunities for applications in IoT. Besides, to be capable enough to handle the total number of connected devices in the IoT, cognitive technologies and contextual intelligence are essential. Although there is a rapid growth in IoT, security and privacy issues are major concerns. These concerns raise significant questions in the IoT domain relating to privacy protection, which includes protocols, information organization, processes and protocols, and technology necessities. The definition of privacy varies for different applications. Security risks can arise when an IoT device is associated with your desktop or laptop. The risk of your data getting leaked while it is collected and transmitted to the IoT device can increase as a result of poor measures for security. In a consumer network arrangement where the devices are connected, this network is further connected with other systems. Any security vulnerabilities in the IoT device could be damaging to the consumer's network. These vulnerabilities are capable of attacking other systems and damaging them. Sometimes unauthorized people might take advantage of the security vulnerabilities to create risks to physical safety. In the IoT system, devices are organized through various hardware and software, so the probability of data leakage through illegal exploitation is high.

The extensive use of IoT applications has led to a significant consequence. For instance, in the deployment stage, the need to track the energy efficiency in the IoT systems is required. Besides that, security, problem-solving techniques, and interoperability with software applications also need attention. No feature can be avoided when considering the advantages of another [1]. This integration gives a different perspective for researchers of various fields to examine the challenges in the IoT system. However, novel security challenges are also introduced due to the distributive character of IoT devices. These characteristics present numerous issues in IoT devices. Besides, if the large volume of data generated by the IoT devices is not transmitted and analyzed steadily, then a significant confidentiality violation may occur. For huge and insufficient systems, the use of existing privacy mechanisms such as access control, authentication, encryption, and network and application security is challenging as each system will have its inbuilt vulnerabilities.

The numerous devices in IoT that are connected generate and exchange data such as system, network, hardware, process, and application data. Machine learning can take advantage of this data to direct and improve the functioning of the system. The large amount of data generated by IoT can be utilized by machine learning that can help to scrutinize the connections and services in IoT. This in turn will increase the security of users' data, owners, and collectors. Machine learning can improve operations such as authentication, access control, data aggregation, and regulatory compliance. Machine learning can keep data more private and protected by reducing the sharing of contextual raw data along with different components of IoT.

Machine learning techniques are known to understand the operational performance of the entire structure and its devices rather than just understanding the operational behavior of a specific device. Machine learning techniques can offer intelligence to systems to detect any irregular behavior and take action promptly.

This technique possibly will alleviate the effect of the attack and, in turn, guide a way to learn for the anticipation of upcoming events of comparable attacks based on a strong understanding of the current causes [2].

This chapter aims to highlight how machine learning can be used to overcome the security challenges in an IoT structure. In Section 6.2, we provide a general idea of the IoT system. We discuss the three-layer architecture with the potential vulnerabilities and security risks in each layer of the IoT system. Section 6.3 presents the most promising machine learning algorithms for securing the IoT system along with its advantages and disadvantages. We also present a summary table of the review of machine learning algorithms. In Section 6.4, we provide the conclusions derived from this survey. In the following section, we discuss the related works in IoT security to highlight the differences between the previous surveys and this survey.

6.1.1 WHY IoT SECURITY IS SO CHALLENGING

Despite the fact that IoT highlights attributes that are by far there in other models, IoT is greatly considered to have a diverse structure and unique challenges with respect to security and privacy.

A troublesome highlight of IoT devices could be that the devices can be detected distantly in an obtainable or impromptu organization. This leads to a heap of opportunities for a straightforward combination of the actual world and the digital world. The outcome is enhanced effectiveness, precision, and financial advantage, notwithstanding decreased human interference. Besides, IoT turns into an example of a broad class of actual-digital frameworks when sensors are added. The following points could encourage research for the unique challenges in IoT.

Size of the devices – given that the current security models do not consider the connection of a billion IoT objects, managing the size of the system will be challenging. Moreover, due to restricted resources, inadequate memory and energy storage could be utilized. These limitations induce the need to implement unique security procedures that can provide services with less storage load on the devices along with robust designs.

Smooth and uninterrupted interactions between the humans and the system could be another troublesome highlight of IoT devices. The IoT devices can now learn and simultaneously maintain the human requirements at the same time due to the latest trends in artificial intelligence and machine learning. Therefore the data will be exchanged between the IoT and the humans. A security breach can occur if any unauthorized user gets access to his data. Hence access control and privacy are important features in IoT.

IoT is an intricate network of numerous different types of objects that are interconnected with each other. Due to these varied types of objects, different types of standards and protocols will also exist. Researchers in the past have stated the connection between IoT devices and their neighboring objects [3]. On the other hand, the IoT network is considered to be exceptionally diverse and dynamic; therefore, the devices may go through impulsive changes, which can result in unexpected alterations in communications.

6.1.2 RELATED WORK

Numerous researchers have led studies on IoT security and have provided convenient points for the current safety measures. However, there are very few existing surveys that have mainly focused on machine learning-based applications for IoT security.

The execution of protection measures against a particular risk is subject to state-of-the-art attacks formed by intruders to discover a path around the current solutions. Besides, knowing which strategies are reasonable, ensuring that the IoT structure is a test in light is important due to the diverse amount of IoT applications [4]. Thus, creating efficient techniques for securing IoT devices is the need of the hour [4, 5].

Machine learning techniques are an amazing approach for studying data to learn about regular and irregular activities as per how IoT components and devices work together. In the IoT system, the data collected from every component can be used to establish regular samples of communication, which can help to discover any irregular behavior in the beginning process. Also, machine learning techniques may perhaps be significant in anticipating novel threats that are frequently transformations of earlier threats, since they can brilliantly foresee potential incomprehensible threats by gaining from the currently used models [6]. Figure 6.2 gives an overview of how machine learning can be used to get a prediction model from a training dataset. Subsequently, the IoT frameworks are obliged to comprise a change by simply encouraging protected communications among devices to safeguard insight empowered by machine learning and deep learning techniques for powerful and secure systems [2].

In recent years, researchers have examined the characteristic susceptibility of IoT devices along with the interconnection between the digital world and the real world [7]. The challenges are classified as those of privacy, trust, identification, and access control. Properties such as unreliability, integrity, confidentiality, and identification are required for data and network security. More solutions are classified as an encryption algorithm and mechanism, safety measures for communication, and protecting sensor data [8]. The level of security required depends on the different applications. Researchers have found that the challenges in IoT take a new dimension, which can be tedious to resolve with traditional methods. The emphasis is on the security approach that depends on a centralized architecture. This in turn makes the IoT application more complicated. Therefore to deal with security issues in IoT, distributed approaches can be used [9]. Another study states that the traditional security countermeasures and the scalability issues which arise are due to the high number of interconnected IoT devices [10]. Concerning security, information

FIGURE 6.2 Use of machine learning to get a prediction model.

secrecy, privacy, and integrity should be ensured. Authorization and authentication are just as important to prevent any unauthorized user from accessing the system. While in privacy requirements, protecting confidential data is of utmost importance as devices could manage sensitive information [11].

The IoT system can be considered as an integration of many layers; therefore, the security issues such as privacy protection do not just belong to a single layer [6]. Kai Zhao et al. have focused on the security issues related to the perception layer in the three-layer IoT structure.

The IoT network is susceptible to various threats that can destroy the system even with so many security mechanisms in place. Thus researchers have found another system, namely Intrusion Detection Systems (IDSs), to overcome these issues [12]. IDS can be used to monitor the working of a host or a network. It can alert the system administrator if any security threat is detected, which is why it is considered to be a vital tool for the safety of conventional systems.

The IDSs are classified based on certain aspects such as detection technique, IDS placement approach, security threat, and validation strategy [12]. Another research focuses on permissible concerns and narrow techniques to understand if the IoT structure can deal with its safety requirements [13].

The authors in Ref. [14] examined safety and protection in a dispersed IoT system. The scientists have listed difficulties that need attention and the precedence of dispersed methods as far as safety and protection are concerned [15]. The distributed IoT approach also permits various IoT devices to communicate regardless of a central system [16].

Nonetheless, in comparison to other studies, this study provides an exhaustive evaluation of the latest improvement in machine learning techniques in terms of security for IoT. We discuss several challenges in the three-layer architecture for the IoT system and the potential machine learning algorithms to overcome the shortcomings in the IoT security context.

6.2   OVERVIEW OF THE IoT SYSTEM

An overview of the IoT system is provided in this section. Many different existing structures describe the IoT system. For example, in Ref. [17] the authors have presented architecture with three layers. In Ref. [18] the authors have made use of service-based architectures. In Ref. [19] the authors have presented five- and seven-layer architectures. In our survey, we use the three-layer architecture as shown in Figure 6.3, and the potential vulnerabilities and security risks in each layer of the IoT system are presented.

6.2.1   PERCEPTION LAYER

Also known as a sensor layer is responsible for categorizing and classifying things and collecting data from them. Sensors like RFID and 2D barcode are attached to the objects to collect data. The type of sensors used depends on the requirement of applications. These sensors collect information related to location, environmental changes, motion, vibration, etc. Nonetheless, the attackers replace the sensors with

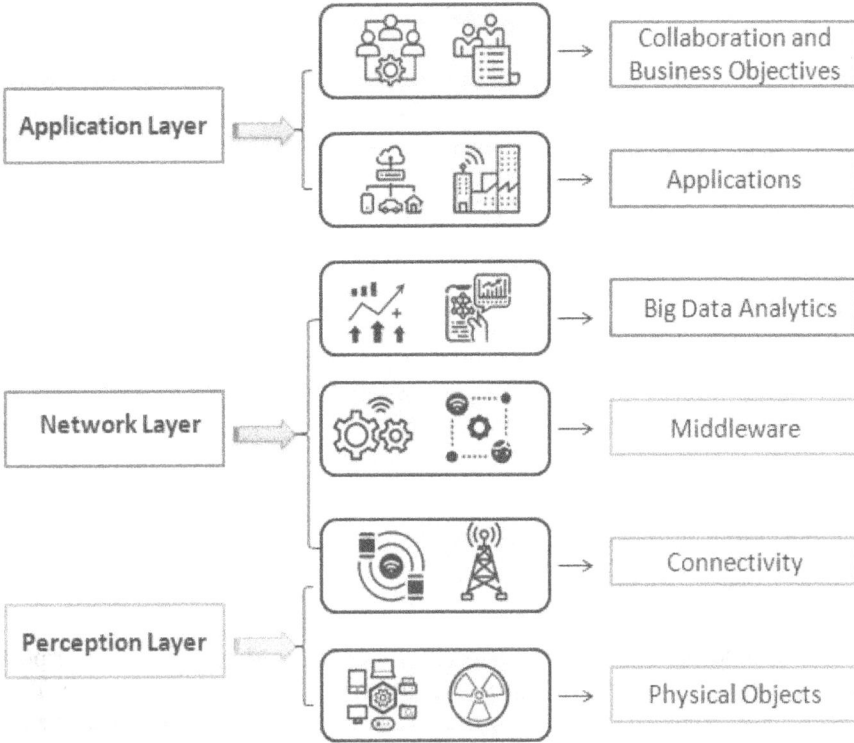

FIGURE 6.3   IoT architecture.

their own considering it as an easy target. As a result, the sensors are a common target for such threats [20-22]. Eavesdropping, which is one of the common security threats, is an illicit real-time attack where the attacker tries to intercept any confidential communications sent over an unsecured channel.

Node Capture is another dangerous attack in the perception layer. The attacker tries to manage the key node, for example, a gateway node. This key is stored in the memory [23]. In another type of attack, fake data is given as an input to the system by the attacker so that the transmission of real data is prevented. In a Replay attack, the attacker eavesdrops and tries to capture the real data sent. The same data is then sent to the receiver [24].

6.2.2   NETWORK LAYER

It is also known as the transmission layer. It acts as a link connecting the perception layer and the application layer. The data gathered from the devices is carried via a wired or wireless medium and transmitted to the sensors by this layer. The network layer is considered to be highly sensitive to attacks, and the major security issues are related to integrity and authentication. A common security threat could be the Denial of Service (DoS) Attack, wherein authenticated users are prevented from using the network resources [25]. Another type of attack is the man-in-the-middle (MITM) attack, wherein an intruder privately monitors transmission among the sender and the receiver. The MITM attacker can interrupt, modify, or even replace the communication traffic (this distinguishes a MITM from a simple eavesdropper). Besides, the victims believe the channel to be protected and are unaware of the intruder [26]. In another type of network layer attack, the data stored on the cloud or on any storage devices could be attacked and modified. This type of attack is known as a storage attack.

6.2.3 Application Layer

Applications that use the IoT structure are characterized in the application layer. IoT applications include intelligent homes, intelligent cities, and intelligent health. IoT is responsible for providing services and these services vary for different applications as it depends on the data collected by the sensors. In Ref. [27] the authors have listed various issues related to smart homes such as weak computational power and less storage. One of the common security threats is Cross-Site Scripting, which is a type of insertion threat. The intruder embeds a script on the client-side. Cross-Site Scripting allows an intruder to add a client-side script and modifies the application in an unauthorized manner [28]. In Malicious Code Attack a piece of code in the system can cause undesired effects and damage the system. This code can be scripted in any component of the system, which can cause serious effects to the system. This code can activate on its own or may require the user's attention and cannot be controlled by an antivirus. Due to a substantial amount of data caused by the numerous connected devices, issues such as data loss and network disturbance can occur. The ability to deal with mass data and data processing is the need of the hour.

6.3 POTENTIAL METHODS IN MACHINE LEARNING FOR SECURING IoT SYSTEMS

In recent years, learning algorithms have been used extensively. The recent progression of learning algorithms is determined as a result of the improvement of the latest methods and the accessibility of a large amount of data, along with the materialization of lesser estimating cost algorithms [29]. Machine learning has progressed impressively in recent years, beginning from a small-scale practice to greater applications in the real world [30].

Usually, the main aim of a learning algorithm is to advance the execution in order to accomplish a given assignment from experience, in assistance with training and learning. While learning intrusion detection, it is important to categorize the behavior as regular or irregular. We can accomplish advancement in execution by merely advancing the categorization accuracy. The training samples from which the algorithms learn are a combination of regular behavior. Algorithms for learning can be broadly categorized as supervised learning, unsupervised learning, and reinforcement learning (RL).

The classification or prediction model in supervised learning methods is formed based on learned mapping [31]. Its results are produced by verifying the input parameters. Furthermore, these techniques define the interaction between the input and the output parameters. Consequently, in supervised learning, the algorithms are trained by using the learning models, which are then used to foresee and categorize the new input at the initial stages [32]. The exceptional developments in supervised learning in recent years have been implemented in profound networks. The functional input is calculated by multi-layer networks [31].

6.3.1 MACHINE LEARNING TECHNIQUES FOR SECURING IoT SYSTEMS

Potential machine learning algorithms are discussed in this section. We have provided a brief on their advantages, disadvantages, and applications in terms of IoT security. Machine learning algorithms are broadly classified as supervised machine learning and unsupervised machine learning methods (see Figure 6.4).

6.3.1.1 Supervised Machine Learning

In supervised learning, the algorithm is trained on a dataset, i.e. there is input data along with its results. The information is split into two sets. The training dataset is utilized to train the network and test dataset which is utilized to predict the outcomes. The most common supervised machine learning algorithms are decision trees (DTs), support vector machines (SVMs), Bayesian algorithms, k-nearest neighbor (KNN), random forest (RF), association rule (AR) algorithms, and ensemble learning (EL) [33]. These algorithms are discussed in detail.

6.3.1.1.1 Decision Trees

Decision tree classifiers can predominantly identify and categorize attributes according to their value. A DT consists of nodes, which are the attributes. The branches represent the value of the node in the sample set. Beginning at the source node the samples will be classified based on the attribute value. The source node is the one that ideally separates the training set [14]. The difficulty while creating the DT lies in choosing the best attribute for the node. Different studies have used the information gain [34] and the Gini index [35] as the splitting criteria.

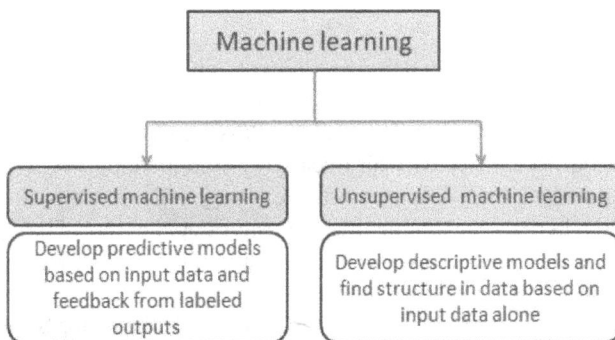

FIGURE 6.4 Classification of machine learning.

The fundamental idea is to choose the attribute which gives maximum information gain during the splitting of each node.

Most of the DT classifiers primarily consist of two cycles, namely building and classification [36]. When constructing a DT, the source node is selected according to the attribute that best splits the training data. The goal is to assign a source node in such a way that it lessens the gap between the classes and the training data. The process is followed for each subtree until the leaf nodes are found. Once the tree is constructed, in the classification process, the training samples with unidentified classes are classified beginning from the source node of the constructed tree during the building process. Subsequently, the path is followed by acquiring the inner nodes of the tree until the leaf node is obtained. Eventually, the predicted classes of the training samples are obtained [34].

The authors in Ref. [34] have given a brief on the key points to simplify the construction of DTs. To begin with, the size of the tree is reduced by applying pre-pruning and post-pruning techniques. This is followed by streamlining the area of the classes. And lastly, the algorithm to improve the search is applied. In the following steps, redundancy is eliminated throughout the search process, which then reduces the data features. Eventually, a set of rules are formed from the tree structure.

One of the drawbacks of DT-based classifiers are that they require a great storage space due to the way the tree is constructed. Also, DT-based classifiers are simpler if only a few trees are concerned. Nevertheless, in some applications, the model for classifying samples is complex and the computational complexity is high. This requires constructing huge trees with many nodes.

Applications such as intrusion detection make use of the DT classifier as its primary classifier or in combination with other machine learning classifiers [37, 38]. According to an earlier study, the researchers used a fog-based system to secure IoT devices. A DT classifier was used to examine the network traffic to identify any mistrustful sources [39].

6.3.1.1.2 Support Vector Machines

The working principle of SVM is to create a hyperplane in an N-dimensional area. This hyperplane uniquely classifies the data points. The dimension of the hyperplane is subject to the number of features. The hyperplane can categorize and divide in a way that the gap between the hyperplane and the neighboring sample points is more for each class [40].

SVM is highly preferred as it is known for its ability to generalize and it is fit for datasets with many feature attributes but fewer sample points [41, 42]. SVMs were created to classify the linearly separable classes in a two-dimensional plane. It also comprises different classes with linearly discrete data points. An outstanding hyperplane is created by SVMs that provide a very high scope by increasing the gap between the hyperplane and the neighboring sample points [43, 44]. SVMs are known for scalability and their capability to execute synchronized intrusion detection. The training samples are also updated dynamically.

Figure 6.5 gives an overview of how to create hyperplanes for SVM to classify the data. The hyperplanes a, b, and c shown in the figure can correctly categorize two things. According to the learning criteria for SVM, it will select the hyperplane c,

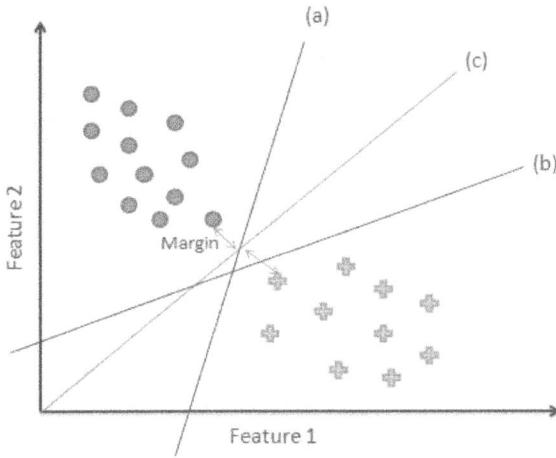

FIGURE 6.5 SVM data classification.

which will increase the distance between the data and in turn maximize its ability to generalize. As a result, SVM will be able to categorize any input data correctly in comparison to other classifiers, even if it is situated near the hyperplane. The diverse applications of SVM vary in terms of security and it is mostly used in intrusion detection and authentication [45-47]. SVMs are used in various applications as it provides high performance. It is also used to resolve issues related to categorizing nonlinear datasets [48, 49]. Previously, SVMs were used for the security of smart grid applications. In Ref. [50], researchers have studied the detection of any malicious activity in the smart grid. This study proved that the use of machine learning algorithms to secure smart grids is effective as compared to conventional techniques.

6.3.1.1.3 Bayesian Theorem-Based Algorithms

The basic idea of the Bayes theorem is the possibility of an event to happen based on the past occurrences associated with the event [51]. Bayes theorem can efficiently analyze the possibility of an attack on the network using past information. Naïve Bayes classifier is the most frequently used machine learning algorithm. It is a type of supervised learning classifier widely used because of its simplicity [52].

The Naïve Bayes classifier computes the posterior probability and makes use of the Bayes theorem to forecast the likelihood of a given dataset. It examines the dataset for unidentified samples and provides and provides a particular tag to these samples. It can be used in intrusion detection to categorize the traffic as regular or malicious. A set of attributes are used by the Naïve Bayes classifiers to categorize the network congestion. The attributes used are connection interval, protocol, and status flag. These attributes could depend on each other but they are worked on separately by the Naïve Bayes classifier. These attributes can be used to predict network traffic as malicious or regular. For that reason, it is called "naïve." The Naïve Bayes classifier finds its applications in intrusion detection over the network [53, 54] and anomaly detection [55, 56]. The striking features of the Naïve Bayes classifier are straightforward execution, simplicity, and sturdiness to unsuitable attributes and require a smaller training set and can be applied to both single and multiple class categorization. Nonetheless, the classifier may not group the required signals from the connections and communications among the attributes. For precise categorization of the attributes, the relations between the attributes are important [57].

6.3.1.1.4 k-Nearest Neighbor

KNN classifiers are used to resolve issues related to classification and regression. It makes use of the Euclidean distance to calculate the distance [58-60]. In simpler words, the KNN algorithm presumes that similar things are close to each other. Figure 6.6 demonstrates the arrangement in KNN. Here the data samples are categorized. The green circles in the figure depict irregular behavior, while the blue circles depict regular behavior. The unknown sample depicted in red is to be categorized as regular or irregular behavior. This unknown sample will be classified based on the votes from its nearest neighbors. The class of this unknown sample will be decided by the highest votes from its nearest neighbors.

As shown in Figure 6.6, if k = 1, i.e. the classification depends on one neighbor, then the classifier will assign it as a regular behavior. If k = 2, i.e. it depends on two neighbors, then the classifier will assign it as a regular behavior. This is so because the closest neighbors are blue, which depicts regular behavior. If k = 3 or k = 4, this will be classified as an irregular behavior. This is for the reason that the closest three and four circles are green. To find the best and most favorable value for k for any given set of samples, it is very important to test many different values of k [61–63]. Given the fact that KNN is an easy classification algorithm and is also feasible for vast data samples [64, 65], the most favorable value for k will differ based on the type of datasets. Thus finding the optimal k value could be a complex task [66, 67].

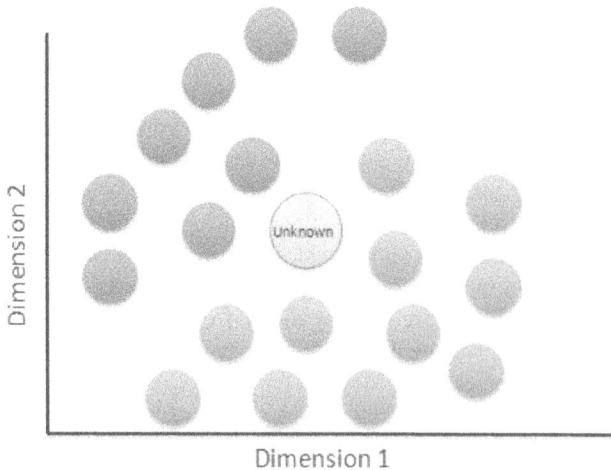

FIGURE 6.6 KNN working principle.

6.3.1.1.5 Random Forest

Random forest is a combination of numerous DTs that are grouped for a strong and accurate prediction model that gives better outcomes [68, 69]. Thus, a RF comprises several DTs that are constructed at random and are skilled to vote for a class. The final output is chosen based on maximum votes [68]. Random forest classifier is primarily formed using DTs but both these algorithms vary considerably.

Usually, after the dataset is provided to the network, DTs devise a set of rules which are used to categorize the new input. To develop a subset of rules, RF makes use of DTs. Therefore, the result of classification is the average of the outcomes. The RF classifier goes past the feature selection process, and it needs only a few inputs [24]. Nevertheless, using RF for applications that require huge training sets may not be feasible as it will require creating many DTs. Anomaly detection and intrusion detection make use of RF classifiers [70, 71].

A study [72] has proved that, for smaller datasets, the RF classifier gives better outcomes as compared to other supervised earning classifiers. In this study, the RF classifier was trained with the attributes received from the network to rightly identify the IoT devices. The authors have concluded that the RF algorithm can be used to rightly identify any unofficial devices [73].

6.3.1.1.6 Association Rule Algorithms

The working principle of AR algorithms is to find an unidentified variable by examining the connections between different variables in a set of samples [74]. For instance, consider two variables, A and B, in a training set T. The AR algorithm will find the interconnection between these variables and examine the connection between them to develop a working model. This model is then used to calculate the class of the latest samples.

Association rule algorithms make use of the repeated variable set. These are commonly used variables that exist together in a training set [31]. In Ref. [75], the authors have studied the TCP/IP relations and the type of threats using AR algorithms. To identify the type of attack, different variables like name, sender's port, receiver's port, and IP of the source were studied. In Ref. [76], the authors have presented an AR algorithm that worked well against intrusion detection. The use of fuzzy ARs resulted in a greater rate for detection and a lesser false positive rate. Association rule methods are not frequently used in comparison to other machine learning techniques; therefore more research is required to verify the use of ARs for IoT security. Association rule algorithms are not as effective as they enlarge quickly [77]. Therefore, efficiency is a major drawback of AR algorithms.

6.3.1.1.7 Ensemble Learning

Ensemble learning is a machine learning technique that is considered to be the most promising. The algorithm gives a combined output based on several machine learning methods to advance the categorization process. Ensemble learning methods target to merge consistent or diverse classifiers to get the ultimate outcome [78]. In an early development stage in machine learning, each method has its pros and cons for a particular application or for a particular dataset. In Ref. [79] the study shows that the optimal method varies based on the applications. But the basic learning technique utilized depends on the information. As we know that the type of data differs for different applications, the optimal method for one application may not be as good for another application. As a result, combining different classifiers is proved to be the best solution to increase efficiency. Ensemble learning makes use of different methods; therefore, it lessens inconsistency. Combining different classifiers has proved to be effective in EL, and hence it can get accustomed easily to a problem [80]. Since EL makes use of several classifiers rather than a single classifier, the time complexity of such systems will be high [81, 82]. Ensemble learning finds its applications in malware, anomaly, and intrusion detection [83–86].

6.3.1.2   Unsupervised Machine Learning

Here, we discuss the regular unsupervised machine learning approaches (i.e. k-means clustering and principal component analysis [PCA]) and their advantages, disadvantages, and applications in IoT security.

6.3.1.2.1   *k-Means Clustering*

This technique is based on unsupervised machine learning. The working principle of this method is to find the clusters in the training set. The algorithm generates k clusters. Features with the same samples will belong to the same clusters. The algorithm works by assigning sample points to different clusters based on the given features [87]. It works in an iterative fashion to assign the sample points and finally generate the outcomes. The dataset and the k clusters are given as the input to the algorithm. To begin with, the centroids are projected and each sample point is assigned to its nearest cluster. This principle is based on the squared Euclidean distance. Once the sample points are allotted to the clusters, the centroids are calculated again by finding the mean of samples allotted to that particular cluster. This process will occur iteratively as long as no samples are pending [88, 89]. The disadvantage of the k-means clustering is to select the value of k. This method is utilized for anomaly detection by differentiating regular and irregular behavior [90, 91]. In Ref. [92] the study proposed by the researchers uses the method of k-means using DTs for anomaly detection. Nonetheless, k-means is proved to be less efficient than the other machine learning algorithms.

6.3.1.2.2   *Principal Component Analysis*

PCA is a method used for feature reduction. It can be used for changing huge variable sets into smaller meaningful sets. The reduced set preserves all the necessary information. The PCA method reduces the interrelated features into unrelated features. These features are called principal components [93-95]. Thus, PCA can be majorly utilized for selecting features to identify concurrent intrusion detection for IoT systems. In a past study, the researchers have presented an idea that combining PCA with other machine learning classifiers gives efficient results concurrent IoT systems [96-98].

We have discussed the most promising machine learning techniques for securing IoT devices along with the advantages, disadvantages, and applications in IoT security. The use of machine learning techniques is limited in cases where the data is less [99, 100]. Table 6.1 presents a summary and use of machine learning methods for IoT security.

TABLE 6.1
Summary of Machine Learning Algorithms for IoT Security

Algorithms	Working Principle	Applications
Decision trees	Techniques based on decision trees develop a prediction model to learn from training samples. The pre-trained model is then used to predict the class of the new sample DT is simple and easy to use	Intrusion detection and irregular traffic sources
Support vector machines	A splitting hyperplane is formed in the feature dimension. The gap between the hyperplane and neighboring points is increased It is popularly known for generalization capability and is suitable for data with large feature attributes but smaller sample points	Malware and intrusion detection malware and attacks in smart grids
Naïve Bayes	Used to calculate the posterior probability. Works on Bayes' theorem to calculate the probability that a particular feature set of unlabeled samples fits a specific label. It is a simple model	Network intrusion detection
k-nearest neighbor	It categorizes a new sample depending on its nearest neighbor's votes.	Detection of intrusions and anomalies
Random Forest	RF allows several DTs to be created and joined together, which results in an accurate model. It is robust and needs lesser input parameters	Detection of intrusion anomalies, DDoS attacks and unauthorized IoT devices
Association rule	Examines the connection between a training dataset to find the relations between the samples, which are then used to create a model. The new samples class is predicted using this model	Detection of intrusion
Ensemble learning	The outcomes of various machine learning algorithms are combined to create a combined outcome for improved performance It minimizes inconsistency and is vigorous to over-fitting	Anomalies, malware, and intrusion detection
k-means clustering	It is a type of unsupervised learning. It identifies clusters based on similarities in attributes. The algorithm generates k clusters. It does not require labeled data; therefore it can be utilized for anonymization	Private data anonymization in an IoT system
Principle component analysis	It is a method used for feature reduction. It can be used for changing huge variable sets into smaller meaningful sets It can reduce dimensionality, which in turn will reduce the complexity	Model attributes are reduced to detect real-time systems in IoT

6.4 MACHINE LEARNING APPLICATION IN INDUSTRY

Machine learning and IoT are widely used in different industries such as urban cities, smart homes, and healthcare. In modern manufacturing industries, the manufacturers make use of machine learning applications in IoT by keeping track of the quality of the products and their building process [101, 102]. This is achieved by making use of high-level automation systems. For example, in the steel manufacturing industry, the manufacturers of steel utilize the data to analyze and understand the working of the system. Input data such as images and videos are used to form the structured data such as the quality of the product, which will give an idea of how the outcome of the product will be [103].

Table 6.2 summarizes the machine learning applications in IoT. A majority of the industry applications make use of machine learning to convert the raw data to a distinct outcome.

Energy, water, and gas industries conserve resources by predicting their usage and it can be dynamically allocated. In the past, these processes were carried out in different ways but did not generate accurate results. Machine learning and IoT have greatly improved this process by keeping track of all the devices and have further helped in the dynamic allocation and load balancing [104–106].

TABLE 6.2
Machine Learning Applications in Industry

Industries	Services	Devices and Sensors	Analysis and Outcomes
Energy, water, gas utilities	• Synchronized data collection • Predicting demand supply • load balancing • Dynamic allocation	• Meters used for utilities	• Significant cost and resource savings by users connected over this network
Modern manufacturing	• Production line mechanization • Monitoring remotely • Diagnostics with floor sensors • Tools management	• IoT devices staged on objects and embedded in machines • Controllers or gateways • Cameras	• Optimal scheduling of production lines • Anomaly and emission detection • Reduced cost and improved quality
Healthcare	• Consulting doctor remotely • Old age care • Management of disease • Wellness and fitness programs	• Personal devices • Wearable medical devices • Mobile phones	• Synchronized disease management • Reduced cost
Insurance	• Property damage prediction • Remote assessment and inspection • Collection of user data	• Sensors that describe the state and usage of the insured unit	• Development of newer insurance models
Consumer goods and retail	• Consumer preferences • Synchronized and precise data collection • Maintaining inventory	• Supply chain analytics improved with concurrent data • Synchronized user profiling • Analytics to extract context from raw sensor data	• Targeted branding and advertising • Development of newer applications
Transportation	• Instantaneous tracking of shipping vehicles • Managing and tracking units	• Sensors, RFID tags • Mounted gateway devices	• Improved service levels • Reduced costs

In the manufacturing industries, manpower is reduced by making use of IoT systems with cameras and controllers. Any irregular activity can be detected and alerted from a prompt action against unauthorized parties. Moreover, future predictions can improve the efficiency and productivity of the systems. Machine learning and IoT are largely advanced in the health care industry. Different wearable devices keep track of the patient, which helps the doctors in the personalized and accurate analysis. The same process also applies to the insurance industry [107, 108]. Personal devices can help gather data that can be further analyzed for suitable situations.

In the retail industry, predicting customer needs has become easy. The data can be gathered from the Internet by making use of different applications and the sensors can be placed on the stores.

Sensors attached to the vehicles can be used to keep track of shipping and transportation. These applications in the transportation industry can help in the efficient allocation of buses based on their timings. This dynamic allocation of limited resources can help in reducing the cost of the operation.

6.5   FUTURE WORK

Regardless of all the efforts, numerous specialized protection challenges have stayed unaddressed in IoT. These difficulties can hinder the development of privacy-preserving technologies. With regards to data privacy and policies, machine learning system in the IoT framework can use information from various categories like security arrangements, recorded information, client information, and access information to normalize protection approaches and information practices, and investigate them to guarantee the data is as per guidelines, arrangements, or clients and frameworks necessities.

Dynamic access control is accountable for controlling and directing security in IoT. To recognize unauthorized entities in the framework, machine learning calculations can analyze links between different components, find data with regards to the relations between them, like informal organization cooperation, and utilize this extra information to improve IoT security and presentation.

A large number of gadgets are associated with IoT systems over the world. Therefore standardization and interoperability are two important aspects. They have divergent qualities and use diverse information protection techniques. Standardization can diminish the gaps among conventions and decrease system complexity.

Compliance with guidelines is a necessity for any innovation, and IoT isn't an exemption. Because of the volume and variety of protocols in IoT, the traditional mechanisms for compliance are not viable with the present necessities. Before long, it will be unavoidable that these cycles will be mechanized, and machine learning will have a basic impact intending to this objective. Machine learning will help them work smarter and quicker.

Future research could use the insight capacity of machine learning techniques to plan security strategies that can adequately fulfill different security compromises under various activity modes inside a predefined application.

6.6 CONCLUSION

Machine learning applications create modern visions and ideas. These ideas are given to the IoT systems to examine changes and to upgrade the systems. The quick development of IoT needs a powerful and innovative approach for security and confidentiality. Since various structures with different problem-solving techniques and inadequate assets interrelate in an IoT structure, the necessity to accomplish this task is complex. To protect the data in IoT, exhaustive use of machine learning methods is being used popularly. This is because a huge amount of data is accessible and advancements in machine learning.

In this study, we identify and classify the sources of data in IoT and discuss the use of these data resources in machine learning for strong privacy management. We provide a complete review of the potential uses of machine learning techniques. These techniques are then differentiated in terms of their applications, advantages, and disadvantages in IoT security.

By examining the machine learning techniques and providing its basic features, along with the IoT applications, we have provided a basis for researchers to identify the basic modules of IoT applications and exploit relevant techniques according to their requirements.

REFERENCES

1. S. Ray, Y. Jin, A. Raychowdhury, "The changing computing paradigm with internet of things: a tutorial introduction," IEEE Design & Test, vol. 33, no. 2, pp. 76–96, April 2016.
2. M. A. Al-Garadi, A. Mohamed, A. K. Al-Ali, X. Du, I. Ali M. Guizani, "A survey of machine and deep learning methods for Internet of Things (IoT) security," IEEE Communications Surveys & Tutorials, vol. 22, no. 3, pp. 1646–1685, third quarter 2020.
3. M. A. Zarandi, R. A. Dara, E. Fraser, "A survey of machine learning-based solutions to protect privacy in the internet of things," Computers & Security, vol. 96, p. 101921, 2020.
4. P. P. Ray, "A survey on Internet of Things architectures," Journal of King Saud University – Computer and Information Sciences, vol. 30, no. 3, pp. 291–319, 2018.
5. G. Rathee, S. Garg, G. Kaddoum, B. J. Choi, "Decision-making model for securing IoT devices in smart industries," IEEE Transactions on Industrial Informatics, vol. 17, no. 6, pp. 4270–4278, June 2021.
6. K. Zhao, L. Ge, "A survey on the internet of things security," in 2013 Ninth International Conference on Computational Intelligence and Security, pp. 663–667, Emeishan, China. 2013.
7. A. R. Sfar, E. Natalizio, Y. Challal, Z. Chtourou, "A roadmap for security challenges in the Internet of Things," Digital Communications and Networks, vol. 4, no. 2, pp. 118–137, 2018.
8. Y. Meng, J. Li, "Data sharing mechanism of sensors and actuators of industrial IoT based on blockchain-assisted identity-based cryptography," Sensors, vol. 21, p. 6084, 2021.
9. D. E. Kouicem, A. Bouabdallah, H. Lakhlef, "Internet of things security: a top-down survey," Computer Networks, vol. 141, pp. 199–221, 2018.
10. S. Sicari, A. Rizzardi, L.A. Grieco, A. Coen-Porisini, "Security, privacy and trust in Internet of Things: the road ahead," Computer Networks, vol. 76, pp. 146–164, 2015.
11. K. M. Sadique, R. Rahmani, P. Johannesson, "Towards security on Internet of Things: applications and challenges in technology," Procedia Computer Science, vol. 141, pp. 199–206, 2018.

12. B. B. Zarpelão, R. S. Miani, C. T. Kawakani, S. C. de Alvarenga, "A survey of intrusion detection in Internet of Things," Journal of Network and Computer Applications, vol. 84, pp. 25–37, 2017.
13. R. H. Weber, "Internet of Things – new security and privacy challenges," Computer Law & Security Review, vol. 26, no. 1, pp. 23–30, 2010.
14. R. Roman, J. Zhou, J. Lopez, "On the features and challenges of security and privacy in distributed internet of things," Computer Networks, vol. 57, no. 10, pp. 2266–2279, 2013.
15. E. Fazeldehkordi, O. Owe, J. Noll, "Security and privacy in IoT systems: a case study of healthcare products," 2019 13th International Symposium on Medical Information and Communication Technology (ISMICT), pp. 1–8, 2019.
16. J. Granjal, E. Monteiro, J. S. Silva, "Security for the Internet of Things: a survey of existing protocols and open research Issues," IEEE Communications Surveys & Tutorials, vol. 17, no. 3, pp. 1294–1312, third quarter 2015.
17. A. H. Ngu, M. Gutierrez, V. Metsis, S. Nepal, Q. Z. Sheng, "IoT middleware: a survey on issues and enabling technologies," IEEE Internet of Things Journal, vol. 4, no. 1, pp. 1–20, Feb. 2017.
18. H. Geng, "Networking protocols and standards for internet of things," in Internet of Things and Data Analytics Handbook, pp. 215–238, Wiley, California, 2017.
19. J. Lin, W. Yu, N. Zhang, X. Yang, H. Zhang, W. Zhao, "A survey on Internet of Things: architecture, enabling technologies, security and privacy, and applications," IEEE Internet of Things Journal, vol. 4, no. 5, pp. 1125–1142, Oct. 2017.
20. H. Suo, J. Wan, C. Zou, J. Liu, "Security in the Internet of things: a review," in Proceedings of the 2012 International Conference on Computer Science and Electronics Engineering (ICCSEE), pp. 648–651, Hangzhou, China. 23–25 March 2012.
21. D. Kozlov, J. Veijalainen, Y. Ali "Security and privacy threats in IoT architectures," in Proceedings of the 7th International Conference on Body Area Networks, pp. 256–262, Oslo, Norway. 24–26 February 2012; Brussels, Belgium: ICST (Institute for Computer Sciences, Social-Informatics and Telecommunications Engineering).
22. X. Xiaohui. "Study on security problems and key technologies of the Internet of things," in Proceedings of the 5th International Conference on Computational and Information Sciences (ICCIS), pp. 407–410, Shiyan, China. 21–23 June 2013.
23. M. V. Bharathi, R. C. Tanguturi, C. Jayakumar, K. Selvamani. "Node capture attack in wireless sensor network: a survey," in Proceedings of the 2012 IEEE International Conference on Computational Intelligence & Computing Research (ICCIC), pp. 1–3, Coimbatore, India. 18–20 December 2012.
24. D. Puthal, S. Nepal, R. Ranjan, J. Chen, "Threats to networking cloud and edge datacenters in the internet of things," IEEE Cloud Computing, vol. 3, no. 3, pp. 64–71, May–June 2016.
25. S. Prabhakar, "Network security in digitalization: attacks and defence," International Journal of Research in Computer Applications and Robotics, vol. 5, no. 5, pp. 46–52, May 2017.
26. M. Conti, N. Dragoni, V. Lesyk, "A survey of man in the middle attacks," IEEE Communications Survey & Tutorials, vol. 18, pp. 2027–2051, 2016.
27. B. Ali, A. I. Awad, "Cyber and physical security vulnerability assessment for IoT-based smart homes," Sensors, vol. 18, pp. 817, 2018.
28. S. Gupta, B. B. Gupta, "Cross-Site Scripting (XSS) attacks and defense mechanisms: classification and state-of-the-art," International Journal of System Assurance Engineering Management, vol. 8, pp. 512–530, 2017.
29. E. Bertino, N. Islam, "Botnets and internet of things security," Computer, vol. 50, no. 2, pp. 76–79, 2017.
30. C. Kolias, G. Kambourakis, A. Stavrou, J. Voas, "DDoS in the IoT: Mirai and other botnets," Computer, vol. 50, no. 7, pp. 80–84, 2017.

31. M. I. Jordan, T. M. Mitchell, "Machine learning: trends, perspectives, and prospects," Science, vol. 349, no. 6245, pp. 255–260, 2015.
32. J. Franklin, "The elements of statistical learning: data mining, inference and prediction," The Mathematical Intelligencer, vol. 27, no. 2, pp. 83–85, 2005.
33. W. Du, Z. Zhan, "Building decision tree classifier on private data," in Proceedings of the IEEE international conference on Privacy, security and data mining-Volume 14, pp. 1–8. 2002: Australian Computer Society, Inc.
34. S. B. Kotsiantis, I. Zaharakis, P. Pintelas, "Supervised machine learning: a review of classification techniques," Emerging Artificial Intelligence Applications in Computer Engineering, vol. 160, pp. 3–24, 2007.
35. J. R. Quinlan, "Induction of decision trees," Machine Learning, vol. 1, no. 1, pp. 81–106, 1986.
36. J. Schmidhuber, "Deep learning in neural networks: an overview," Neural Network, vol. 61, pp. 85–117, 2015.
37. K. Goeschel, "Reducing false positives in intrusion detection systems using data-mining techniques utilizing support vector machines, decision trees, and Naive Bayes for off-line analysis," in SoutheastCon, pp. 1–6, 2016: IEEE.
38. G. Kim, S. Lee, S. Kim, "A novel hybrid intrusion detection method integrating anomaly detection with misuse detection," Expert Systems with Applications, vol. 41, no. 4, pp. 1690–1700, 2014.
39. S. Alharbi, P. Rodriguez, R. Maharaja, P. Iyer, N. Subaschandrabose, Z. Ye, "Secure the internet of things with challenge response authentication in fog computing," in 2017 IEEE 36th International on Performance Computing and Communications Conference (IPCCC), pp. 1–2. 2017: IEEE.
40. S. Tong, D. Koller, "Support vector machine active learning with applications to text classification," Journal of Machine Learning Research, vol. 2, no. Nov, pp. 45–66, 2001.
41. V. Vapnik, The Nature of Statistical Learning Theory. Springer Science & Business Media, Springer, New York, NY, 2013.
42. A. L. Buczak, E. Guven, "A survey of data mining and machine learning methods for cyber security intrusion detection," IEEE Communications Surveys & Tutorials, vol. 18, no. 2, pp. 1153–1176, 2015.
43. H.-S. Ham, H.-H. Kim, M.-S. Kim, M.-J. Choi, "Linear SVM-based android malware detection for reliable IoT services," Journal of Applied Mathematics, vol. 2014, 10 pages, Article ID 594501, 2014.
44. L. Lerman, G. Bontempi, O. Markowitch, "A machine learning approach against a masked AES," Journal of Cryptographic Engineering, vol. 5, no. 2, pp. 123–139, 2015.
45. W. Hu, Y. Liao, V. R. Vemuri, "Robust support vector machines for anomaly detection in computer security," in ICMLA, pp. 168–174, 2003.
46. Y. Liu, D. Pi, "A Novel Kernel SVM algorithm with game theory for network intrusion detection," KSII Transactions on Internet & Information Systems, vol. 11, no. 8, pp. 4043–4060, 2017.
47. C. Wagner, J. François, T. Engel, "Machine learning approach for IP-flow record anomaly detection," in International Conference on Research in Networking, pp. 28–39. 2011: Springer.
48. A. Heuser, M. Zohner, "Intelligent machine homicide," in International Workshop on Constructive Side-Channel Analysis and Secure Design, pp. 249–264, Springer, Berlin & Heidelberg, 2012.
49. G. D'Agostini, "A multidimensional unfolding method based on Bayes' theorem," Nuclear Instruments and Methods in Physics Research Section A: Accelerators, Spectrometers, Detectors and Associated Equipment, vol. 362, no. 2–3, pp. 487–498, 1995.
50. I. Ullah, Q. H. Mahmoud, "An intrusion detection framework for the smart grid," 2017 IEEE 30th Canadian Conference on Electrical and Computer Engineering (CCECE), 2017, pp. 1–5.

51. M. Ozay, I. Esnaola, F. T. Y. Vural, S. R. Kulkarni, H. V. Poor, "Machine learning methods for attack detection in the smart grid," IEEE Transactions on Neural Networks and Learning Systems, vol. 27, no. 8, pp. 1773–1786, 2016.

52. G. E. Box, G. C. Tiao, Bayesian Inference in Statistical Analysis. John Wiley & Sons, Wisconsin, Madison, 2011.

53. M. Panda, M. R. Patra, "Network intrusion detection using Naive Bayes," International Journal of Computer Science and Network Security, vol. 7, no. 12, pp. 258–263, 2007.

54. S. Mukherjee, N. Sharma, "Intrusion detection using Naive Bayes classifier with feature reduction," Procedia Technology, vol. 4, pp. 119–128, 2012.

55. S. Agrawal, J. Agrawal, "Survey on anomaly detection using data mining techniques," Procedia Computer Science, vol. 60, pp. 708–713, 2015.

56. M. Swarnkar, N. Hubballi, "OCPAD: one class Naive Bayes classifier for payload based anomaly detection," Expert Systems with Applications, vol. 64, pp. 330–339, 2016.

57. A. Y. Ng, M. I. Jordan, "On discriminative vs. generative classifiers: A comparison of logistic regression and Naive Bayes," Advances in Neural Information Processing Systems, vol. 14, pp. 841–848, 2002.

58. P. Soucy, G. W. Mineau, "A simple KNN algorithm for text categorization," in Proceedings IEEE International Conference on Data Mining, pp. 647–648, 2001: IEEE.

59. Y. Liao, V. R. Vemuri, "Use of k-nearest neighbor classifier for intrusion detection1," Computers & Security, vol. 21, no. 5, pp. 439–448, 2002.

60. A. O. Adetunmbi, S. O. Falaki, O. S. Adewale, B. K. Alese, "Network intrusion detection based on rough set and k-nearest neighbour," International Journal of Computing and ICT Research, vol. 2, no. 1, pp. 60–66, 2008.

61. C.-F. Tsai, Y.-F. Hsu, C.-Y. Lin, W.-Y. Lin, "Intrusion detection by machine learning: a review," Expert Systems with Applications, vol. 36, no. 10, pp. 11994–12000, 2009.

62. L. Li, H. Zhang, H. Peng, Y. Yang, "Nearest neighbors based density peaks approach to intrusion detection," Chaos, Solitons & Fractals, vol. 110, pp. 33–40, 2018.

63. A. R. Syarif, W. Gata, "Intrusion detection system using hybrid binary PSO and K-nearest neighborhood algorithm," in 2017 11th International Conference on Information & Communication Technology and System (ICTS), pp. 181–186. 2017: IEEE.

64. Z. Deng, X. Zhu, D. Cheng, M. Zong, S. Zhang, "Efficient kNN classification algorithm for big data," Neurocomputing, vol. 195, pp. 143–148, 2016.

65. M.-Y. Su, "Real-time anomaly detection systems for Denial-of-Service attacks by weighted k-nearest-neighbor classifiers," Expert Systems with Applications, vol. 38, no. 4, pp. 3492–3498, 2011.

66. H. H. Pajouh, R. Javidan, R. Khayami, D. Ali, K.-K. R. Choo, "A two-layer dimension reduction and two-tier classification model for anomaly-based intrusion detection in IoT backbone networks," IEEE Transactions on Emerging Topics in Computing, vol. 7, pp. 314–323, 2019.

67. W. Li, P. Yi, Y. Wu, L. Pan, J. Li, "A new intrusion detection system based on KNN classification algorithm in wireless sensor network," Journal of Electrical and Computer Engineering, vol. 2014, 8 pages, Article ID 240217, 2014.

68. L. Breiman, "Random forests," Machine Learning, vol. 45, no. 1, pp. 5–32, 2001.

69. D. R. Cutler et al., "Random forests for classification in ecology," Ecology, vol. 88, no. 11, pp. 2783–2792, 2007.

70. J. Zhang, M. Zulkernine, "A hybrid network intrusion detection technique using random forests," in The First International Conference on Availability, Reliability and Security, pp. 8–269. 2006: IEEE.

71. Y. Chang, W. Li, Z. Yang, "Network intrusion detection based on random forest and support vector machine," in 2017 IEEE International Conference on Computational Science and Engineering (CSE) and Embedded and Ubiquitous Computing (EUC), vol. 1, pp. 635–638. 2017: IEEE.

72. R. Doshi, N. Apthorpe, N. Feamster, "Machine learning DDoS detection for consumer internet of things devices," arXiv Preprint arXiv:1804.04159, 2018.

73. Y. Meidan et al., "Detection of unauthorized IoT devices using machine learning techniques," arXiv Preprint arXiv:1709.04647, 2017.

74. R. Agrawal, T. Imieliński, A. Swami, "Mining association rules between sets of items in large databases," ACM Sigmod Record, vol. 22, no. 2, pp. 207–216, 1993.

75. H. Brahmi, I. Brahmi, S. B. Yahia, "OMC-IDS: at the cross-roads of OLAP mining and intrusion detection," in Pacific-Asia Conference on Knowledge Discovery and Data Mining, pp. 13–24, Springer Science and Business Media LLC, Springer, Berlin, Heidelberg, 2012.

76. A. Tajbakhsh, M. Rahmati, A. Mirzaei, "Intrusion detection using fuzzy association rules," Applied Soft Computing, vol. 9, no. 2, pp. 462–469, 2009.

77. S. Kotsiantis, D. Kanellopoulos, "Association rules mining: a recent overview," GESTS International Transactions on Computer Science and Engineering, vol. 32, no. 1, pp. 71–82, 2006.

78. M. Woźniak, M. Graña, E. Corchado, "A survey of multiple classifier systems as hybrid systems," Information Fusion, vol. 16, pp. 3–17, 2014.

79. P. Domingos, "A few useful things to know about machine learning," Communications of the ACM, vol. 55, no. 10, pp. 78–87, 2012.

80. C. Zhang, Y. Ma, Ensemble Machine Learning: Methods and Applications. Springer, Boston, MA, 2012.

81. L. E. Santana, L. Silva, A. M. Canuto, F. Pintro, K. O. Vale, "A comparative analysis of genetic algorithm and ant colony optimization to select attributes for an heterogeneous ensemble of classifiers," in 2010 IEEE Congress on Evolutionary Computation (CEC), pp. 1–8, 2010: IEEE.

82. N. M. Baba, M. Makhtar, S. A. Fadzli, M. K. Awang, "Current issues in ensemble methods and its applications," Journal of Theoretical and Applied Information Technology, vol. 81, no. 2, p. 266, 2015.

83. D. Gaikwad, R. C. Thool, "Intrusion detection system using bagging ensemble method of machine learning," in 2015 International Conference on Computing Communication Control and Automation (ICCUBEA), pp. 291–295. 2015: IEEE.

84. A. A. Aburomman, M. B. I. Reaz, "A novel SVM-kNN-PSO ensemble method for intrusion detection system," Applied Soft Computing, vol. 38, pp. 360–372, 2016.

85. R. R. Reddy, Y. Ramadevi, K. Sunitha, "Enhanced anomaly detection using ensemble support vector machine," in 2017 International Conference on Big Data Analytics and Computational Intelligence (ICBDAC), pp. 107–111. 2017: IEEE.

86. S. Y. Yerima, S. Sezer, I. Muttik, "High accuracy android malware detection using ensemble learning," IET Information Security, vol. 9, no. 6, pp. 313–320, 2015.

87. H. H. Bosman, G. Iacca, A. Tejada, H. J. Wörtche, A. Liotta, "Ensembles of incremental learners to detect anomalies in ad hoc sensor networks," Ad Hoc Networks, vol. 35, pp. 14–36, 2015.

88. J. A. Hartigan, M. A. Wong, "Algorithm AS 136: A k-means clustering algorithm," Journal of the Royal Statistical Society. Series C (Applied Statistics), vol. 28, no. 1, pp. 100–108, 1979.

89. A. K. Jain, "Data clustering: 50 years beyond k-means," Pattern Recognition Letters, vol. 31, no. 8, pp. 651–666, 2010.

90. G. Münz, S. Li, G. Carle, "Traffic anomaly detection using k-means clustering," GI/ITG Workshop MMBnet, 2007.

91. M. H. Bhuyan, D. K. Bhattacharyya, J. K. Kalita, "Network anomaly detection: methods, systems and tools," IEEE Communications Surveys & Tutorials, vol. 16, no. 1, pp. 303–336, 2014.

92. A. P. Muniyandi, R. Rajeswari, R. Rajaram, "Network anomaly detection by cascading k-Means clustering and C4. 5 decision tree algorithm," Procedia Engineering, vol. 30, pp. 174–182, 2012.

93. P. Laskov, P. Düssel, C. Schäfer, K. Rieck, "Learning intrusion detection: supervised or unsupervised?" in International Conference on Image Analysis and Processing, pp. 50–57. 2005: Springer.

94. H.-b. Wang, Z. Yuan, C.-d. Wang, "Intrusion detection for wireless sensor networks based on multi-agent and refined clustering," in WRI International Conference on Communications and Mobile Computing, vol. 3, pp. 450–454. 2009: IEEE.

95. Q. Li, K. Zhang, M. Cheffena, X. Shen, "Channel-based Sybil detection in industrial wireless sensor networks: a multi-kernel approach," in IEEE Global Communications Conference (GLOBECOM 2017), pp. 1–6. 2017: IEEE.

96. M. Xie, M. Huang, Y. Bai, Z. Hu, "The anonymization protection algorithm based on fuzzy clustering for the ego of data in the Internet of Things," Journal of Electrical and Computer Engineering, vol. 2017, 10 pages, Article ID 2970673, 2017.

97. S. Wold, K. Esbensen, P. Geladi, "Principal component analysis," Chemometrics and Intelligent Laboratory Systems, vol. 2, no. 1–3, pp. 37–52, 1987.

98. S. Zhao, W. Li, T. Zia, A. Y. Zomaya, "A dimension reduction model and classifier for anomaly-based intrusion detection in internet of things," in Dependable, Autonomic and Secure Computing, pp. 836–843. 15th International Conference on Pervasive Intelligence & Computing, 3rd International Conference on Big Data Intelligence and Computing and Cyber Science and Technology Congress (DASC/PiCom/DataCom/CyberSciTech), 2017 IEEE.

99. X. Zhu, Z. Ghahramani, J. D. Lafferty, "Semi-supervised learning using Gaussian fields and harmonic functions," in Proceedings of the 20th International conference on Machine learning (ICML-03), pp. 912–919. 2003.

100. X. J. Zhu, Semi-Supervised Learning Literature Survey, University of Wisconsin-Madison Department of Computer Sciences, Madison, WI, 2005.

101. O. Y. Al-Jarrah, Y. Al-Hammdi, P. D. Yoo, S. Muhaidat, M. Al-Qutayri, "Semi-supervised multi-layered clustering model for intrusion detection," Digital Communications and Networks, vol. 4, no. 4, pp. 277–286, 2018.

102. S. Rathore, J. H. Park, "Semi-supervised learning based distributed attack detection framework for IoT," Applied Soft Computing, vol. 72, pp. 79–89, 2018.

103. V. Mnih et al., "Human-level control through deep reinforcement learning," Nature, vol. 518, no. 7540, p. 529, 2015.

104. R. S. Sutton, A. G. Barto, Reinforcement Learning: An Introduction (No. 1). MIT press, Cambridge, MA, 1998.

105. M. A. Aref, S. K. Jayaweera, S. Machuzak, "Multi-agent reinforcement learning based cognitive anti-jamming," in Wireless Communications and Networking Conference (WCNC), pp. 1–6. 2017: IEEE.

106. S. Machuzak, S. K. Jayaweera, "Reinforcement learning based anti-jamming with wideband autonomous cognitive radios," in IEEE/CIC International Conference on Communications in China (ICCC), pp. 1–5, 2016: IEEE.

107. G. Han, L. Xiao, H. V. Poor, "Two-dimensional anti-jamming communication based on deep reinforcement learning," in IEEE International Conference on Acoustics, Speech and Signal Processing (ICASSP), pp. 2087–2091. 2017: IEEE.

108. Y. Gwon, S. Dastangoo, C. Fossa, H. T. Kung, "Competing Mobile Network Game: embracing antijamming and jamming strategies with reinforcement learning," in IEEE Conference on Communications and Network Security (CNS), pp. 28–36. 2013: National Harbor, MD.

7 IoT-Enabled Heart Disease Prediction Using Machine Learning

Vaishali Baviskar
GH Raisoni College of Engineering and Management
Savitribai Phule Pune University, Pune, India

Divya Srivastava and Madhushi Verma
Bennett University, Greater Noida, India

Pradeep Chatterjee
TML Business Services Ltd, Pune, India

Sunil Kumar Jangir
University Institute of Engineering,
Chandigarh University, Mohali, India

Manish Kumar
Mody University of Science and
Technology, Lakshmangarh, India

CONTENTS

7.1 INTRODUCTION

Health is the fundamental need of any individual. According to World Health Organization (WHO) statistics, 24% of deaths in India and one-third of all global deaths are due to heart diseases. Out of the 31% of global deaths, nearly 85% were due to heart attack and stroke (World Health Organization, 2017). Heart disease is the most hierarchical reason for death all over the world. According to the WHO survey, heart-related disease is responsible for taking 16.9 million lives every year, which is 21% of all global deaths. Heart disease killed 1.6 million Indians in 2016, making India a leading part of this survey. According to the global burden of disease report in 2016, heart disease is one that spoils not only the person but also the financial stability of a person. The WHO's estimates suggest that India has lost up to 230 billion from 2005 to 2016 due to heart-related disease, so there is a need to predict heart-related diseases.

The heart is a primary part of the body that controls the functioning of all other organs. Heart failure can lead to a dangerous situation. For heart consultants, it is complicated to predict heart disease at the right time. The heart disease diagnosis through conventional medical history mainly has not been considered reliable in many views. Non-invasive methods like IoT-based machine learning (ML) techniques, which are reliable and efficient, can be used for classifying fit people and those having heart disease. These techniques for prediction can aid the medical field. Apart from dietary control and a healthy lifestyle, the diagnosis at the right time is also critical. Age, cholesterol, gender, high blood pressure, smoking, obesity, family history, physical inactivity, poor diet, diabetes, alcohol intake, and heredity are various causes of heart diseases (Nashif et al., 2018).

There are different heart diseases, such as coronary heart diseases, angina pectoris, congestive heart failure, cardiomyopathy, congenital heart disease, arrhythmias, and myocarditis. Primary causes for heart disease are extra body fat, tobacco, unhealthy food, overweight, excessive sugar, and symptoms are pain in the chest and arm (Khourdifi and Bahajm, 2019).

Developments in information and communication technologies have led to the introduction of the Internet of Things (IoT). In the recent health care surroundings, IoT technology has been well received by doctors and patients as it is useful in several medical areas like healthcare management, patient health monitoring, and patient information. Nowadays, IoT technology is being popularly used in the healthcare industry. IoT is the essential upcoming technology and acquiring attention in healthcare applications. However, IoT-based heart attack detection systems raise privacy and safety concerns. In healthcare, Body Sensor Network (BSN) is one of the significant technologies used for monitoring a patient who is examined using wireless, tiny-powered, and lightweight sensor nodes. Data privacy of a patient is the security risk associated with this innovative technology used in healthcare applications. Recording health data using smart systems is proliferating nowadays. The utilization of Wi-Fi sensors in smart systems for collecting data and distributing it to the patients to make them aware of their health is the major challenge. IoT devices produce massive data in the healthcare environment, and to handle and get easy access to this vast volume of data, cloud computing can be used.

In this domain, cloud-based applications play the leading role. In this research, all these aspects are covered to implement a smart and secured healthcare monitoring system. Firstly, an IoT-based healthcare monitoring system is proposed using various BSNs. In the next stage, efficient deep learning (DL) algorithms are used for disease prediction based on the sensor network's input parameters.

There are certain issues and challenges prevailing in the healthcare domain which can be tackled using IoT. The growing prices of healthcare and the accessibility of innovative special health devices are the IoT's visualization elements in the associated healthcare (Rghioui and Oumnad, 2018). The physicians are not in direct contact with each patient, and certain problems which are generated due to this are as follows:

    a. *Adherence Monitoring:* Physical consultants do not have the proper access to check each patient, whether their patients follow the prescribed treatment. It is reasonably expected that the absence of linkage and connection causes the threat of hospitalization and, as an outcome, raises the financial load for patients as well as their families.

    b. *Inadequate and Prospective Time:* Because of the limited time and substantial growing population, physicians cannot give patients adequate time. Due to the short time, physicians cannot check daily routine, sleep time, exercise follow-up, and time-to-time diet, which are essential factors for predicting and diagnosing heart disease.

IoT applications have been found to be beneficial for dealing with such issues and helpful in monitoring chronic diseases like congestive heart failure and diabetes, which need special attention.

7.2 METHODOLOGY OF HEART DISEASE PREDICTION

Heart disease prediction at early stages can be a lifesaving system for people living in remote areas. Due to the advent of remote-sensing technologies that take advantage of the IoT-enabled devices, sensing of body parameters related to heart diseases has become possible. Parameters like body temperature, blood sugar levels, blood pressure, electrocardiogram-related signals, etc., can be sensed in rural areas, and data can be sent to the cloud for processing. Due to these areas' remote nature, there is a limited possibility of a doctor being physically present to examine the patient. However, the patient's physical parametric data can be sensed and presented to the doctor via the cloud, and the doctor can suggest treatments whenever necessary (Sreejith et al., 2016).

A sensing circuitry and a prediction engine are required to sense different body parameters used for heart disease prediction (Venkatesan et al., 2018). The sensing circuitry contains more body sensors like ECG, heart rate, temperature, blood pressure, etc., while the prediction engine consists of software components to process this data and find the final output heart disease class, whether it is normal or abnormal (Swapna et al., 2018). A sample system that identifies heart diseases based on the input parameters can be observed in Figure 7.1. Here, the input data is taken from hardware sensors and is given to different rule bases for analysis. For instance, the symptoms rule base is used for analyzing patient's symptoms and mapping them with the general symptoms of heart disease, the diabetes rule base is used to analyze the input data and finds out the presence of diabetes, and similarly, the cholesterol rule base is used to find out the existence of cholesterol levels in the body of a patient.

All these rules are combined to determine the probability of heart disease in the patient's body by analyzing the parameters' readings. Once the heart disease probabilities are evaluated, then ECG sensor readings are extracted. These readings are

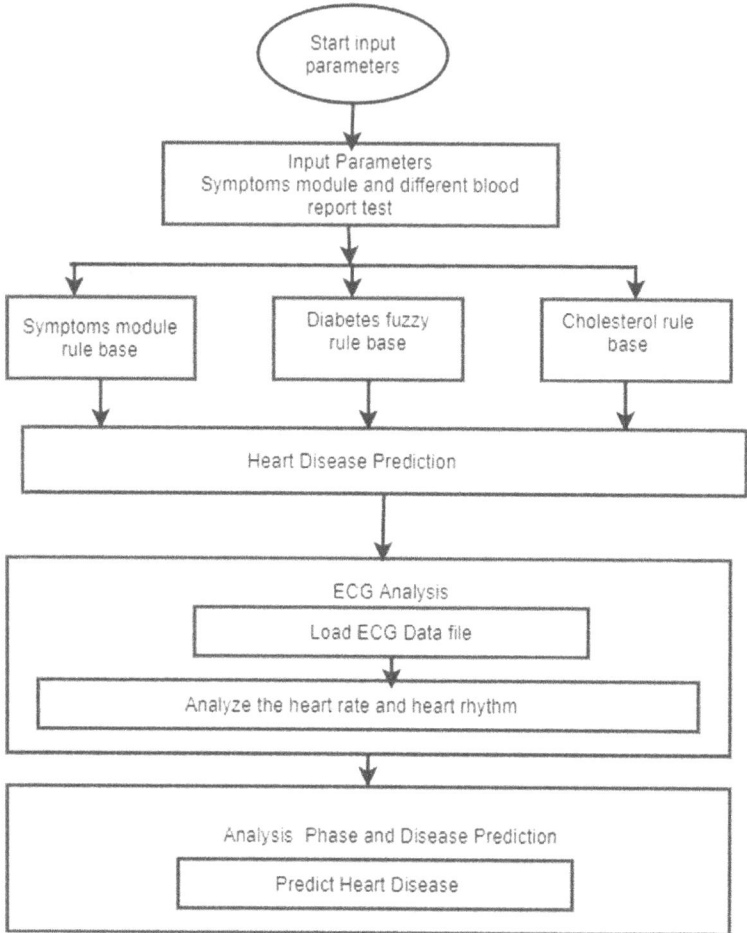

FIGURE 7.1   A heart disease prediction system.

analyzed to discover the heart speed and rhythms. The heart rate is measured over a temporal span of more than 30 minutes, and any abnormalities in the same are evaluated. The heart rhythms are evaluated using the ECG waveforms (Luo et al., 2017). Sample ECG waveforms for a normal person and a person suffering from heart disease can be observed in Figure 7.2. From the typical ECG waveform, it can be observed that the P, PR, Q, S, ST, QRS, etc., wave intervals are uniformly distributed. Nevertheless, from the abnormal ECG waveforms, it can be observed that there are changes in the ST waveforms (they get suppressed or elevated), there is a sharp rise in the T-point, there are rapid changes in the ECG waves, and there are longer R-R intervals (bradycardia). Such observations and more are enough to analyze if a person is going through any pain due to heart disease (Zhanpeng et al., 2009).

When the ECG sensor data, patient symptoms, diabetes, cholesterol sensors, and other parameters are combined, a final prediction of the person's heart health can

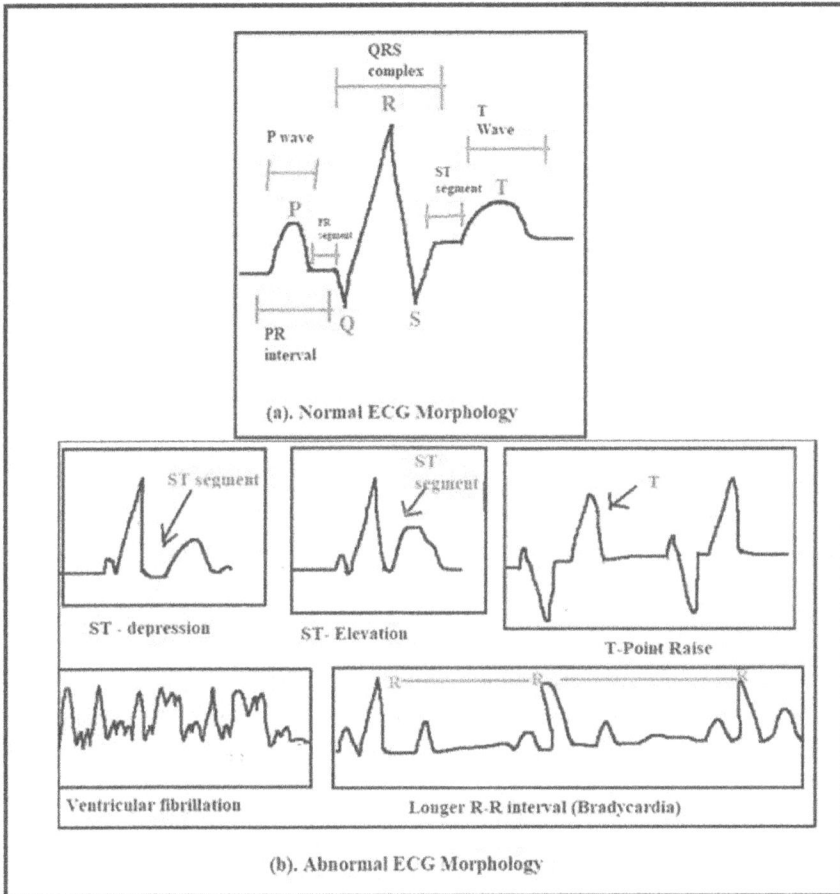

FIGURE 7.2   Normal and abnormal ECG waveforms.

be made. To improve the accuracy of this prediction, there is a need to design a high-performance classifier. Neural network-based classifiers are preferred over others because they can analyze data patterns more effectively than non-neural counterparts. It is observed that convolutional neural networks and deep convolutional neural networks can produce high-accuracy classification systems when combined with effective patient parameter extraction techniques. The results obtained from the classifier can be used for post-processing to analyze or predict any future heart diseases that the person might have (Abdeldjouad et al., 2020). For example, if the person's symptoms, diabetes, and ECG analysis results are normal, but the value of cholesterol increases over time, it can be predicted that the person might suffer from cardio-related disorders.

Moreover, if everything except diabetes is normal, even then the person is susceptible to heart diseases, as diabetes is the leading cause of any disease in the body. A mutually inclusive relationship between these parameters is the main pattern that is analyzed during heart disease prediction. To predict the presence of disease, equation 7.1 is applied to evaluate a heart disease prediction threshold.

$$H_{DTh} = MAX\left(\left\|\langle ECG|BP|Glucose\rangle\right\|\right), \tag{7.1}$$

where H_{DTh} is the heart disease prediction threshold, and *ECG, BP, and Glucose* are the values of the respective parameters. The ‖ indicates the variance of the signals over a given temporal period, while the MAX & | indicates the maximum variance in the given values over a period (Shah et al., 2020).

Thus, using this analysis, heart disease can be predicted using different input parameters. This chapter describes a system that can sense different human body parameters remotely and then send them over the cloud to an automated classification system. The classification system is based on a ML classifier that uses the Cleveland dataset to predict heart diseases. Once the data is sensed, then a classifier can be built by combining recurrent neural networks and long short-term memory (LSTM) systems that can classify the inputs into different heart disease classes with more than 95% accuracy. Due to the high accuracy of heart disease detection, the system can be used in present environments. The system's design and deployment are also described in this chapter, wherein different sensor specifications and their usage instructions are mentioned in detail. Once the sensing is done, the data is given to a lossless compressor to reduce the data size. This reduction in data size ensures that a smaller number of bytes are required for communication, thereby assisting the system to work virtually, even remotely where network connectivity is low. Also, the IoT server design is done such that the communication responses have a one-byte size so that the system response is fast, and the system can work even under 2G conditions. The designed system is also tested under different real-time environments, and their performance evaluation is also presented in this chapter. Readers would be able to deploy the proposed system with complete network design and hardware design after referring to this chapter.

7.3 SENSORS USED FOR HEART DISEASE PREDICTION

The medical purpose's electronic equipment uses various sensors for analysis, i.e. converting electrical signals into digital values. Sensors increase and grow the medical device intelligence which monitors the vibrant signs and additional health elements. An expanding and aging population accelerates the innovative and various kinds of medical devices, including different sensors used for patients. Healthcare experts need reliable, real-time, and exact diagnostic outcomes provided by equipment to track the patient. Medical equipment is to be designed that is reliable, has improved performance, and is of lower cost. Low cost, portable, less power consumed sensors are used in medical device design, mostly in non-invasive, portable, smaller equipment like heart-rate monitors, pulse meters, blood-glucose meters (Singh et al., 2019). Figure 7.3 shows various phases for cardiac disease prediction.

In today's era, sensors are being used in medical devices for diagnosis. Various researchers use the sensors like pressure, ECG, blood pressure, temperature, and humidity sensors for disease prediction (Ganesan and Sivakumar, 2019; Jahangir et al., 2019 Nashif et al., 2018).

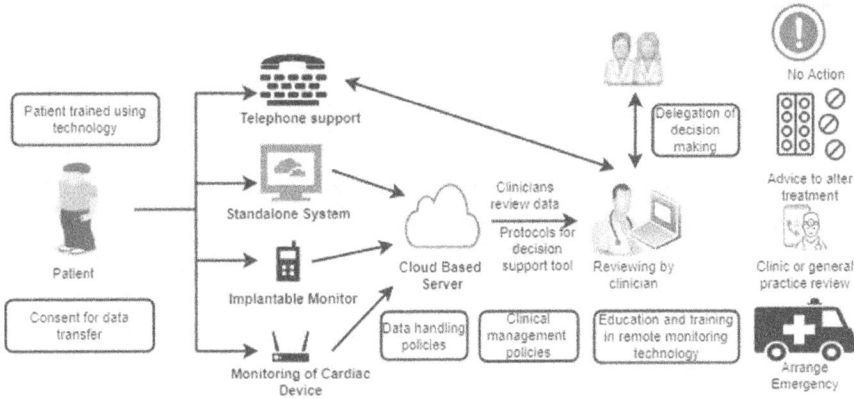

FIGURE 7.3   Various phases for heart disease detection.

7.4   HEART DISEASE PREDICTION IN REAL-TIME USING INTERNET OF THINGS

The synchronized monitoring of multi parameters is combined with various sensors that work together to get the essential data without noise troubles. The implemented system is built to achieve a high accuracy of prediction by decreasing and minimizing human interferences. It is built on IoT diagnostic scheme for remotely placed heart patients to measure blood pressure, heart rate, body temperature, and various new parameters using biomedical sensors. The number of the composed dataset can be kept and diagnosed as well as predicted for diseases (Brahmbhatt et al., 2020).

To predict heart diseases in real-time, various sensors are needed to be interfaced with the system. These sensors read the device parameters and then pass these parameters to the server. The server analyses these parameters and finally predicts heart diseases. The real-time detection system consists of wearable gadgets that can sense heart rate, cholesterol levels, and blood pressure and take feedback from the user in terms of the survey. These gadgets are interfaced using Bluetooth or near field communication (NFC) with the server (Shankar et al., 2020). The server keeps track of all this data for each user and performs classification-based analysis to evaluate if the person is having any heart-related issues or not. The sample architecture of such a system can be observed in Figure 7.4, wherein a communication interface between the patient and the doctor is depicted.

From the system, it can be observed that the overall health data of the user is read using a variety of sensors, as mentioned in the previous section. Each of these sensors is connected to a Bluetooth-powered smartphone. The smartphone collects further user-specific data like previous medicine reminders, location, etc. All this data is aggregated and given to the cloud server. The cloud server stores each of this information into a structured query language (SQL) table, which can be observed from Table 7.1. Here the input data from different sources are merged, and the merged data is analyzed from user to user.

All these readings are given to the classifier, which is deployed on the server. Based on these readings, a classifier is trained to get the final status of the person.

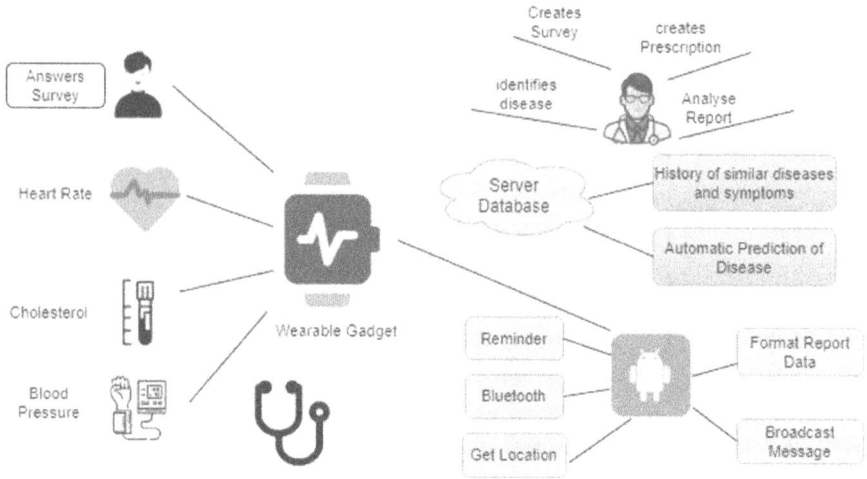

FIGURE 7.4   A real-time heart disease detection system.

The main advantage of using cloud-based processing systems is merging all these information sources and obtaining a continuously learning classification system. For instance, the patient's knowledge in the given Table 7.1 can be used to classify diseases for other patients based on their input information. In the next section, a study on different classifiers used for heart disease detection has been presented. Study suggests that convolutional neural networks outperform other classification systems to predict heart disease.

Another example of heart disease prediction using patient implanted sensors can be observed from Figure 7.4, wherein a Bluetooth-powered device is connected directly to the patient's body (Vinayaka and Gupta, 2020). Based on this sensor's readings, the doctor's mobile phone transfers the data to a central hospital cloud. The cloud can perform signal processing, feature extraction, online training, and finally classification.

7.5   CLASSIFIERS USED FOR HEART DISEASE PREDICTION

To effectively predict heart diseases in the IoT environment, there are multiple DL and ML algorithms available. Every algorithm out of these utilizes different kinds of DL and ML algorithms (Tougui et al., 2020). Table 7.3 showcases the various DL and ML classifiers used for heart disease classification.

In M. Ganesan and Dr N. Sivakumar (Ganesan and Sivakumar, 2019), a new IoT- and cloud-based healthcare app was developed to diagnose and monitor severe diseases. The classifier was trained utilizing the benchmark dataset. The identification of the presence or absence of disease is done during the testing phase.

AKM Jahangir Alam Majumder et al. proposed multiple sensors systems using a smart IoT, which gives an early warning of disease risk. It continuously collects the data from the user and sends it to the android phone via Bluetooth using the Body

TABLE 7.1

Sample of Data Stored on the Server

User Identification	BP (S)	BP (D)	Heart Rate	Glucose Levels	ECG Data	Time-Stamp	Condition
1	120	80	75	120		20-10-2020 5:30 P.M.	Normal
1	150	90	85	115		21-10-2020 6:30 P.M.	Normal
1	190	125	95	105		22-10-2020 6:30 P.M.	Abnormal
1	195	135	105	90		23-10-2020 6:30 P.M.	Abnormal
2	120	80	75	120		21-10-2020 6:30 P.M.	Normal
2	120	75	80	116		22-10-2020 6:30 P.M.	Normal
2	125	85	80	120		23-10-2020 6:30 P.M.	Normal
2	125	95	90	125		24-10-2020 6:30 P.M.	Normal

Area Sensor (BAS) system. The data investigation and processing took place to view the user's real-time plots of impending cardiac arrest. An IoT device with little power consumption communication model was developed (Majumder et al., 2019), which collects normal body temperatures and heart rate using smartphones. Here ML and signal processing techniques were used to analyze sensor data and to predict high-accuracy cardiac arrests. A smartphone-based wearable device was implemented for heart rate detection and used a combination of ECG and body temperature. A heart rate analysis was done on the mobile platform where the patient could view the body temperature and real-time ECG signal plots. Architecture for this system can be observed in Figure 7.5 as follows.

In Mohm Ayoub Khan (Khan, 2020), IoT background was projected for heart-related disease prediction using a Modified Deep Convolutional Neural Network (MDCNN) algorithm. The health parameters like ECG and blood pressure were monitored utilizing a heart monitoring device and smartwatch attached to the user. The system performance was investigated by equating the proposed algorithm with existing Deep Neural Networks (DNNs). MDCNN performed better compared to other methods.

Shadman Nashif et al. (Nashif et al., 2018) proposed accurate heart disease prediction using ML algorithms in WEKA, a Java-based Open Access platform for data mining. The proposed algorithm achieved an accuracy of 97.53% using SVM with 10-fold cross-validation. The system of real-time monitoring of patient method was established using Arduino by identifying parameters like a heartbeat, body temperature, humidity, and blood pressure. The patient's real-time video streaming was

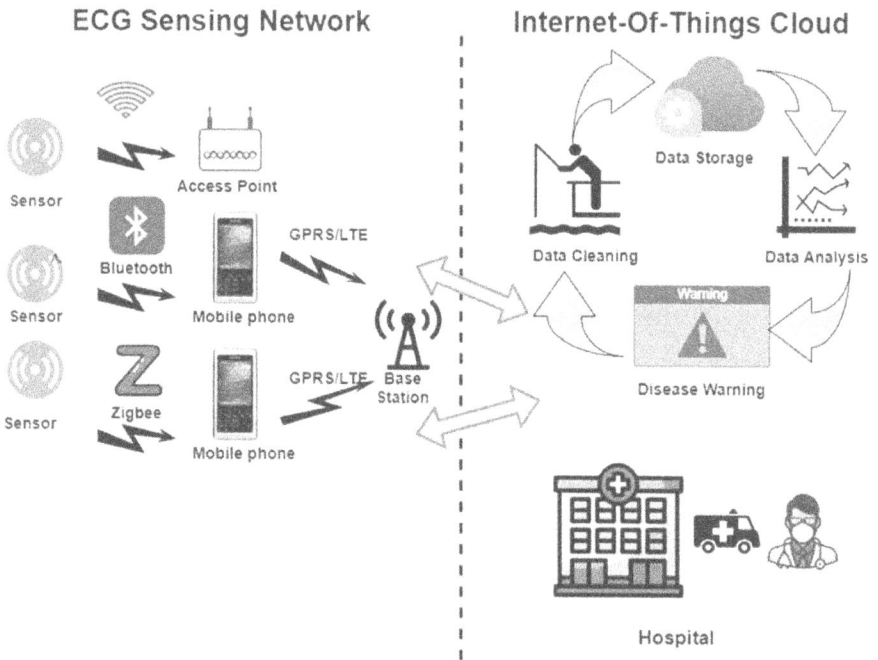

FIGURE 7.5 A sample IoT enabled ECG processing system.

Accuracy

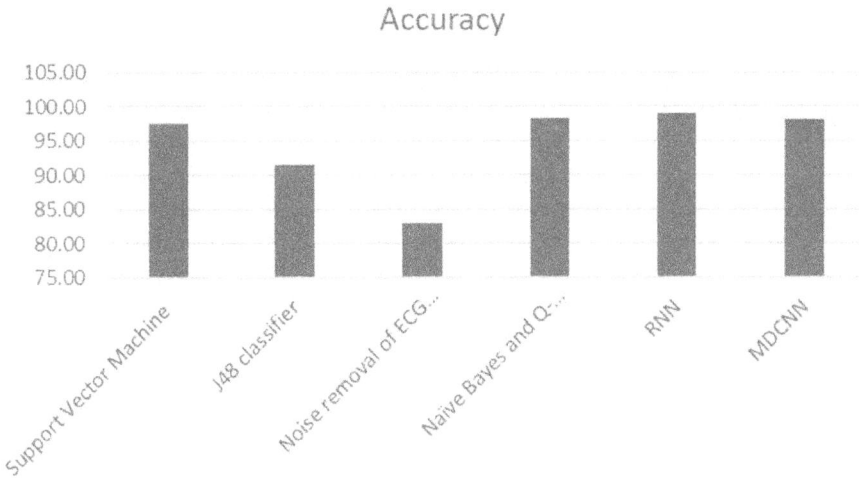

FIGURE 7.6 Accuracy of various ML and DL algorithms.

observed, and the prescribed physical consultation was notified using GSM technology, if any parameter exceeded the threshold value. A mobile application was developed, which stored the doctor's and patient's history (Yang et al. 2016). A preprocessing technique was proposed, which improved the accuracy of the classification of ECG signals.

Figure 7.6 shows an accuracy comparison of various ML and DL algorithms used for heart disease prediction. Comparison is shown between support vector machine, J48, Decision tree, Naïve Bayes, RNN, and MDCNN. Among these RNN has shown maximum accuracy (Jiang et al., 2017). So, for further experimentation we have tried a hybrid algorithm of RNN and LSTM.

From Tables 7.2 and 7.3, it is inherent that the DNN as a hybrid classifier can categorize the input data with a higher accuracy of 96%. Thus, RNN and LSTM can be termed as highly accurate heart disease prediction systems for IoT (Sharma, 2019).

TABLE 7.2

Comparative Analysis of Machine Learning and Deep Learning Classifiers Using Various Sensors

Sr no.	Author	Year	Dataset Used with Sensors	Algorithm Used	Accuracy
1	Shadman Nashif et al.	2018	Cleveland and Statlog Heart Disease dataset (Swapna et al., 2013) using a body temperature sensor, heartbeat, humidity, blood pressure	Support Vector Machine	97.5%
2	M. Ganesan and Dr. N. Sivakumar	2019	Heart disease dataset using ECG sensor, pulse sensor, temperature sensor	J48 classifier	91.48%

(Continued)

TABLE 7.2 *(Continued)*

Comparative Analysis of Machine Learning and Deep Learning Classifiers Using Various Sensors

Sr no.	Author	Year	Dataset Used with Sensors	Algorithm Used	Accuracy
3	A. K. M. Jahangir Alam Majumder et al.	2019	Heart disease dataset from the UCI Repository. A pulse sensor and a temperature sensor	Noise removal of ECG signal with decision tree	83.3%
4	Dr. Yogesh Kumar et al. (Khatal and Sharma)	2020	Real-time monitoring patient dataset. Temperature sensor (LM35), ECG sensor, Heart Rate sensor, Raspberry Pi, GS module	Naïve Bayes and Q-learning(Hybrid) classifier	98.3%
5	Khatal and Sharma	2020	1. Heart disease dataset from the UCI Repository 2. Real-time dataset from IoT device Temperature sensor (LM35), ECG sensor, Heart Rate sensor	RNN	Better than others shown graphically
6	Mohm Ayoub Khan	2020	Three datasets used: 1. Heart disease dataset from UCI Repository, 2. Framingham dataset from Kaggle, and 3. Public health and sensor data. A heart monitoring device and smartwatch are used to monitor the blood pressure and ECG of the patient	MDCNN	98.2%

TABLE 7.3

Comparison of Different ML and DL Classification Methods for Evaluation of Heart Diseases

Classification Method	Feature Extraction Method x	Classification Accuracy (%)
Variable learning rate CNN (Sreejith et al., 2016)	R-peak detection with non-linear transform	92.70%
CNN with PCA (Shah et al., 2020)	Denoising filters, AlexNet, Pan-Tomkins	92%
PCA applied with ANN (Raman et al., 2020)	Electrocardiogram	88.50%
Neuro-Fuzzy (Raman et al., 2020)	Heart rate variability over a span of time	94.20%
SVM (Raman et al., 2020)	Heart rate variability over a span of time	96%
Decision Tree (Zubair et al., 2016)	Heart rate variability over a span of time	81%
GMM (Zubair et al., 2016)	Heart rate variability over a span of time	90.50%
SVM (Zubair et al., 2016)	Heart rate variability over a span of time	86.50%
5 Layer CNN (Li et al., 2017)	Heart rate variability over a span of time	84.10%
5 Layer CNN-LSTM (Li et al., 2017)	Heart rate variability over a span of time	90.90%
AdaBoost (LS) (Kiranyaz et al., 2016)	Clinical Non-linear patient-specific features	90%
AdaBoost (ML) (Kiranyaz et al., 2016)	Clinical Non-linear patient specific features	85.80%
AdaBoost (ML_diag) (Kiranyaz et al., 2016)	Clinical non-linear patient-specific features	85.60%
AdaBoost (NDDF) (Kiranyaz et al., 2016)	Clinical non-linear patient-specific features	85.22%

(Continued)

TABLE 7.3 *(Continued)*

Comparison of Different ML and DL Classification Methods for Evaluation of Heart Diseases

Classification Method	Feature Extraction Method x	Classification Accuracy (%)
AdaBoost (Perceptron) (Kiranyaz et al., 2016)	Clinical non-linear patient-specific features	85.22%
AdaBoost (Pocket) (Kiranyaz et al., 2016)	Clinical non-linear patient-specific features	84.55%
AdaBoost (Stumps) (Kiranyaz e al., 2016)	Clinical non-linear patient-specific features	87.57%
SVM (RBF) (Kiranyaz et al., 2016)	Clinical non-linear patient-specific features	83.33%
SVM (Poly3) (Kiranyaz et al., 2016)	Clinical non-linear patient-specific features	83.38%
FSC (Kiranyaz et al., 2016)	Clinical non-linear patient-specific features	87.75%
DT (Isin and Ozdalili, 2017)	Discrete wavelet transforms	92.64%
Hybrid algorithm (Baviskar et al., 2021)	Clinical patient data – Cleveland data	96%

In our study, we have implemented the system as shown in Figure 7.7

Here, we have used a pulse sensor, ECG sensor, temperature sensor, and oxygen saturation sensors to capture the body parameters. The results are recorded using the Arduino Uno kit. GUI interface is designed to access these parameters. To access the data and for security, cloud services may be used. Arduino Uno is the controller board to which the different analog sensors are connected through

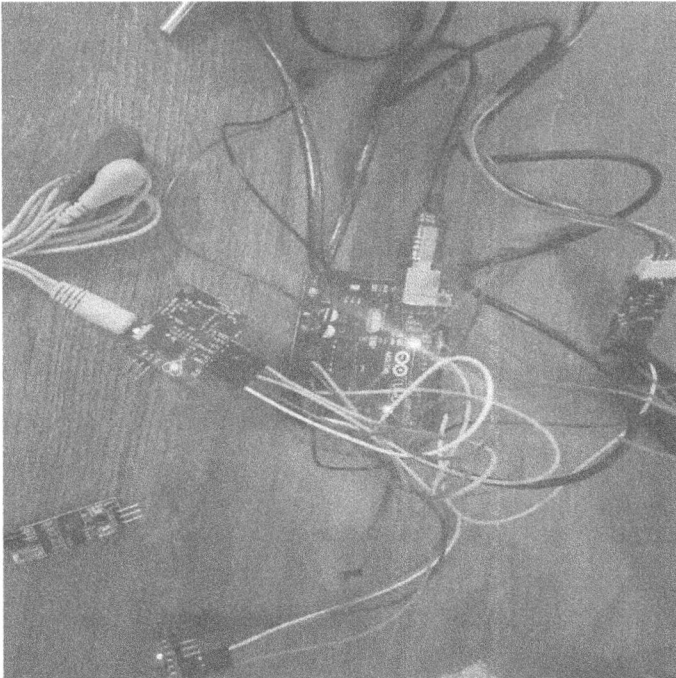

FIGURE 7.7   IoT device for heart disease prediction.

analog pins. Data from sensors are uploaded to the cloud and processed further for generating prediction results.

7.6   CONCLUSION

We have described various applications of IoT for heart disease prediction. Heart disease is predicted using various non-invasive methods like electrocardiogram and body parameters data captured with the help of sensors. By applying signal processing techniques to ECG signals, heart disease can be predicted, which is too complicated and long-term. IoT-based methods capture the patient data at an early stage, and using various ML and DL algorithms heart disease can be predicted. On the existing Cleveland heart disease dataset, high accuracies have been achieved using ML and DL algorithms. In the future, real-time data would be captured using the prototype shown in Figure 7.7, and a hybrid DL algorithm would be applied to get accurate heart disease prediction in the IoT-controlled environment.

REFERENCES

Abdeldjouad F. Z., Brahami M., and Matta N., "A Hybrid Approach for Heart Disease Diagnosis and Prediction Using Machine Learning Techniques," *Proceedings International Conference on Smart Living and Public Health*, vol. 12157, pp. 299–306, 2020.

Brahmbhatt, Darshan H., and Cowie Martin R., "Remote Management of Heart Failure: An Overview of Telemonitoring Technologies," *Cardiac Failure Review*, vol. 5, pp. 86–92, May 2019.

Ganesan M., and Sivakumar N., "IoT Based Heart Disease Prediction and Diagnosis Model for Healthcare Using Machine Learning Models," in *IEEE International Conference on System, Computation, Automation and Networking, Pondicherry, India*, pp. 1–5, 2019.

Isin A., and Ozdalili S., "Cardiac Arrhythmia Detection Using Deep Learning," *Procedia Computer Science*, vol. 120, pp. 268–275, 2017.

Jiang C., Song S., and Meng M. Q.-H., "Heartbeat Classification System Based on Modified Stacked denoising Autoencoders and Neural Networks," in *IEEE International Conference on Information and Automation (ICIA), IEEE*, pp. 511–516, 2017.

Khan, M. A., "An IoT Framework for Heart Disease Prediction Based on MDCNN Classifier," *IEEE Access*, vol. 8, pp. 34717–34727, 2020.

Khatal, Sunil S., and Sharma Yogesh Kumar, "Analyzing the Role of Heart Disease Prediction System Using IoT and Machine Learning," *International Journal of Advanced Science and Technology*, vol. 29, no. 9, pp. 2340–2346, May 2020.

Khourdifi, Youness, and Bahajm Mohamed, "Heart Disease Prediction and Classification Using Machine Learning Algorithms Optimized by Particle Swarm Optimization and Ant Colony Optimization," *International Journal of Intelligent Engineering and Systems*, vol. 12, no. 1, pp. 242–252, 2019.

Kiranyaz S., Ince T., and Gabbouj M., "Real-Time Patient-Specific ECG Classification by 1-D Convolutional Neural Networks," *IEEE Transactions on Biomedical Engineering*, vol. 63, no. 3, pp. 664–675, March 2016.

Li D., Zhang J., Zhang Q., and Wei X., "Classification of ECG Signals Based on 1D Convolution Neural Network," in *IEEE 19th International Conference on E-Health Networking, Applications and Services, Healthcom*, Dalian, China, pp. 1–6, 2017.

Luo, Kan, Li Jianqing, Wang Zhigang, and Cuschieri Alfred, "Patient-Specific Deep Architectural Model for ECG Classification," *Journal of Healthcare Engineering*, vol. 2017, Article ID 4108720, p. 13, 2017.

Majumder A. K. M Jahangir Alam, ElSaadany Yosuf Amr, Young Roger Jr, and Ucci Donald R., "An Energy Efficient Wearable Smart IoT System to Predict Cardiac Arrest," *Hindawi Advances in Human-Computer Interaction*, vol. 2019, Article ID 1507465, p. 21, 2019.

Nashif, Shadman, Raihan Rakib Md., Islam Rasedul Md., and Imam Mohammad Hasan, "Heart Disease Detection by Using Machine Learning Algorithms and a Real-Time Cardiovascular Health Monitoring System," *World Journal of Engineering and Technology*, vol. 6, no. 4, pp 854–873, November 22, 2018.

Raman M., Sharma V. K., Hiranwal S., and Bairwa A. K., "Efficient Method for Prediction Accuracy of Heart Diseases Using Machine Learning," in *Proceedings of International Conference on Communication and Computational Technologies. Algorithms for Intelligent Systems*, pp. 113–121, 2020.

Rghioui, Amine, and Oumnad Abdelmajid, "Challenges and Opportunities of Internet of Things in Healthcare," *International Journal of Electrical and Computer Engineering (IJECE)*, vol. 8, no. 5, pp. 2753–2761, October 2018.

Shah D., Patel S., and Bharti S. K., "Heart Disease Prediction Using Machine Learning Techniques," *SN Computer Science*, vol. 1, p. 345, 2020.

Shankar V., Kumar V., Devagade U., Karanth V., and Rohitaksha K., "Heart Disease Prediction Using CNN Algorithm," *SN Computer Science*, vol. 1, p. 170, 2020.

Sharma Yogesh Kumar, "Health Care Patient Monitoring Using IoT and Machine Learning," *International Organization of Scientific Research Journal of Engineering*, vol. 6, no. 3, pp. 68–73, March 2019.

Singh Vikramjit, Gupta Amit, Sohal J. S., and Singh Amritpal, "Multi-Scale Fractal Dimension to Quantify Heart Rate Variability and Systolic Blood Pressure Variability: A Postural Stress Analysis," *Fluctuation and Noise Letters*, vol. 18, no. 4, 1950019, p. 16, 2019.

Sreejith S., Rahul S., and Jisha R. C., "A Real Time Patient Monitoring System for Heart Disease Prediction Using Random Forest Algorithm," *Advances in Intelligent Systems and Computing*, vol. 425, pp. 485–500, 2016.

Swapna G., Rajendra Acharya U., VinithaSree S., and Suri J. S., "Automated Detection of Diabetes Using Higher Order Spectral Features Extracted from Heart Rate Signals," *Intelligent Data Analysis*, vol. 17, no. 2, pp. 309–326, March 2013.

Swapna G., Soman Kp, and Vinayakumar R., "Automated Detection of Diabetes Using CNN and CNN-LSTM Network and Heart Rate Signals," *Procedia Computer Science*, vol. 132, pp. 1253–1262, 2018.

Tougui I., Jilbab A., and El Mhamdi J., "Heart Disease Classification Using Data Mining Tools and Machine Learning Techniques," *Health and Technology*, vol. 10, pp. 1137–1144, 2020.

Venkatesan C., Karthigaikumar P., Paul A., Satheeskumaran S., and Kumar R., "ECG Signal Pre-Processing and SVM Classifier-Based Abnormality Detection in Remote Healthcare Applications," *EEE Access*, vol. 6, pp. 9767–9773, 2018.

Vinayaka S., and Gupta P. K., "Heart Disease Prediction System Using Classification Algorithms," *Proceedings Communications in Computer and Information Science*, vol. 1244, pp. 395–404, 2020.

World Health Organization, 2017, Newsroom, Fact sheets, Detail, Cardiovascular diseases (CVDs), accessed May 17, 2017, https://www.who.int/news-room/fact-sheets/detail/cardiovascular-diseases-(cvds)

Yang Zhe, Zhou Qihao, Lei Lei, Zheng Kan, "An IoT-Cloud Based Wearable ECG Monitoring System for Smart Healthcare," *Journal of Medical Systems*, vol. 40, p. 286, 2016.

Zhanpeng J., Sun Y., and Cheng A. C., "Predicting Cardiovascular Disease from Real-Time Electrocardiographic Monitoring: An Adaptive Machine Learning Approach on a Cell Phone," in *Annual International Conference of the IEEE Engineering in Medicine and Biology Society*, pp. 6889–6892, 2009.

Zubair M., Kim J., and Yoon C., "An Automated ECG Beat Classification System Using Convolutional Neural Networks," in *6th International Conference on IT Convergence and Security, Prague, Czech Republic*, pp. 1–5, 2016.

8 Internet of Everything, the Future of Globalization
A Comprehensive Study

Prashant Hemrajani, Amisha Kirti Gupta,
Manoj Kumar Bohra, and Amit Kumar Bairwa
Manipal University Jaipur
Jaipur, India

CONTENTS

8.1   INTRODUCTION

The impact of technology is evident in the world today. A colossal use of fast-developing mechanisms is being observed in every part of human life. In the past half of a decade, we have noticed the evolution of everyday objects into something akin to machines with abilities. We have seen smart televisions, vehicles, fridges, etc. but in the time to come, it will also extend to objects like clothing, mats, and so on. Before we would even realize it, it would become an integral part of our lives.

This brings us to the rise of the Internet of Things (IoT), the intensive network of everyday "things," making them capable of receiving, sending data, and communicating over the Internet. Many industrial activities are preparing to switch over to the new era with IoT steering the ship across the seas of evolution.

DOI: 10.1201/9781003145004-8

8.2   THE INTERNET OF EVERYTHING (IOE)

Mankind has begun to propel the process of adaptation to the changes in the planet of technology and the Internet of Everything (IOE) is no exception. The IOE is the next step to the rapid growth of IoT, now forming connections among "people, process, data and things." Many industrial and other activities are undergoing innovative changes to bring about high efficiency, real-time statistics and data analysis, and the ability to direct with understanding.

In this study, we will look over the opportunities of IOE in the real world, the key applications including smart roads, smart transport, smart power grids, and smart living standards.

The advent of the Internet of Things (IoT) has the potential to bring about changes like never before, bringing about foundational developments in the way industries handle their production, as well as the process by which the consumer market interconnect with these companies. The IOE expands the IoT concept by adding links to data, people, and (business) processes. It is said to consist of "four pillars," namely "people, process, data, and things," and is layered over the concept of IoT with one pillar: "things." These links can be people-to-people (P2P), machine-to-people (M2P), and machine-to-machine (M2M) [1, 2].

The "pillars" of the IOE are as follows:

1. *People:* Connections among people have reached new heights. No matter where you are in the world, you can remain connected. As the concern shifts toward IOE, this will change for even more valuable and innovative ways.
2. *Data:* Collecting, gathering, analyzing, and combining information is an integral part of the development of IOE. As it progresses, data collection will play an important role in how systems make smart decisions.
3. *Things:* Everyday "objects" are made more aware and more informative about their surrounding using the Internet, connecting it via sensors, consumer devices, and so on.
4. *Process:* All the aforementioned aspects – people, data, things – are connected together with processing so that every piece of information is delivered at the right time and at the right place [1].

8.3   RFID TECHNOLOGY

It has a big impact on the IOE. Chip-less RFID technology has opened the doors for RFID to enter new domains like tracking of items. This uses electromagnetic ID for encoding of data which in turn reduces cost. It can identify objects, keep track of the information collected from its surroundings [3].

Two tag systems are involved:

1. First is knowns as active reader tag
2. Second is known as passive reader tag

RFID technology is mainly used in healthcare, agriculture, and nation's security systems (Figure 8.1) [4, 5].

P2M

M2M Connections P2P

Network System

Data Centre in the Virtual World

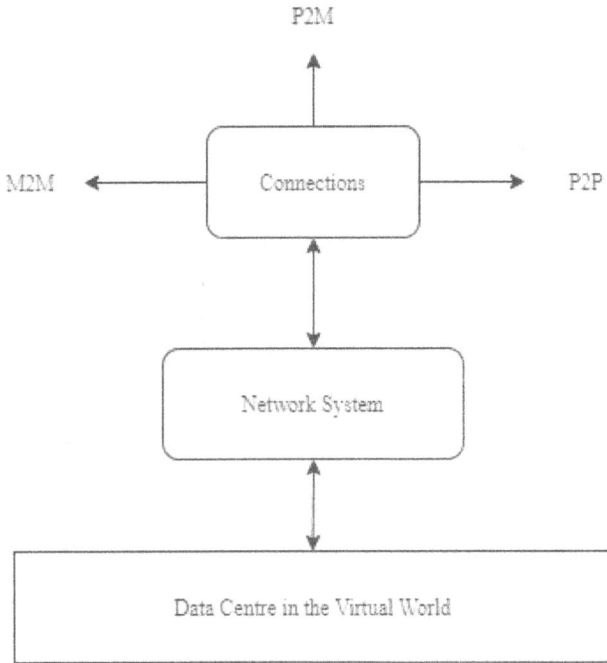

FIGURE 8.1 IOE Architecture.

8.4 PEOPLE-TO-PEOPLE (P2P), MACHINE-TO-PEOPLE (M2P), AND MACHINE-TO-MACHINE (M2M) CONNECTIONS

1. *People-To-People (P2P):* When information is passed on from one computer to another via the Internet, synonymous with the passing of information between people, it is called People-to-People connections. It is a collaborative kind of connection where interaction takes place with the help of devices [6]. This happens through, for instance, social networks.

2. *Machine-to-People (M2P/P2M):* People-to-machine and machine-to-people is a type of connection where humans interact with machines to form better judgments regarding data and analytics [7]. For instance, an application that provides you information about the nearby hotels and restaurants, aiding you to select one. Then, we have the accuracy of weather forecasts depending on smart devices [6].

3. *Machine-to-Machine (M2M):* Lastly, we have machine-to-machine connections. Machine-to-machine (M2M) tackles any kind of technology, such as sensors and actuators, that makes it possible for networked devices to exchange information, interpret the information, and perform actions without human assistance through the Internet or other networks [7]. ATMs, for example, get the authorization to allow the user to withdraw or deposit cash. Another example is wireless local area networks (Figure 8.2).

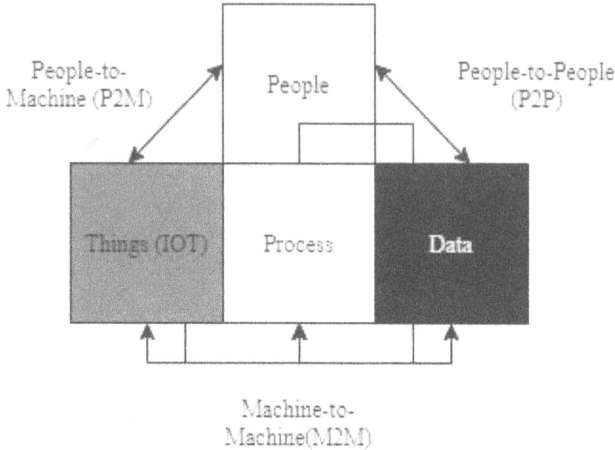

FIGURE 8.2   The pillars of IOE. (From Ref. [16].)

The term "Internet of Things" (IoT) was coined by Kevin Ashton of the American Multination Company, Procter & Gamble, in 1999 [8]. He believed then that Radio frequency identification (RFID) is essential for the development of IoT [9]. The roots are believed to go back to the Massachusetts Institute of Technology's (MIT) Auto-ID Centre. The ratio of things and people dramatically grew from 0.08 in 2003 to 1.84 in 2010 [10].

Now, to connect what is left to be connected, we are now advancing toward IOE. As said before, after the combination of people, data, things, and processes, the Internet will have something akin to invincible power to simulate mind-blowing experiences in industries and even in our everyday lives. The advent of this concept was made possible by the idea of empowering systems through ubiquitous Internet, meaning that it should exist everywhere and for everyone, the usage and functioning of big data and artificial intelligence (AI) [11].

Cisco made a statement in 2013 that IOE was aimed to produce 19 trillion dollars' worth of Stake in the following decade, i.e. 2013–2022. The reason for this is that IOE creates opportunities for various sectors and organizations to achieve feats like never before with a new face of connectivity. This would lead to an increase in global economic growth, reliability in public services like healthcare, and a surplus of productivity [12]. With this in mind, let us look at the opportunities of IOE.

8.4.1   Fog Computing and Cloud Computing

Fog Computing refers to the decentralized framework wherein data, cloud, storage, etc., are located between the source of the data and the cloud. It earns its name "fog" as a cloud that is nearer to the source (surface in comparison) as it brings the cloud closer to where the source of the data lies, the place where it was created.

The applications involved are both – what could be run cloud and what could be run in other smart devices, all forming a bridge of connection through the Internet of

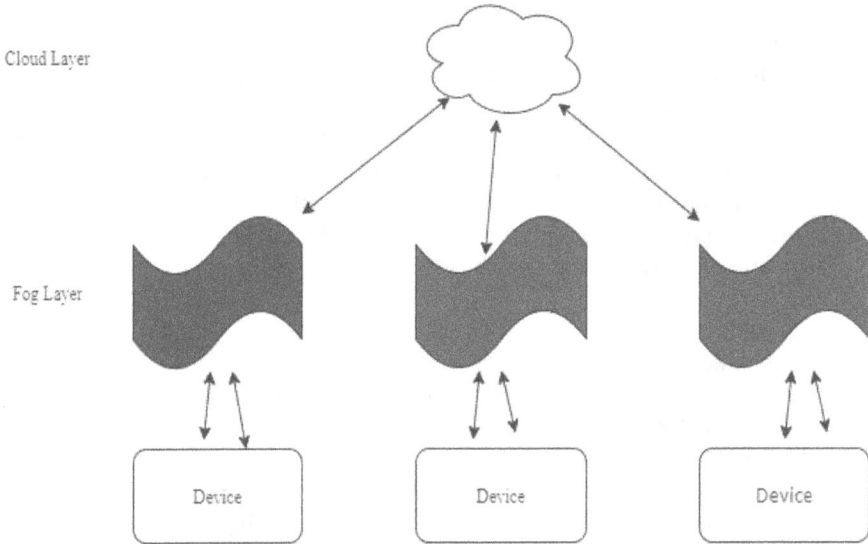

Cloud Layer

Fog Layer

Device Device Device

FIGURE 8.3 Fog computing.

Things. This comes with the execution time being shortened and problems related to scalability and reliability. If we need a higher level of performance in the IOE paradigm, we need methods to combine the advantages of cloud computing with devices run with the help of the concept of IoT. Now, in a highly dynamic environment, fog computing is more preferred. It is believed to be able to cater to an exponential increase in devices and what really is demanded by IoT. Hence, it would be an important asset in the next step, which is the IOE (Figure 8.3) [13].

8.5 OPPORTUNITIES AND APPLICATIONS OF THE INTERNET OF EVERYTHING

Population growth has put a critical restraint on the kind of quality of services being provided to the masses, and even the availability of these. Water provisions, electricity, transportation, sanitation, and other amenities are becoming harder to be distributed economically to the public in cities. Traffic congestions are a major problem on roads related to public transportation and adjusting information according to the congestions and traffic light timings is equally strenuous. We need optimization and that is where sensors in traffic lights, vehicles, etc., come in [14].

Another thing to note is that traffic monitoring in Smart Cities is something that does not have any scope for delay in communication from the place of data collection to the cloud data center and then back to the starting point. Fog computing is a concept that has been introduced to reduce processing of the traffic situation in question and hence brings in the sense of proximity between the user and the cloud services [15].

It is said that the use of IOE will become almost necessary to deploy the functioning of smart cities in the future, including Smart Grids. The mining industry would also have the contribution of IOE for the betterment of the sector in monitoring and improving safety [16].

In the coming days, with the help of IOE, cities will have an advantage from being connected. These will progress to become "Smart Cities" with the processing of "Big Data," involving, for instance, the development of better highways, sensors of road damages, growth of agriculture, healthcare, and education [16].

There are some expectations attached to IOE as follows:

a. It should cover the needs of everyone and everything. It should scale all the requirements spanning from urban and rural areas to underwater and even outer space. This can be made possible by taking in mind various communication and transmission distances.
b. It should be intelligent enough to make its own independent decisions, predictions, and analysis based on the data collected from its surroundings. The data collected is massive from the span of scalability mentioned above.
c. Combining what has been discussed above – scalability and intelligence – IOE encompasses a diverse classification of applications including geographical diversity (based on geographical areas), stereoscopic diversity (based on spatial positions), business diversity (based on social utilities), and technology diversity (based on technologies of Information and Communication Technology (ICT)) [11].

Sensors are an important part of any IoT- and IoE-based device. It can be understood in the way of explaining the human body. Whenever our body performs a command from the brain, it has a lot of sensors working together at the same time to support the action. We have our eyes, taste buds, nose, ears, and our sense of touch all working simultaneously. Sensors are found in many automatic systems today: health care, mobile devices, computing systems, and much more. Sensor fusion, as the term suggests, means the fusion of various sensors. Great changes were brought about with the use of MEMS-based inertial sensors [17].

These days, at an individual level, you can see the effects of IOE when you use your mobile phones to adjust the cooling or heating of your room, alarm clocks that are customized according to the user with the help of the Internet, and even adjusting the security settings with respect to the circumstances. Checking up on the conditions of the pandemic is another use of IOE in the present time [18].

Service providers are interested in the usage of IOE for feedback information such as the most used features and least used features. This helps in better chances of making their service more user-friendly and more appealing to the users (Figure 8.4) [19].

Smart Grid: The need for Smart Grids had risen almost 20 years ago as a result of inadequacies in the provision of power, resulting in blackouts [20]. Smart grids have brought about the solution for an alternative for energy production and have the option for optimization based on the data gathered

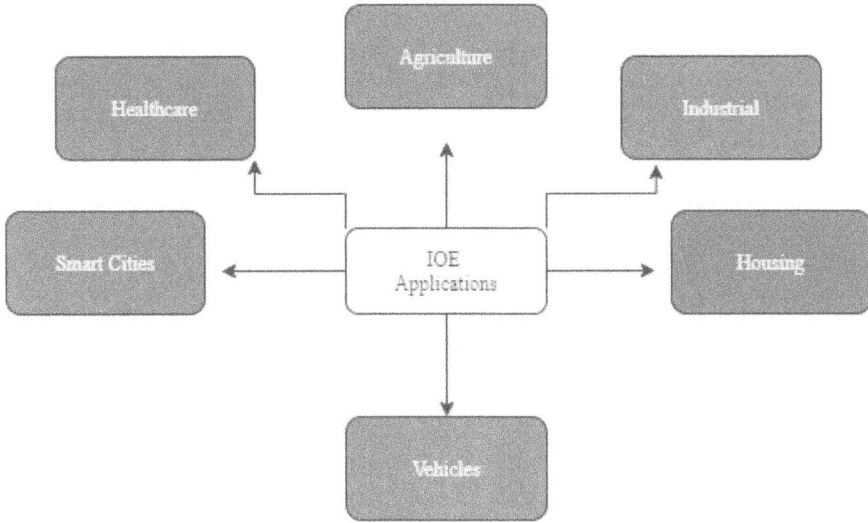

FIGURE 8.4 Applications of IOE.

by the open system. With IOE, network connection is made possible; power apparatus makes data processing and reading possible; speed is increased in both ways, and all of this cultivates better performance and efficiency [21, 22].

The advantages of Smart Grids in this century are [23]:

1. Consumers take part actively
2. Demand response and distributed generations [21]
3. Accommodation is provided for storage
4. The need for qualitative service is fulfilled
5. Efficient when operated and there is the optimization of utilization

In order to achieve this, Smart Grids comprise of elements like phasor measurement units (PMUs) [20], capable of sending and receiving massive amounts of data per minute. The advent of Internet Protocol Version 6 (IPv6) [24] makes it possible for billions of devices to be connected at the same time in theory, thus overcoming the possible limitation [25]. It is necessary to keep using better methods for big data management as the costs increase with passing years (Figure 8.5) [26].

FIGURE 8.5 Smart grids.

Smart Cities: As many new infrastructures are being created as we speak to support the growing standards of living and the growing population, these projects are being formed on the foundation of Information and Communication technologies [27]. For the best results, data is gathered by sensors and IT systems [24, 28].

In Ref. [21], a case study was mentioned to explain how Smart Cities would integrate a better living for the citizens. Here, a person had gone for the service provided by a medical facility wherein the doctor was monitoring their condition. The doctor is notified of the person fainting through an alarm and an ambulance is made to bring the person safely to the medical center. The ambulance uses a smart driving application to reach their home immediately. This study portrays how a Smart City brings together smart health, smart transportation, and smart public safety service.

Medical centers also have a chronic concern of waiting time. It is something directly related to the hospital's efficiency since every second matters in healthcare. More waiting time leads to precious time lost by the patient and the hospital in turn loses the patient and, as a result, its reputation. The development in the ICT sector and IOE is now under progress to eliminate this factor entirely to enable monitoring of patients by the staff in the medical centers [29].

One of the notable facts about the IOE is how it will decentralize businesses. Of course, it does not mean that now industries would not have a centralized footing. In fact, this gives them an opportunity to expand their footing into more activities besides what they were formally majoring at [30].

In Ref. [31], the IOE is described as the "IOT gateway on an enormous scale."

An automatic network consists of functions of optimization, configuration, healing, etc., on its own. The measure and vision were taken up by IBM in 2001 and were aimed at the development of an organized framework with the ability to manage itself. This endeavor was proposed to overcome the hurdle of the complexity of network systems.

As mentioned, it manages many things on its own including decision making and adjusting itself to the changes in network systems. This progress in technology and device innovations requires the usage of gateways. And while the current use of cellular technology might be apt for applications, collisions of packets would only increase with time and then degradation in efficiency would inevitably follow. Moreover, the current trends are unsuitable for quality of service (QoS) support and lead to the consumption of a good deal of amount of power.

CARS with IOE has produced coined "Sensing as a Service" (SenaaS). The example given in [32] is that of a radio station relying on CARS to help cover road accidents and transmit information about traffic situations. This proves to be pretty feasible in terms of commercial aspects. Then we have the instance of smartphones having sensors to monitor the environment, health-related conditions, and transportation (Figure 8.6).

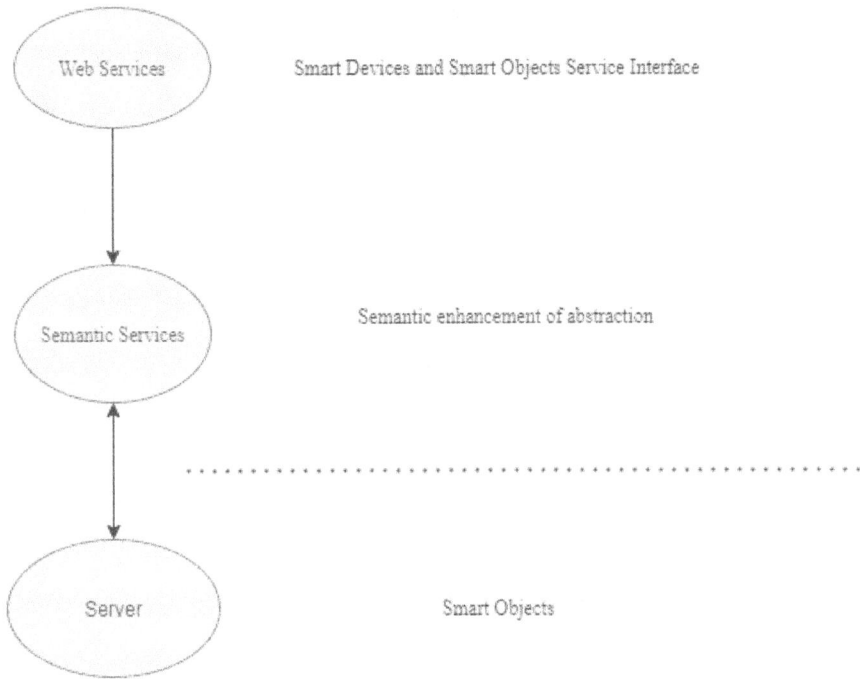

FIGURE 8.6   Sensor as a Service (SenaaS).

8.5.1   CHALLENGES FACED BY IOE

With every step forward, there come hurdles to pass through. IOE is no exception, with security being a major debate. As the data collection becomes higher and higher, keeping track of delicate frameworks becomes difficult. The drive to guarantee the security and privacy of the user and to keep their information confidential should take the lead focus. Mobility is another factor to take another consideration since the fast-developing world around is mobile and dynamic. The future Internet needs to be resourceful, highly connective, and reliable [24].

Security: This includes many aspects like ensuring that the user matches the identity given, is able to go along with the task in their minds, the fact that the data was sent out is the same as the received, and not having any intruders in the network [24].

In [33], there are seven principles proposed for privacy by design (PbD):

a. Features of security or privacy should be proactive instead of reactive
b. The features of security/privacy should be placed by default and should remain so
c. These features should be part of the design, implanted into the embedded system

d. The features should be integrated with positive-sum as functionality
e. Protection should be provided from one end to the other when it comes to security or privacy
f. Transparency and visibility are a must
g. These features should be in the boundary of respect of the users

As talked about before, smart grids have set a new benchmark for energy efficiency and reliability. But there are some security issues related to it. IoT and IOE have transformed the way the grid functions in today's time, and as an effect of this use, we have a major risk of cyber-attacks [34]. These are potentially harmful since they could lead to a full breakdown of the system. It is said to be of two types, namely active and passive attacks, and for reasons such as hacking, monitoring of traffic, eavesdropping along with ransomware attacks, and Denial of Service (DoS) [35].

Smart Cities have some grim realities associated with cyber security [36, 37]:

a. Increased use of IOE and related technology would make data vulnerable
b. These concerns bring in the fact that there are not sufficient aides in the department of cyber security

These can become an alarming consequence in the present time.

8.5.2 BLOCKCHAIN TECHNOLOGY

For higher scalability and efficient problem solving, blockchain technology can be used. This would ensure the protection of security against malicious attempts to intrude data centers and abusers of privacy.

It also helps in the identification of various access levels to put through limitations on unofficial implementations.

Two functions of the mechanism of blockchain:

1. Generation of transaction by the user in the system
2. Blocks are required to be in sequence and should remain unchanged [5]
3. **The Question of "Anytime" and "Anywhere":** There are many parts of the world that fall under remote areas and hence are hard to scale to and reach. In such geographical regions, it is hard to plant nodes. Consequently, the provision of IOE projects is not cost-efficient. Even if IOE projects are deployed, the communication between nodes will not remain to be strong enough to bring about connections "anytime" and "anywhere."
4. **How to Keep It Going:** Another main issue comes in when we think about constantly providing power to nodes. Battery-powered nodes would easily run out and hence, the process would be easily broken. This is especially true when it comes to remote areas. It is important for developing schemes that are sustainable and efficient.
5. What is thought to be the greatest power of IOE can be its ultimate shortcoming too. In order to pull off such an intensive Internet framework,

continuous computation of big data is necessary. To make every project node adapt to the local algorithms is a tedious task [11].

6. In the fast-changing world, in order to make economic progress, organizations would have to adapt with a high level of awareness about the changes happening every day. They will require equipment, new proposals for running their business, and new models [22].

7. **Sensors** are another important aspect of IOE. But it faces some key challenges as well.

   a. High consumption of power
   b. Transmission of accurate data collected by the sensors
   c. There may be a break in connections which could lead to delay in communication
   d. Security of data in sensors; it is prone to attacks and can be hacked (Figure 8.7) [38]

8.6 SECURITY OF IOE WITH BIG DATA ANALYTICS

Big data, mentioned several times in this chapter, is considered to be a huge volume of data and data variety and using that data in efficient ways for processing information for impactful decision making. Big data essentially narrows down to collecting data on a continuous basis, related to environment and circumstances. And the majority of it is stored in the cloud platform. Although thought to be secure, there are major concerns regarding this. The accuracy of generation of big data with IOE is extremely high. Intruders, hackers, and cyber attackers may harm the integrity of the data hence produced, altering the data leading to disruption of peace and order [39].

8.6.1 THE FUTURE OF GLOBALIZATION

Keeping the opportunities and challenges in mind, IoE is a revolutionary concept driven to connect almost anything to the Internet, not only electrical devices but any object which was not earlier controlled by any virtual idea. When you think about

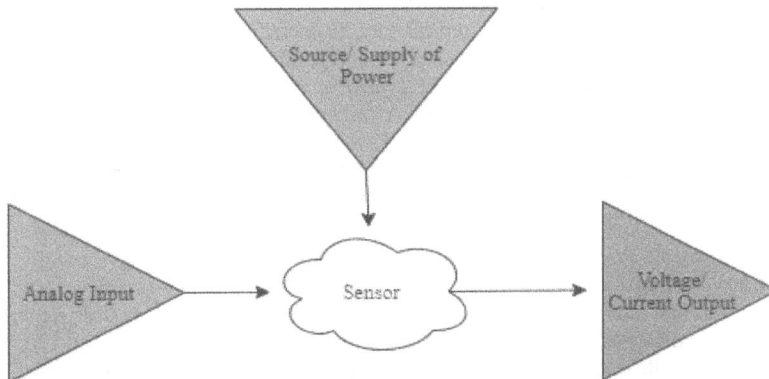

FIGURE 8.7 Sensors. (From Ref. [39].)

the level of the technology being theorized and the same even being brought into practice, it is a massive step toward bringing the world even closer. If it was earlier being referred to as a "village," then now it can be said that the proximity of the world at large has set into motion changes in the very essence of how we perceive globalization. It should be seen not as a "one-man army" but as an army for the whole world. In other words, it is not a stand-alone system but is an infrastructure to have services integrated on a global level [38].

8.7   CONCLUSION

The Internet has already evolved dramatically in the past twenty years. The onset of the era of the virtual world, wherein almost every person has some sort of access to the Internet, our economy and society now largely depend on the development of industries like the ones who handle cloud computing, big data, IoT and the next step, IOE. With each new development, we see new demands for information. The definition of IOE can be seen and traced back to Cisco, wherein the new environmental system will be connected in all ways.

But as we enter into the next decade, there are bigger hurdles to cross and even bigger things to achieve. Inventors, scientists, engineers, and everyone else involved in the world of science have continuously persevered to welcome the new wave of technology where "everything," or rather, "anything" that comes to one's mind, would have a way to stay connected. IOE is not simply a concept of one day; rather it has the power to change the world as we know it. It has huge opportunities and applications in the real world in the present itself, which gives a strong hint of its bright future. Many more opportunities would open up as it acceptance and new ways to adapt start to flow in. This doesn't mean, of course, that it would remain unchallenged. Any innovation has to stay rooted despite the odds to become better and better as each day passes. Security, coverage, battery provision, etc., are to name a few but there would be a time when even these would be overcome.

REFERENCES

1. Evans, David. "The internet of everything: How more relevant and valuable connections will change the world." Cisco Internet Business Solutions Group (IBSG) (2012).
2. Raj, Anu, and Shiva Prakash. "Internet of everything: A survey based on architecture, issues and challenges." *2018 5th IEEE Uttar Pradesh Section International Conference on Electrical, Electronics and Computer Engineering (UPCON).* IEEE, 2018.
3. Srinivas, K., Jabbar, M. A., and Neeraja, K. S. "Sensors in IoE: A review." *International Journal of Engineering and Technology (UAE)* 7 (2018): 158–160.
4. Dey, Shuvashis, Jhantu Kumar Saha, and Nemai Chandra Karmakar. "Smart sensing: Chipless RFID solutions for the internet of everything." *IEEE Microwave Magazine* 16.10 (2015): 26–39.
5. Kaur, Maninder Jeet, Sadia Riaz, and Arif Mushtaq. "Cyber-physical cloud computing systems and internet of everything." *Principles of Internet of Things (IoT) Ecosystem: Insight Paradigm.* Springer, Cham, 2020. 201–227.
6. Farhan, Laith, et al. "A concise review on Internet of Things (IoT)-problems, challenges and opportunities." *2018 11th International Symposium on Communication Systems, Networks & Digital Signal Processing (CSNDSP).* IEEE, 2018.

7. Ismail, Sura F. "IOE solution for a diabetic patient monitoring." *2017 8th International Conference on Information Technology (ICIT).* IEEE, 2017.
8. Ashton, Kevin. "That 'internet of things' thing." *RFID Journal* 22.7 (2009): 97–114.
9. Magrassi, Paolo. "Why a universal RFID infrastructure would be a good thing." *Technology* 16 (2002): 0038.
10. Evans, Dave. "The internet of things: How the next evolution of the internet is changing everything." *CISCO White Paper* 1 (2011): 1–11.
11. Liu, Yalin, et al. "Unmanned aerial vehicle for internet of everything: Opportunities and challenges." *Computer Communications* (2020): 66–83.
12. Bradley, Joseph, et al. "Internet of everything (IoE): Top 10 insights from Cisco's IoE value at stake analysis for the public sector." *Economic Analysis* (2013): 1–5.
13. Shojafar, Mohammad, and Mehdi Sookhak. "Internet of everything, networks, applications, and computing systems (IoENACS)." *International Journal of Computers and Applications* 27 (2020): 213–215.
14. DeNardis, Laura. *The Internet in Everything: Freedom and Security in a World with No Off Switch.* Yale University Press, United States, 2020.
15. Schatten, Markus, Jurica Ševa, and Igor Tomičić. "A roadmap for scalable agent organizations in the internet of everything." *Journal of Systems and Software* 115 (2016): 31–41.
16. Miraz, Mahdi H., et al. "A review on Internet of Things (IoT), Internet of everything (IoE) and Internet of nano things (IoNT)." *2015 Internet Technologies and Applications (ITA).* IEEE, 2015.
17. Vaya, Dipesh, and Teena Hadpawat. "Internet of Everything (IoE): A new era of IoT." *ICCCE 2019.* Springer, Singapore, 2020. 1–6.
18. Kashyap, Ramgopal. "Applications of wireless sensor networks in healthcare." *IoT and WSN Applications for Modern Agricultural Advancements: Emerging Research and Opportunities.* IGI Global, Hershey, Pennsylvania, 2020. 8–40.
19. Demirkan, Haluk, et al. "Innovations with smart service systems: Analytics, big data, cognitive assistance, and the internet of everything." *Communications of the Association for Information Systems* 37.1 (2015): 35.
20. Yinger, Robert J., and Ardalan E. Kamiab. "Good vibrations." *IEEE Power and Energy Magazine* 9.5 (2011): 22–32.
21. Yun, Miao, and Bu Yuxin. "Research on the architecture and key technology of Internet of Things (IoT) applied on smart grid." *2010 International Conference on Advances in Energy Engineering.* IEEE, 2010.
22. Khan, Hasnaat. "Cyber Security Challenges in Smart Grids." (2020). https://engrxiv.org/ua3wp/
23. Matusiak, Bożena, Anna Pamuła, and Jerzy S. Zieliński. "New idea in power networks development. Selected problems." *Przegląd Elektrotechniczny* 87.2 (2011): 148–150.
24. Jara, Antonio J., Latif Ladid, and Antonio Fernandez Gómez-Skarmeta. "The Internet of Everything through IPv6: An analysis of challenges, solutions and opportunities." *Journal of Wireless Mobile Networks Ubiquitous Computing, and Dependable Applications* 4.3 (2013): 97–118.
25. Zieliński, Jerzy S. "Internet of Everything (IoE) in smart grid." *Przegląd Elektrotechniczny* 91.3 (2015): 157–159.
26. Pfisterer, Dennis, Mirjana Radonjic-Simic, and Julian Reichwald. "Business model design and architecture for the internet of everything." *Journal of Sensor and Actuator Networks* 5.2 (2016): 7.
27. Caragliu, Andrea, Chiara Del Bo, and Peter Nijkamp. "Smart cities in Europe." *Journal of Urban Technology* 18.2 (2011): 65–82.
28. Komninos, Nicos, Marc Pallot, and Hans Schaffers. "Special issue on smart cities and the future internet in Europe." *Journal of the Knowledge Economy* 4.2 (2013): 119–134.

29. Vlacheas, Panagiotis, et al. "Enabling smart cities through a cognitive management framework for the internet of things." *IEEE Communications Magazine* 51.6 (2013): 102–111.

30. "What the Internet of Everything really is – a deep dive." Available at https://www.i-scoop.eu/internet-of-things-guide/internet-of-everything/, 2020.

31. Kang, Byungseok, Daecheon Kim, and Hyunseung Choo. "Internet of everything: A large-scale autonomic IoT gateway." *IEEE Transactions on Multi-Scale Computing Systems* 3.3 (2017): 206–214.

32. Abdelwahab, Sherif, et al. "Enabling smart cloud services through remote sensing: An internet of everything enabler." *IEEE Internet of Things Journal* 1.3 (2014): 276–288.

33. Singh, Parminder, et al. "Blockchain and fog based architecture for Internet of Everything in smart cities." *Future Internet* 12.4 (2020): 61.

34. Usak, Muhammet, et al. "Health care service delivery based on the Internet of things: A systematic and comprehensive study." *International Journal of Communication Systems* 33.2 (2020): e4179.

35. Mohanty, Saraju P. "Security and privacy by design is key in the Internet of Everything (IoE) era." *IEEE Consumer Electronics Magazine* 9.2 (2020): 4–5.

36. A. O. Otuoze, M. W. Mustafa, and R. M. Larik, "Smart grids security challenges: Classification by sources of threats." *Journal of Electrical Systems and Information Technology* 5.3 (2018): 468–483.

37. Kesar, Shalini. "Cybersecurity and smart cities." *Societal Challenges in the Smart Society* 217 (2020): 217–224.

38. TM Forum Inform. 2020. Infographic: The Top 20 Internet Of Everything Challenges – TM Forum Inform. Available at: https://inform.tmforum.org/features-and-analysis/2016/07/infographic-top-20-internet-everything-challenges, 2020.

39. Karthiban, Mr K., and Jennifer S. Raj. "Big data analytics for developing secure internet of everything." *Journal of ISMAC* 1.02 (2019): 129–136.

9 A Review of Human–Robot Interaction for Automated Guided Vehicles Using Robot Operating Systems

Prashant Hemrajani, Tanishka Mohan,
Manoj Kumar Bohra, and Amit Kumar Bairwa
Manipal University Jaipur
Jaipur, India

CONTENTS

9.1 INTRODUCTION

As time passes, the world advances into a new phase of technological developments. These advances in technology strive to make the quality of life better for all Human Beings. While developments are happening all around the world surreal innovations have also taken place in the area of Automated Guided Vehicles (AGVs). These ubiquitous machines are replacing human operators in warehouses, industries, terminals, hotels, and so many other applications. They are improving work efficiency and are more flexible and versatile than traditional methods. AGVs have been providing industrial material handling capabilities in intra-logistic applications for decades, simply following predefined paths for navigation, mapping, and sensors for indicating directions.

AGVs are battery-powered machines that generally work using GEL-based batteries. Their software framework is based on the Robot Operating System (ROS).

DOI: 10.1201/9781003145004-9

ROS is an interface that gathers software frameworks for creating robot software. A number of sensors are involved in its working; for example, the proximity sensor helps in the detection of movement, which directly contributes to making it start or stop. Photosensors help in the detection of objects in the vehicle's pathway. The speed controlling is usually dealt with by fuzzy inference systems. Lane paths, signal paths, or signal beacons are used for navigating in an optimized manner and predicting danger in the path. Visual-based sensors are used for vehicular guidance and provide a lookout of its surroundings. The control system helps in automatic accurate pathfinding, motion control, localization, and mapping parallelly. The progress of navigation technology has contributed significantly to the growing utilization of AGVs. The transfer device sends a signal to the control device, after which the driving device receives a signal, which further moves the vehicles [1, 2] in the desired direction.

Human–Robot Interaction (HRI) helps in guiding the execution of a task by an AGV. They help in instilling a sense of safety and dependence for its proper functioning. These are high-level intelligent systems that are able to extrapolate from a set of basic motions to the completion of a task without the user having to demonstrate all aspects of said task. It requires the definition of actions that are both easily programmed by humans and easily comprehended and disambiguated by the AGV and its sensors. It works on an algorithm that enables it to figure out its trajectory to reach the final destination, and the target location is given by the user using GUI. Navigation is performed via odometry (measurements of vehicle motion), global positioning satellite system. It is greatly improved by simultaneous localization and mapping (SLAM) [3] based on front and rear light detection and ranging (LIDAR) scans. It must be provided with a comprehensive variety of sensors, comprising range, proximity, touch, vision, sound, temperature, and so on, owing to the unstructured format of anthropic domains and the somewhat unexpected movements of people.

AGVs of today are equipped with powerful decision-making technology. There is a trade-off between previously and dynamically generated paths and this means that the vehicle has to make a decision based on safety concerns and speed. While progress is being made on all fronts, the AGVs are also becoming more proficient in carrying out human tasks as well as ergonomically challenging tasks in industries. In the future, a parallel increase in sensor processing capabilities and hardware robustness will result in people and robots sharing a common workspace in a harmonic manner.

9.2 THE ROBOT OPERATING SYSTEM (ROS)

Robot Operating System or ROS is an open-source software framework that is used for designing robotics applications. In complex environments whether it be navigating an AGV or grasping objects or preventions of accidents, ROS provides tremendous amounts of versatility and notably demonstrates its strength. It is not an actual operating system (meta operating system), but it provides a set of tools and functionalities like the abstraction of hardware, low-level control of gadgets, or deliverance of information between process and package management on a diversified computer cluster. The ROS can be modified, however, as per the task's requirement.

In robotics, ROS comes loaded with system drivers and interface packages with different embedded systems (sensors and devices). It comes with ready-to-deploy packages which can be used for various algorithms and functions such as SLAM and adaptive Monte Carlo Localization (AMCL). SLAM is basically the combination of complex algorithms that helps in plotting the exact position in the stated environment by collecting data every few seconds and aligning accordingly. Using SLAM, the machine can locate itself and construct a virtual map. Using an inertial measurement unit (IMU), sensors can also use visual information, non-visible datasets, and fundamental positional data.

ROS is a loosely coupled device built, where every process is associated with a NODE, and each and every NODE is accountable for a task. Everything in ROS is in the form of packages, and the packages should always be inside the workspace. It helps to bundle the code in a format that is easy to manage. Messages are data structures that are loaded by NODES with bits of information. With the help of messages passing through a logical channel called TOPIC, NODES form a connection between each other. Using the publish or subscribe model, each NODE transmits or gets data from the other node. In robotics research and development, the ultimate purpose of ROS is to facilitate code reuse so that you can find a built-in package framework. The entire process is explained in Figure 9.1 through a block diagram. For displaying active NODE in the running system, a command line tool named Rosnode in ROS is brought into use. For listing it, Rosnode list command is used.

The system allows users to use landmarks to identify paths, and the simulated AGV is capable of following these paths [4]. For the navigation and localization purposes of the engine, the ROS navigation stacks are brought into use [5]. The control architecture employed for the navigation system (Figure 9.2) involves a number of independent Linux operating systems as independent modules.

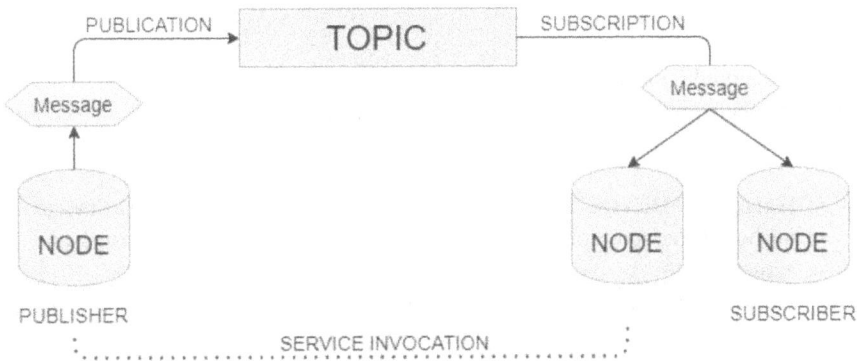

FIGURE 9.1 Basic working of ROS.

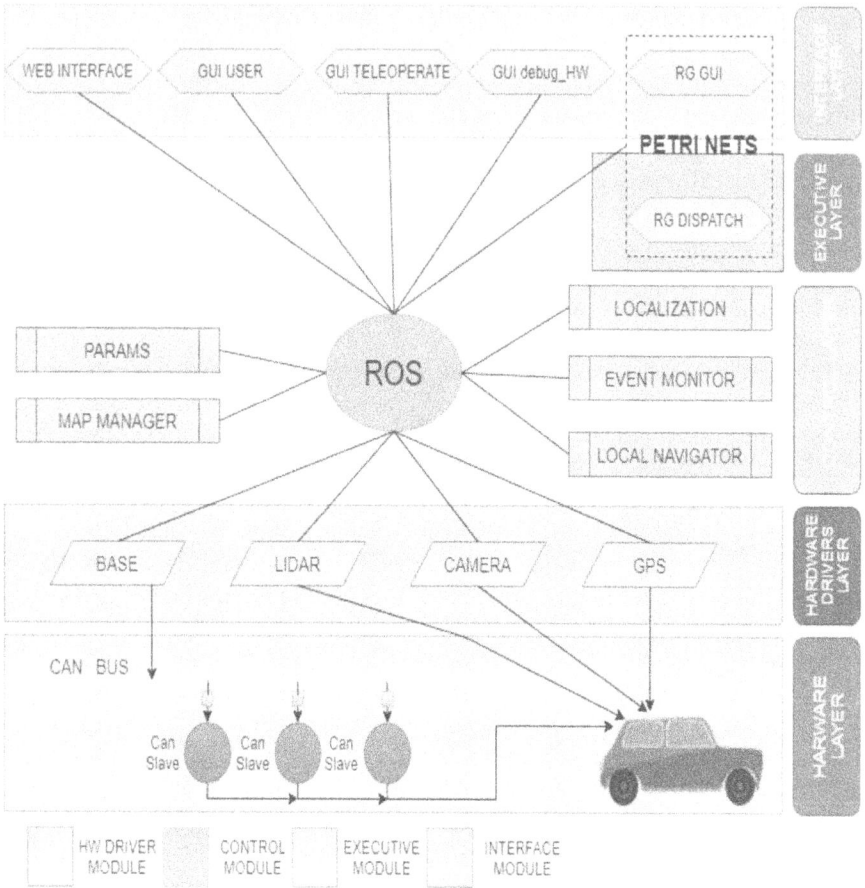

FIGURE 9.2 Car software architecture. (Reference taken from [6].) (*Abbreviations:* ROS-Robot Operating System; GUI- Graphical user interface; HW- Hardware; RG- RoboGraph; GPS - Global Positioning System; LIDAR- Light detection and ranging.)

These modules exchange data using the ROS-provided [6] inter-process communication mechanism. On the basis of functionality, modules can be categorized into four sets:

- **Hardware Driver:** All processes that monitor the hardware devices on board the automobile is included in the hardware driver (sensors and actuators).
- **Control Layer:** The basic navigation algorithms are performed by the control layer: constructive control or preventing obstacles (local navigator), localization (localization), route planning (map manager), and perception (event monitor) that process sensor information to identify simple incidents [7].

- **Executive Layer:** In order to carry out the current behavior, the executive layer coordinates the series of acts that other modules need to perform. The aim of this study is this layer.
- **Interface Layer:** The interface layer consists of a series of user and web interaction procedures.

9.3   HUMAN–ROBOT INTERACTION (HRI)

In the same cell, robots and humans coexist and share assignments as per their abilities. There may be multiple modes of communication between a human and a robot, but these modes are primarily determined by whether or not the individual and the robot are near to one another. As a consequence of the interplay between the robot and the human, a realistic combination of the two "systems" emerges. The reciprocal weight of the human and robot compatible features influences this equilibrium. Because their activity is not restricted to a narrow area, robots with built-in detectors often provide a flexible platform. The interaction between a human and a robot can mainly be categorized into two types:

- **Remote Interaction:** The human and the robot are a certain distance that is separated in space also or even if it is temporary. They are not co-located [8].
- **Proximate Interaction:** Humans and robots are co-located [8].

In a ROS environment, a smart decision-making process that enables human-robot work allocation is developed and incorporated [9]. This ensures the HRI collaborative work is in a successive manner. The contact between a person and a robot is often done through a depth sensor and a software tool for the gesture handler. Typically, these encounters involve physically articulated robots, and their representation differentiates them significantly from other technological innovations [10]. With the help of HRI, humans are able to bring about their in-life experience, i.e. with other human beings, into the mechanism of human–robot interface as well. In Figure 9.3, you can very well see the cycle of perception and then what causes the reaction of the action received for HRI.

In the automation world in the upcoming years there have been grand developments in relation to HRI. Often the systems in AGVs work on the regulated human-machine interface in the case of the driver of the car like they have some sort of advanced driver assistance (ADAS) onboard innovations, such as path monitoring, adaptive cruise control, auto parking, forecasting braking, safety systems for pedestrians, and blind-spot alarm systems [10]. Management should devote more time and energy to duties of accountability now that they are free of hard labor. As a result, there will be "less human interference," in the sense that the more the facility is automated and operated by automation, the more its systems communicate without the need for a human to resolve issues, make repairs, levitate objects, or schedule work. In order to account for workload safety concerns, cognitive knowledge could be used to adaptively define the terms of robot interference.

Furthermore, self-driven automobiles need graphical interfaces for them to perceive other traffic users' behaviors and motives, and the automobile will require methods to

PERCEPTION CYCLE

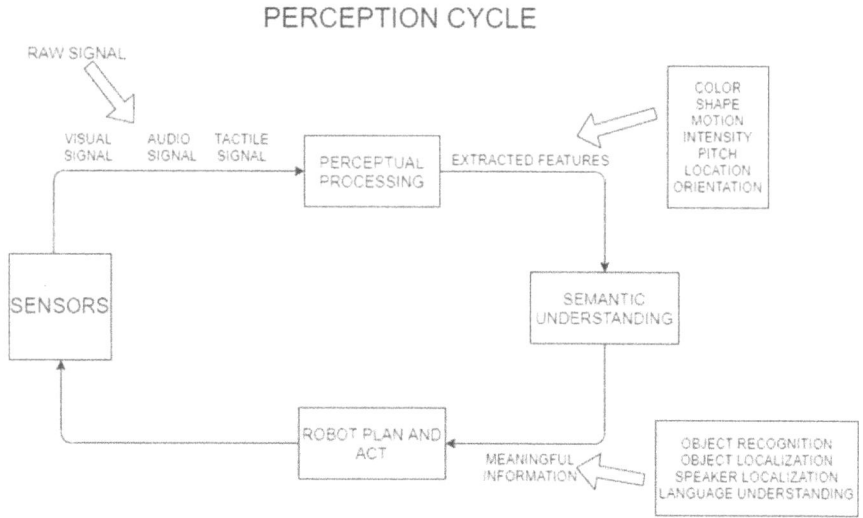

FIGURE 9.3   The perception cycle of HRI.

analyze them and convey their actions accordingly. All of the multiple elements of a robot, from systems to actuators, from sensing to control, should be assessed for the security and reliability of the physical interactions. The reaction to automated vehicles by travelers can be further estimated by the HRI. Pressure sensors and minimal "tactile" sensors deployed along the framework may be required for finer HRI, such as assisting individuals, handing over objects, or shaking hands – all "physical" occurrences.

9.4   AUTOMATED GUIDED VEHICLES

AGVs are portable, fully automated systems employed mainly for transport processes and also for versatile assembly line system solutions. Requirements for AGV systems range from automatic ports where shipments are passed around in warehouses to ship pallets. AGVs are used by hospitals to simplify procedures, such as washing and drug storage, and to move food and other supplies within platforms. Basically, it means that we can transfer any amount of load with the help of AGVs [11]. Today, the function of industrial AGVs requires specified facilities and maps that are easily obtainable to localize them [12]. The traffic direction along the flow path can either be defined as unidirectional or bi-directional [13]. AGVs are sufficiently automated and provide low-cost transport activities [14].

Subjective factors, fatigue, or the weight of the moved commodities have no influence on the performance of a robot AGV. In less time, a single self-driving car will meet the deadlines of one or more individuals. The benefits of AGVs are amplified when used for repetitive tasks in the factory and/or on a large scale. Programming allows for accurate path and task definition for the robot.

AGVs generally follow a defined path that has been set for them in order to perform their tasks. LIDAR is a technique that mainly employs a laser sensor (or distance sensor). The laser sensor point cloud comes with advanced precision distance estimation and is ideal for SLAM map construction. When a LIDAR sensor is mounted on an AGV sends out a series of laser pulses that calculate the distance between obstacles and the automobile. This data is used to develop a complete 360° environmental map of the functional environment, which allows the AGV to navigate around the facilities without the need for any supporting investment. To provide automotive navigation, AGV databases and camera-based vision are frequently used in conjunction with LiDAR sensors.

For new equipment and installation, AGV companies usually consider the four techniques listed below:

- **Magnetic Navigation:** This is usually utilized by light-duty AGVs and uses magnetic tape for guidance. The main benefit of utilizing magnetic tape is that it can be used; however, it can be applied and removed from anywhere as per our demand for the change of route. It also saves money by avoiding the cost of reorganizing the warehouse or storage facility floor.
- **Laser-Guided Navigation:** The vehicle uses triangulation to evaluate its precise location using reflectors situated on the surrounding walls, allowing it to complete the tasks needed in the industrial zone. Unlike magnetic guidance systems, a laser-guided algorithm does not require any floor work.
- **Vision Guided Vehicles:** This platform allows cars to run in either automatic or manual phase, giving them a lot of versatility. The vehicle's machine actively creates a 3D mapping of its regulatory area. These generally use optic sensors.
- **Natural Navigation:** These do not require much of the necessities like the others and are easily installed in the systems already in place. LIDAR is the major technology utilized in this.

Guide path has electronically encoded patterns that are interpreted by onboard sensor devices and signals into numerical signals used by an onboard microcomputer to monitor the direction module, as well as the collection of sensor direction signals, and routing despite disruptions or anomalies in the guidance for consistent and precise travel between the selected sites along the guide path [16]. Figure 9.4 shows the algorithm for implementing the AGV path tracking model in reference with [15].

1. To set the vehicle's route, the Route Planner Module sends a list of reference points to the Internal Controller.
2. The first reference point is read by the Internal Controller.
3. The Kalman filter algorithm is used by the Network Delay Estimator module to determine the next RTT value.
4. The Position Predictor module calculates the vehicle's position at the moment the control information is processed.
5. The Internal Controller estimates the reference velocities of the wheels and assigns them to the car.
6. The vehicle follows the kinematic model's instructions. Only after the actual RTT delay has been performed, the controller's control commands are obtained and implemented in the automobile.

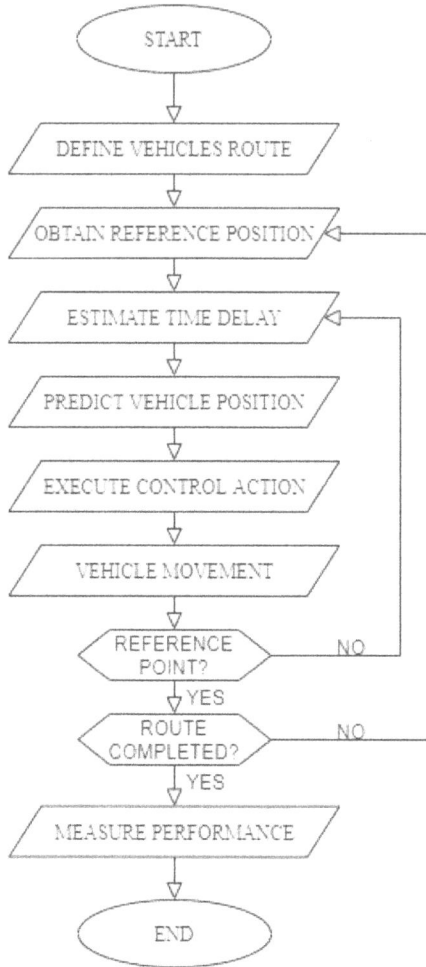

FIGURE 9.4   Flowchart of the AGV path tracking. (Reference taken from [15].)

7. The Internal Controller observes the next point if the vehicle has approached the reference point.
8. If not, the process is continued with the next RTT value estimate.
9. If the car has covered the specified path, the cycle concludes with the computation of performance values using the cost functions defined.
10. If not, the pattern repeats with the next reference point being read.

As you can see, Figure 9.5 represents the structure of a classical AGV model [4]. Demands from the Production Planning and Control (PPC) module power the AGV Control System (ACS) that breaks down the general operations into internal operations.

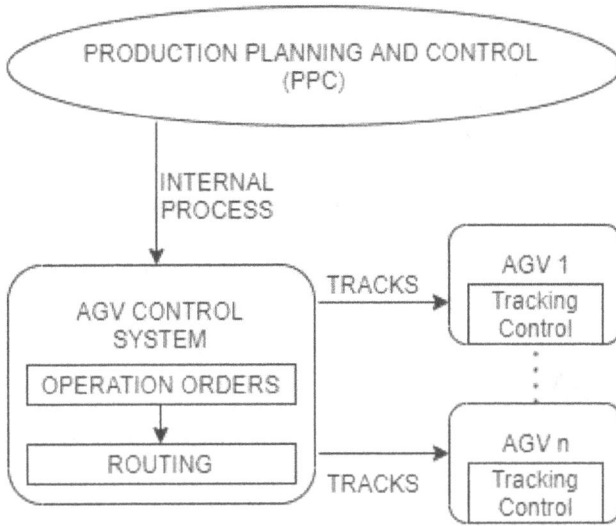

FIGURE 9.5   Structure of a classical AGV model.

In comparison to internal procedures that define the processes necessary to satisfy externally stimulated processes, general processes are externally triggered processes, such as client demands [4]. When the AGV is faced with any issue its automated response is to reach a halt position. Instead of facing this issue, it must be able to [4]:

- prevent hurdles on the course on its own.
- deal with issues where the ACS isn't involved, such as multi-robot scenarios or pick-and-place operations.
- be easier to update and less costly throughout system design and setup through using enhanced trajectories to save time, energy, and assets (e.g. floor abrasion).

These hurdles can only be crossed when the AGVs are able to [4]:

- locate themselves (even if they stray from the predetermined path).
- communicate with one another.
- act out and modify their behavior patterns (role play) to tackle specific issues without the need for core involvement.

9.5   MAJOR ISSUES FACED

To build this co-dependent environment of robots and humans, there are still many issues faced. While constructing of this stable environment the four major issues faced by the robot in [17] are:

1. *Self-Reliance:* The robot must be able to look after itself and not rely on the human to help it out for the completion of its tasks. It must be able to safeguard itself while performing the tasks with efficiency.

2. *Self-Awareness:* This does not suggest that it must be completely human-like; instead it must be competent enough to recognize its threshold and know in which tasks it needs to seek assistance from the human and which are the issues which it can resolve on its own.
3. *Adaptiveness:* The robot must be able to adapt to different environments and different controllers and adjust its functions accordingly in its capacity.
4. *Dialog:* There must be efficient communication between the robot and the operator. It must be able to properly intercept the task allotted to it. Dialog is a two-way thing so conveying of information properly is a necessity.

9.6 INTERNET OF THINGS IN THE WORLD OF AUTOMATED GUIDED VEHICLES

While the AGVs are greatly useful and perform their tasks brilliantly, it comes with their set of challenges. To overcome those issues Internet of Things (IoT) is brought on board. It allows the optimum use and management of AGVs in real-time data capture and analysis, besides remote monitoring and control. Industrial IoT technologies generally produce huge amounts of data in lots of variations which create the difficulties of Big Data. AI-driven analytics must overcome these issues. As you can see in Figure 9.6 [18], these all eventually lead up to the IoT cloud, after which the rest of the processes take place. The IoT-enabled AGV system will produce all statistics

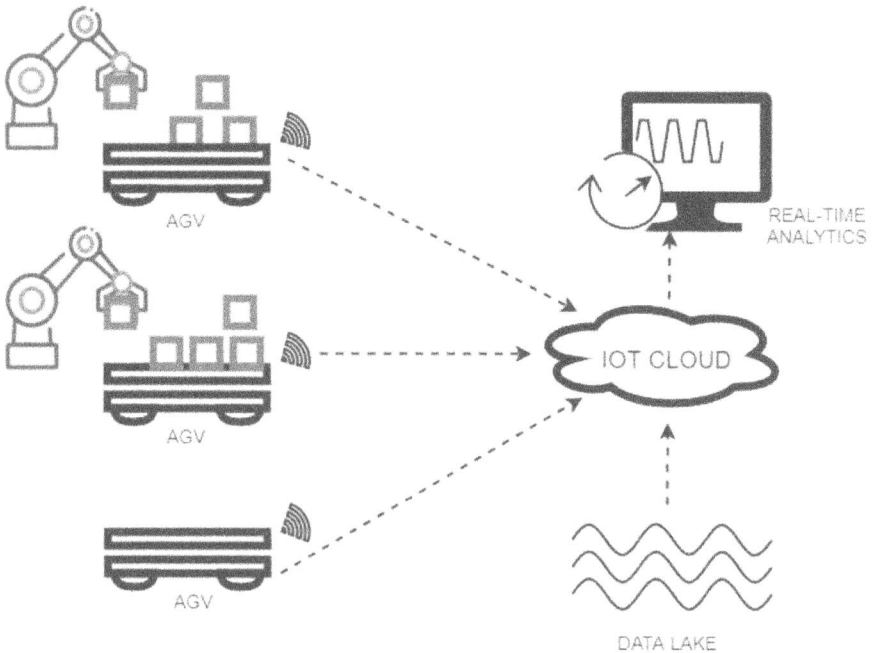

FIGURE 9.6 Remote monitoring, managing, and detecting failures in AGVs connected to the control center through real-time AI-based analytics. (Reference taken from [18].)

related to material transfer in the factory [19]. These ensure trustable operations of AGV applications [20].

In a traditional AGV, computer-based processes and systems are used to allocate work to the automobile, which is then combined with cloud storage and directed along the route [21, 22]. With this all the data would reach the assigned person in time hence avoiding any delays or human errors like handling, carrying, or transportation of the load to the factory [19].

IoT helps out in the field of AGV in:

- **Condition Monitoring:** IoT sensors help in capturing data of variables that contribute to the robustness of the device. It is possible to set sirens for severe battery levels, carriage load, heat, reverberation, etc. Concerns can be quickly solved as the data is gathered in real-time.
- **Maintenance and Remote Servicing:** Data capturing also helps in analyzing the efficiency of the device. The amount traveled, operation time, and the heaviness of the load carried are important variables that need to be tracked.
- **Performance Tracking and Usage:** When we bring IoT on board the cost of maintenance reduces tremendously for the device. The maintenance department can look after the AGVs with the help of IoT and can come on-site only when any problem arises. By these the daily maintenance measures are reduced drastically.

The location of the AGV is normally determined from its velocity and processing speed, with a GPS position error of more than several meters due to external variables, such as floor or rail frictional force [23]. With the assistance of an internet-enabled dashboard, data visualization of real-time AGV factors is possible. When many guidelines and a common approach are followed, AGV innovations provide a brilliant solution for boosting the effectiveness of the associated logistic process [24].

9.6.1 Automated Guided Vehicle in Production Line

With the progression in time, slowly the more and more manpower is being replaced with AGVs. Consumers may be waiting for self-driving cars, but in the industrial sector, AGVs are already being used to boost production and enhance operational flexibility and flow. AGVs that use intelligent communication techniques and are connected to ERP systems and manufacturing execution systems (MESs) are now an increasingly important part of synchronized and make-to-order production. AGVs are expected to make huge returns on investments made on them to completely replace manual labor [24]. The returns can only be expected when all the criteria are being followed in order to minimize the risks. In order to complete the process of pre-implementation, implementation, post-implementation, and improvement and maintenance a basic scheme of critical success factor (CSF) and key performance indicators (KPI) is followed [24]. KPI creates the scope of improvements as through it the company receives feedback.

The basic scheme of CSF in reference to [24] is:

1. Technological
   - Ability to estimate the position in real time
   - Space for AGV angling
   - Maximum speed
   - Guidance system
   - Towing capacity
2. Safety
   - Ergonomic aspect
   - Low regulation
   - Wellbeing
3. Organizational
   - Traffic frequency (heat map)
   - HRM and work standardization
   - Decision-making aspect
   - Order frequency

The basic scheme of KPI in reference to [24] is:

1. Delivery on Time
   - Delay
   - Oversupply – earlier delivery
2. Delivery in Full
   - Incomplete delivery
   - Lost deliveries
   - Damaged items
3. Qualitative KPI
   - Visualization
   - Standardization
   - Ergonomic aspects (RULA, OWAS, NIOSH, etc.)
4. Cost of Delivery
   - Maintenance cost
   - Costs of accidents and repair
   - Safety and ergonomics costs

Assembly lines are also a region where AGVs are commonly used. The workload in this industry is inequitable and fluctuating. As a result, the docking machines must be customized to the requirement [14]. The logistics industry has a plethora of new AGV implementations. The application of an AGV to a mobile pallet stretch wrapping device is one instance.

9.7 CONCLUSION

Where once the factories were not interconnected and the machines had to be operated by an individual, making them completely dependent on each other, now with the help of these technologies they can form real-time connections and enable a stable and efficient working environment. A regulated-device calibrated by a human pilot can be tolerated in industrial applications only. It still is quite a big idea to bring about its use among us individuals in our day-to-day lives. We learned that ROS plays the base in the operation of AGVs in coordination with the HRI. The ROS is responsible for the observations by the various sensors and completes the efficient operation of our automated vehicle. The robot uses algorithms to locate individuals and seek assistance, bearing in mind the burden it places on those it requests (journey path to the assisting destination) as well as the robot's own need for a quick performance evaluation [25].

The AGVs are path specific and their paths are centered on magnet or inductive guide tapes that are applied on or on the surface of the site [26–28]. The AGV adjusts the automated docking pattern in a coordinated manner centered on human movement. Human workers can exist side by side with an AGV in an unfamiliar setting for autonomous docking activities, according to realistic experimental findings [29]. In case a disaster happens, the robotic system should revert to the most human-safe setup possible [30].

We also learned about the importance of the Internet of Things in the operation of AGVs. In short, it helps in the safe and proper handling of materials, improving onsite safety of the workers, saving of all data, and sending it to the relevant authority for proper analysis and increasing productivity while reducing manpower [19]. It helps in low cost and better maintenance of the device in terms of many parameters. There is still research going on in this field, with lots of new aspects being discovered. This is a highly accelerated field and eventually in the upcoming years will play an important role in our life.

REFERENCES

1. Yu, J., Chen, Y., Ouyang, L., Liao, W., & Bi, S. (2016). An image enhancement method for non-uniform illumination with illumination constraints for vision-guided AGV. In *2016 International Conference on Advanced Mechatronic Systems (ICAMechS)*, Melbourne, VIC (pp. 148–153). http://dx.doi.org/10.1109/ICAMechS.2016.7813437
2. Su, S., et al. (Winter, 2020). Positioning Accuracy Improvement of Automated Guided Vehicles Based on a Novel Magnetic Tracking Approach. *IEEE Intelligent Transportation Systems Magazine*, 12:4, 138–148. https://doi.org/10.1109/MITS.2018.2880269
3. Bore, D., Rana, A., Kolhare, N., & Shinde, U. (2019). Automated guided vehicle using robot operating systems. In *2019 3rd International Conference on Trends in Electronics and Informatics (ICOEI)*, Tirunelveli, India (pp. 819–822). http://dx.doi.org/10.1109/ICOEI.2019.8862716
4. Bader, M., Richtsfeld, A., Suchi, M., Todoran, H. G., Kastner, W., Vincze, M. (2015). Balancing centralized control with vehicle autonomy in AGV systems, In *The Eleventh International Conference on Autonomic and Autonomous Systems (ICAS 2015)*, Italy (pp. 1–7).
5. Marder-Eppstein, E., Berger, E., Foote, T., Gerkey, B., & Konolige, K. (May, 2010). The office marathon: Robust navigation in an indoor office environment. In *2010 IEEE International Conference on Robotics and Automation (ICRA)*, Alaska (pp. 300–307).
6. López, J., Sánchez, P., Sanz, R., & Paz, E. (2020). Implementing Autonomous Driving Behaviors Using a Message Driven Petri Net Framework. *Sensors*, 20, 449. https://doi.org/10.3390/s20020449

7. Romera, E., Alvarez, J. M., Bergasa, L. M., & Arroyo, R. (2018). ERFnet: Efficient Residual Factorized ConvNet for Real-Time Semantic Segmentation. *IEEE Transactions on Intelligent Transportation Systems*, 19, 263–272.

8. Goodrich, M. A., & Schultz, A. C. (2008). *Human-Robot Interaction: A Survey.* Now Publishers Inc., Boston – Delft.

9. Tsarouchi, P., Alexandros-Stereos, M., Makris, S., & Chryssolouris, G. (2017). On a Human-Robot Collaboration in an Assembly Cell. *International Journal of Computer Integrated Manufacturing*, 30:6, 580–589. https://doi.org/10.1080/09511 92X.2016.1187297

10. Eyssel, F., Kanda, T., Keijsers, M., Šabanović, S., Belpaeme, T., & Bartneck, C. (2020). *Human-Robot Interaction: An Introduction.* Cambridge: Cambridge University Press.

11. Ullrich, G. (2014). *Automated Guided Vehicle Systems: A Primer with Practical Applications.* Berlin & Heidelberg: Springer-Verlag.

12. Beinschob, P., & Reinke, C. (2015). Graph SLAM based mapping for AGV localization in large-scale warehouses. In *2015 IEEE International Conference on Intelligent Computer Communication and Processing (ICCP)*, Cluj-Napoca (pp. 245–248). https://doi.org/10.1109/ICCP.2015.7312637

13. Gaskins, R. J., & Tanchoco, J. M. A. (1987). Flow Path Design for Automated Guided Vehicle Systems. *International Journal of Production Research*, 25:5, 667–676. https://doi.org/10.1080/00207548708919869

14. Schulze, L., Behling, S., & Buhrs, S. (May, 2008). Automated guided vehicle systems: a driver for increased business performance. In *Proceedings of the International Multiconference of Engineers and Computer Scientists*, Hong Kong (pp. 1275–1280).

15. Lozoya, C., Martí, P., Velasco, M., & Fuertes, J. (2007). Effective real-time wireless control of an autonomous guided vehicle. In *2007 IEEE International Symposium on Industrial Electronics*, Spain (pp. 2876–2881).

16. MacKinnon, A. S., Willemsen, D. J., & Hamilton, D. T. (1985). *U.S. Patent No. 4,530,056.* Washington, DC: U.S. Patent and Trademark Office.

17. Fong, T., Thorpe, C., & Baur, C. (2003). Collaboration, Dialogue, Human-Robot Interaction. In: *Robotics Research.* Berlin, Heidelberg: Springer. 255–266.

18. Cupek, R. et al. (2020). Autonomous Guided Vehicles for Smart Industries – The State-of-the-Art and Research Challenges. In: Krzhizhanovskaya V. et al. (eds) *Computational Science – ICCS 2020. ICCS 2020. Lecture Notes in Computer Science*, vol. 12141. Cham: Springer. https://doi.org/10.1007/978-3-030-50426-7_25

19. Shejwal, Y., & Gera, S. K. (2018). IoT Based Automated Guide Vehicle. *IRJET*, 5: 5, 3256–3259.

20. Kamarudin, K. (2019). Development of IoT Based Mobile Robot for Automated Guided Vehicle Application. *Journal of Electronic & Information Systems*, 1. https://doi.org/10.30564/jeisr.v1i1.1061

21. AlZubi, A. A., Alarifi, A., Al-Maitah, M., & Alheyasat, O. (2020). Multi-Sensor Information Fusion for Internet of Things Assisted Automated Guided Vehicles in Smart City. *Sustainable Cities and Society*, 102539. https://doi.org/10.1016/j.scs.2020.102539

22. Gomathi, P., Baskar, S., & Shakeel, P. M. (2020). Concurrent Service Access and Management Framework for User Centric Future Internet of Things in Smart Cities. *Complex & Intelligent Systems.* https://doi.org/10.1007/s40747-020-00160-5

23. Cheong, H. W., & Lee, H. (2018). Requirements of AGV (Automated Guided Vehicle) for SMEs (Small and Medium-Sized Enterprises). *Procedia Computer Science*, 139, 91–94.

24. Hrušecká, D., Lopes, R., & Juřičková, E. (2018). Challenges in the Introduction of AGVs in Production Lines: Case Studies in the Automotive Industry. *Serbian Journal of Management*, 14. https://doi.org/10.5937/sjm14-18064.

25. Schneier, M., & Bostelman, R. (2015). *Literature Review of Mobile Robots for Manufacturing.* Maryland: US Department of Commerce, National Institute of Standards and Technology.
26. Walenta, R., Schellekens, T., Ferrein, A., & Schiffer, S. (2017). A decentralised system approach for controlling AGVs with ROS. In 2017 IEEE AFRICON, (pp.1436–1441). http://dx.doi.org/10.1109/AFRCON.2017.8095693
27. Bostelman, R., & Messina, E. (2016). Towards development of an automated guided vehicle intelligence level performance standard. In *Autonomous Industrial Vehicles: From the Laboratory to the Factory Floor, ASTM,* West Conshohocken, PA. https://doi.org/10.1520/STP159420150054
28. Jaiganesh, V., Jayashankar, D. K., & Girijadevi, J. (2014). Automated Guided Vehicle with Robotic Logistics System. *Procedia Engineering,* 97, 2011–2021. https://doi.org/10.1016/j.proeng.2014.12.444
29. Song, K.-T., Chiu, C.-W., Kang, L.-R., Sun, Y.-X., & Meng, C.-H. (2020). Autonomous docking in a human-robot collaborative environment of automated guided vehicles. In *2020 International Automatic Control Conference (CACS),* Hsinchu, Taiwan (pp. 1–6). https://doi.org/10.1109/CACS50047.2020.9289713
30. Alami, R., Albu-Schäffer, A., Bicchi, A., Bischoff, R., Chatila, R., De Luca, A., ... & Villani, L. (2006, October). Safe and dependable physical human-robot interaction in anthropic domains: State of the art and challenges. In *2006 IEEE/RSJ International Conference on Intelligent Robots and Systems* (pp. 1–16). IEEE.

10 Analysis of Cascading Behavior in Social Networks and IoT

Jitendra Kumar and Mukesh Kumar
NIT Patna
Patna, India

CONTENTS

10.1 INTRODUCTION

Internet of Things (IoT) is a set of interconnected processing devices having unique identification (UID). IoT has also the capacity to move information over an association without expecting H2H or H2C communication. Nodes or devices are connected to the Internet differently, which is envisioned that an individual may associate with the Internet by many things. Essentially, there will be a ton of things associated with the Internet instead of individuals who are associated. It drives IoT to be improved and the quickest developing advancements in the Internet world.

Complex systems are formed naturally using several communities. In such a system, these communities are usually elaborated as groups having a dense connection

DOI: 10.1201/9781003145004-10

of nodes within themselves and sparse connection of nodes to the nodes of other groups [Chefri 2019]. Similarly, we can characterize the community as indicated by the design of the organization that inside community, individuals are profoundly associated, and across the community, individuals are freely connected [Grivan 2002]. While considering the definition of the community, we can observe that in the overlapping community, a node must be in several communities and when we talk about a non-overlapping community, it is also concluded that nodes must reside in a single community. Several aspects of everyday life deal with the involvement of networks, social platforms, and biological networks. The WWW is just an example of that involvement while observing the structure of networks. Let us consider some disciplines such as Physics, Computer Science, and Social Science. The identification of community is an important task in communication aspects that deal with a subset of vertices of the network. Afterward, these regions help gain insights into the community that is related to itself [Fionda 2017].

In this era, as we know that all are following each other; for example, if someone saw his friend having good sports shoes and if he then adopts to wear them, some of his/her friends may also adopt to wear these sports shoes and so on in the entire community structure. If we observe, a diffusion concept arises here in which nodes are influencing their sibling in the community. A little modification has been considered in terms of advancement that allows diffusing the community by using cascading behavior. Let us consider all the nodes of the community have the same behavior, and after sometimes a new idea or behavior comes toward the community. Initially adopting these new behaviors by the nodes of community structure is unreliable. Hence, to make all the nodes of the community structure reliable, proposed models such as increasing the payoff and key people concepts are used. There is also another aspect where this concept can be applied, that is, IoT-enabled devices. In this scenario, data gathering and data distribution take place at every node or device of that community.

Real-world application of purposed methodology might be seen in fields such as Social Networks, IoT, Medical, Agriculture, Product Advertisement, Blog Networks, Protein Interaction Networks, and Epidemic Spreading.

The remaining sections of this chapter are structured as follows. Section 10.2 addresses the literature survey of cascading behavior and community recognition in the field of informal organization and IoT, while in Section 10.3, we discuss the methodology related to cascading behavior, which includes the two methods: first, "Increase the payoff" and second, "Key people concept." In Section 10.4, we discuss the implementation of the methodology, while Section 10.5 is related to the challenges of this methodology. Section 10.6 deals with the result and discussion part, and lastly Section 10.7 concludes the entire work done.

10.2   RELATED WORK

In the middle of the 20th century, several classical studies were done to establish a basic strategy of research, which is related to the spreading of a new behavior among the nodes or people of a community and studying the facts that facilitated its progress or cascading behavior in that community. Some of these facts deal mainly with a scenario in which the person-to-person or node-to-node influence was elaborated primarily to informational effects. We know that nodes pay attention to the decisions of their neighbors, which provides them indirect information about them to try the innovation. To capture such instructive impacts, there are two most compelling early bits of research work. Ryan has clarified about the appropriation of crossover seed corn among farmers in Iowa that depends on the neighbor in their community [Easley 2010], and Coleman, Katz, and Menzel have explored the selection of antibiotic medicine by specialists or doctors in the United States [Easley 2012]. Ryan and Gross have clarified that when they talked with farmers to decide how and when they would start utilizing crossover seed corn, it was concluded that while the majority of the farmers acquired the knowledge about hybrid seed corn from salesmen, some of them were convinced to use it based on their neighbor's experience. Coleman, Katz, and Menzel further investigated the adoption of a new medicine by specialists at a clinical store. Similarly, some other significant examinations mainly deal with the dispersion of developments. It considers a framework that manages the selection of behavior by direct-benefit impacts instead of informational ones. The examinations depict direct-benefit impacts, in which the spread of innovations like the phone, the fax machine, and email has relied upon the incentives. On which individuals need to speak with friends who have just received the innovation or behavior. There is the presence of malevolent use of publicly supporting frameworks by examining two group turfing sites, namely ZBJ and SDH, Wang et al. [Wang 2012] stated. The investigation of Wang et al. is identified with cascading behavior of community [Wang 2012]. A batch of rewarded scholars is active on the web to populate or control the conventional wisdom [Liu 2016], where spam crusades are facilitated by these groups. The primary logic of some spam campaigns is generally to elect clients having a high degree in their community to populate spam. Since we additionally realize that the more devotees a client has, the more certainly his re-tweets spread. Till now, researchers have developed three primary attributes of social network community structure, among which the first is the little world phenomenon. It can be represented as follows: let us assume we have two nodes, and they are identified with each other using a method in which several different nodes exist in the network or community. This means to say that these intermediate nodes are utilized uniquely to give a communication path. The second one is the power law, which is the distribution of node degrees by following the example of a power function. The third and last are the observed community structures inside an organization [Batagclj 1992].

The Internet of Things guarantees the communication of numerous kinds of frameworks and applications. These frameworks are used for giving progressively intelligent, flexible, and secure services. An enormous assortment of sensors including RFID, IR, and GPS, buildings structures, transport organizations, and utilities are connected through ICT. Let us consider a smart grid IoT framework, for instance. The grid dependability can be deciphered by the capacity of the power grid framework to convey electricity in the form of quantity and quality (being protected or secure) requested by clients. The flexibility of the grid may be treated as the limit of the power grid framework. That directly suffers and recovers from high-impact dangers. Such perils mainly deal with the framework's unwavering quality such as cataclysmic events, extreme environmental circumstances, and malignant physical or computerized assaults.

A community can be described as a group of strongly connected members inside and loosely connected members outside. So, detection of community has a great impact on social networks for understanding the actual structure and predicting the user behavior. In Figure 10.1, a straightforward model is given to explain the distinction among disjoint and overlapping (covering) communities. Nodes that exist in the center have a place with a different group in a disunited setting and are referred to as a disjoint community. It is denoted by community 3. On the other hand, in overlapping communities, the central node is related to two or more networks. Consequently, community recognition can be categorized into two kinds depending on the capacity to catch covering nodes under various communities. For a great illustration, we should refer to the Louvain calculation to allocate every node to just a single community [Blondel 2008] and the Bron-Kerbosch algorithm [Khanfor 2019] to decide overlapping or covering networks.

Cascading failures in various application zones exist for various reasons. It is not restricted to pandemics, individuals, power stations, teleportation organizations, wireless sensor networks, accounts, and organic systems. Segments in these frameworks are frequently, genuinely, or legitimately associated with one another. As we can see when one segment breaks down, different parts should then make up for the broken-down segment. This is why over-burden segments are making them glitch for trading off the domino chain impact. Without legitimate time executives and appropriate alleviation, the control systems regularly create catastrophic consequences for well-being climate and additionally for individual culture. Let us consider an example; in 2003, the Northeast power outage was set off by a transmission line failure and took the form of a gigantic blackout all over the Northeast area. This blackout influenced nearly 55 million individuals and created monetary harm of nearly $6 billion [Minkel 2008]. We can similarly consider another outline of Amazon Web Administrations' power outage in October 2012. This outage was caused by a memory spill in an information assortment. It influenced numerous locales, including Reddit, Foursquare, and Pinterest, and so forth.

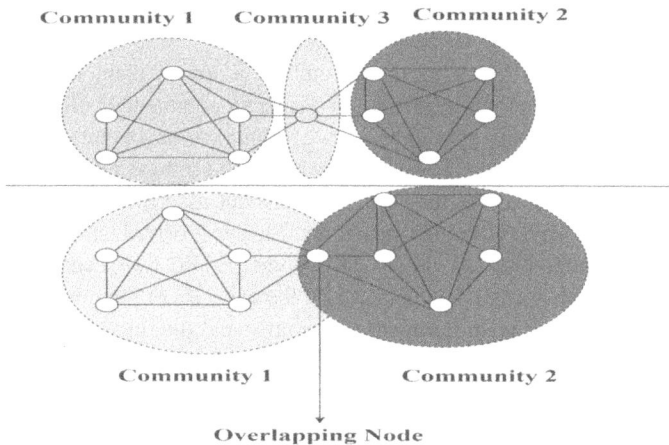

FIGURE 10.1 Structure to differentiate between disjoint and overlapping communities.

10.3 METHODOLOGY USED

10.3.1 About the Data Sets

The ego-Facebook (FB) is a free publicly available dataset at the Snap repository. Facebook data was collected from survey participants using this Facebook app. The dataset includes node features (profiles), circles, and ego networks. The ego-Facebook dataset consists of "circles" (or "friend's lists") from Facebook. This data set mainly has 4039 nodes, 88,234 edges, and 1,612,010 triangles. The purpose of using this data set is to create a social network graph so that after applying the proposed method, it can embrace the new behavior. Some other social network data set can also be used here, but for the sake of simplicity, we have considered a small community graph so that cascading behavior can easily be analyzed.

In this community graph initially, all nodes are denoted by the blank node. Here, a blank node represents a particular behavior of a node in the community. It intends to say that all nodes having the same behavior are living comfortably in their community and generally don't want to adopt a new behavior because of unreliability. This unreliability perception can be resolved using the purposed methodology.

10.3.2 Purposed Method

We have built a model that represents the diffusion of new behavior in the community (Figure 10.2). It is based on the individuals' and the related neighbor's decision. A particular symptom or behavior can start spreading across links of the community partially or completely. However, we are only interested in a complete cascade of the community structure. In this section, we have talked about a particular scenario, and the scenario is, there is a network and on this network, every person has adopted some action or some behavior. So, every person is initially settled in a specific environment and living comfortably with it on this network. After some time, when a new idea or new behavior comes into this network, it creates a decision point for every related node. Here, every related node will have to decide if this new behavior should be adopted or not.

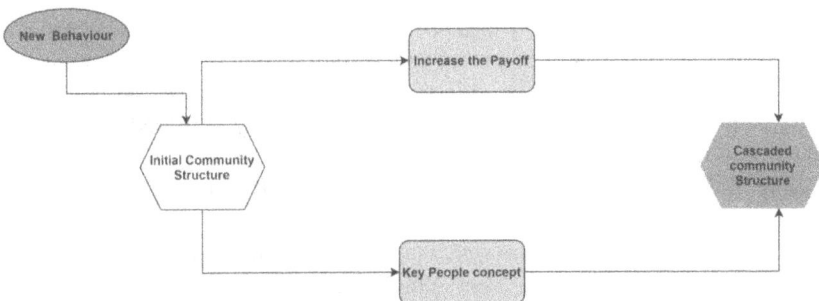

FIGURE 10.2 Cascading procedure.

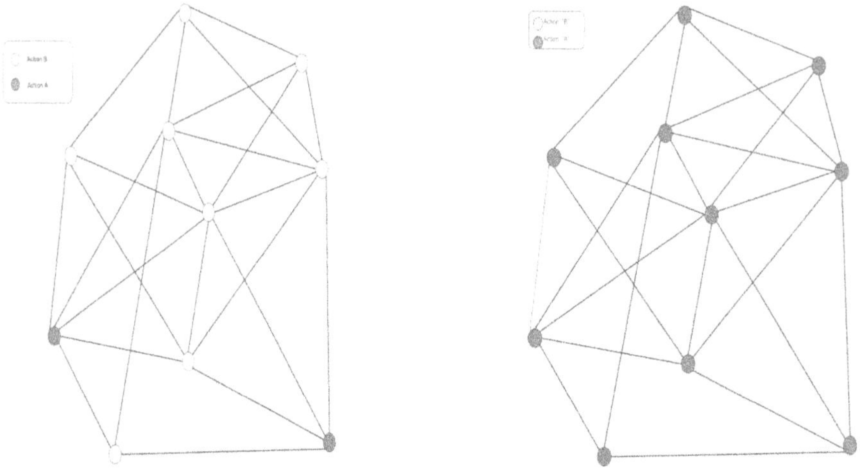

FIGURE 10.3 Initial and cascaded community structure.

In Figure 10.3, all the nodes in the network are blank except for two nodes. So, it has adopted an action; let's call it a blank idea or blank behavior. Two nodes having a gray idea or a gray behavior and they want to diffuse on this network. Here we have looked at the obvious problems with this. What is the problem with this gray idea is to diffuse on this network and make all the nodes gray here? The problem was, as we discussed, that people find it risky to adopt a new idea or new behavior. They are not comfortable shifting to a new thing, especially when all of their friends are doing the old thing. So, no node in this network wants to shift to this gray idea. To overcome such a problem, two methods are proposed that will help to understand the actual concept of diffusion inside the community.

10.4 IMPLEMENTATION

10.4.1 Increase the Payoff

The first solution is to increase the payoff associated with this gray idea so that benefits associated with this gray idea could increase. As the benefit associated with this gray idea increases, more and more people start adopting it. So, we have looked at how we have modeled the entire thing with the help of payoff and several friends. Each idea here let us be the gray one or blank one has a payoff. Based on the payoff and number of friends which they have adopted, the node decides what to do? Hence, by using the proposed method we can increase the payoff associated with this gray idea, and then it will cascade on this network.

We have a community graph in which initially all the nodes have the same behavior B denoted by blank nodes. After some point in time, a new behavior having action A and color gray enters into the community, then cascading of new behavior will be based on the following Algorithm 1.

Algorithm 1 Increase the Payoff

Input: Community structure
Output: Return Num//Num is number of neighbors

```
 1. procedure Find Neighbor (Xi) i = 1, 2… M
 2. Initialize Num = 0
 3. for each1 in neighbors of Graph: do
 4. if G. nodes [each1] ['action'] == c then:
 5. increase the value of Num
 6. end if
 7. end for
 8. Return Num
 9. end procedure
10. procedure Def. Recalculate options (G):
11. dict1 = {}
12. Payoff (A) = a
13. Payoff (B) = b
14. for each in G. nodes: do
15. numA = find neigh (each, 'A', G)
16. numB = find neigh (each, 'B'', G)
17. payoffA = a * numA
18. payoffB = b * numb
19. if payoff A is greater than or equal to payoff B: then
20. dict1 [each] = 'A'
21. else
22. dict1 [each] = 'B'
23. Return dict1
24. end if
25. end for
26. end procedure
```

After increasing the payoff by the above algorithm, we can see the whole graph is cascaded with new behavior.

10.4.2   KEY PEOPLE CONCEPT

[S1]: Let G (V, E): be a community Graph
[S2]: L [a, b]: be a list that initially adopts the new behavior.
[S3]: For all, L [a, b] check is cascade complete or not?
[S4]: If an appropriate bunch of nodes is selected in the list then it may cascade
all the nodes of the community otherwise may not cascade.

The second idea is about choosing the right people. So, if cascading starts with a particular bunch of nodes, then it may be unable to create a complete cascade, but if cascading starts from another bunch of nodes that are all around associated with the rest of the network, then it might end up creating a complete cascade. Let us consider a bike example. Suppose someone wants to buy a new bike and his mother or his siblings are not present who can convince his father. In such a situation, his mother or his siblings work as key people to buy a new bike.

We have a network on which every person has adopted an old idea and then a new idea comes and now wants to diffuse whole networks. In this case, the payoff associated with this new idea should not be changed. Hence, we are not going to change the payoff but we are going to change the people from where this cascade starts and which shows that if we start from some bunch of people, it is unable to create a complete cascade, but if we start from some other bunch of people, it can create a complete cascade. Let us consider another example if we select a sensor having maximum battery lifespan and is more reliable to that network. This means we have to select a coordinator to distribute data to other sensors or computers. In this scenario, cascading gets cascaded appropriately in that network. Similarly, in the opposite way, when we think about a sensor having a minimum life span then cascading looks difficult to every sensor or computer because it may get disrupted in their network.

10.5   CHALLENGES

10.5.1   IMPACT OF COMMUNITIES ON DIFFUSION

In this scenario, we need a community graph consisting of two densely connected communities. Nodes ranging from 0 to 9 are connected in one community, while the remaining nodes ranging from 10 to 19 are densely connected to another community having the probability of 0.5. We can also observe that a sparse link is connected between community 1 and community 2 (see Figure 10.4). When cascading gets

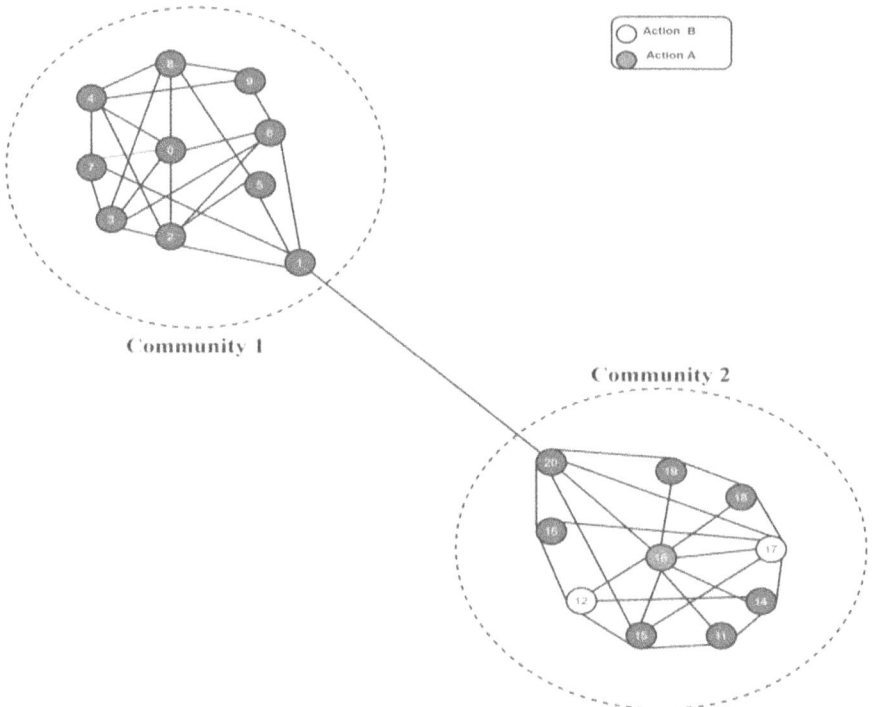

FIGURE 10.4   Impact on the community after diffusion.

started, a condition arises that intends to say that some of the nodes receive behavior A and a portion of the nodes embrace behavior B. Hence, we can conclude here that even after cascading all the nodes in community 1, we cannot cascade all the nodes in community 2 of the same behavior. The impact of communities on diffusion will be based on the following Algorithm 2.

Algorithm 2 Impact of Communities

Input: Community structure
Output: Partially cascaded community structure

```
1.  procedure [S1]: def. Create first community (G). i = 1, 2… M
2.  for i = 0 to 10 do:
3.  G.add node (i)
4.  for i = 0 to 10 do:
5.  for j = 0 to 10 do:
6.  if i< j then
7.  r = random. Uniform (0, 1)
8.  if r < 0.5 then
9.  G.add edge (i, j)
10. end if
11. end if
12. end for
13. end for
14. end for
15. [S2]: def. create second community (G):
16. for i = 11 to 20 do:
17. G.add node (i)
18. for i = 11 to 20 do:
19. for j = 11 to 20 do:
20. if i< j then
21. r = random. Uniform (0, 1)
22. if r < 0.5 then
23. G.add edge (i, j)
24. end if
25. end if
26. end for
27. end for
28. end for
29. [S3]: Add an edge between community 1 and community 2
30. [S4]: apply cascading in the community
31. [S5]: Hence we can conclude here that even after cascading
    all the nodes in community 1 we cannot cascade all the
    nodes in community 2 of the same behavior
33. end procedure
```

Usually, a cascading failure happens when we see that the failure of one segment triggers progressive disappointments of different segments. That results in broad harm to the whole framework or network and even to the climate and the public [Xing 2020]. In the previous decade, the IoT was a concept of gathering individuals and various things or connecting frameworks or systems. IoT is fast developing, planning to change the human culture to smart, productive, efficient, resilient, and helpful in the development of our daily life. We know that IoT is a collaboration of interconnected gadgets. It is also accepted that IoT works cooperatively and is self-governing to perform information assortment, information generation, hand-off data to each other, and cycle data astutely since there are communications and dependencies between various gadgets that exist either in the form of usefulness assault or potential savvy structure. Consequently, cascading failure has become a significant danger to the IoT framework [Zhong 2019, Wu 2019]. It is obvious that when a particular gadget fails, a situation of unforeseen or undesired state arises that might set off to other associated IoT gadgets. In such a situation, this behavior may introduce or animate calamitous cascading failures. Now, for ensuring the capacity of an IoT framework, it is important to comprehend the reasons and actual components of falling cascading behavior. What is the reason for it? Address their belongings in the framework demonstrating and further form flexibility against cascading failure.

10.5.2 CASCADE AND CLUSTER

[S1]: Let G (V, E): be a community Graph
L1 [a, b]: be a list that initially adopts the new behavior and
L2 [a, b]: be a list that contains L1 [a, b]
[S2]: For all, L1 [a, b] in L2 [a, b] check is cascade complete or not?
[S3]: If an appropriate bunch of nodes is selected in the list then it may cascade all the external nodes but it is unable to cascade the cluster having density $\geq (1 - q)$
[S4]: q can be defined as $q = a/(a + b)$
[S5] Even after adopting the new behavior by entire external communicating nodes we are unable to cascade a particular cluster

In this scenario, if the density of the cluster is greater than $(1 - q)$, we can never achieve a complete cascade (Figures 10.5–10.8). Let the payoff associated with behavior A be 2 and the payoff associated with behavior B be 3; then the value of q becomes 2/5, which is equal to 0.4. If the density of the cluster is greater than 0.6, the cluster cannot be cascaded inside the community. So, we can say that there are clusters in every network and the cluster is of density p. If for every node in the cluster every node has at least p fraction of their friends in the same cluster, we call it a cluster of density p and q here is a threshold associated with the new idea. This intends to say that if q fraction of friends or greater than q fraction of friends adopt a particular idea then it will also adopt this new idea. So, the theorem states that if there is a group having a thickness greater than 1 minus q i.e. $(1 - q)$ in this network the then-new idea cannot diffuse inside the cluster. So, here is a group having a thickness greater than 1 minus q i.e. $(1 - q)$, and then a cascade starts from a particular list of nodes. Even if the cascade infects all these external nodes in the network, we have seen that it is unable to defuse inside this cluster.

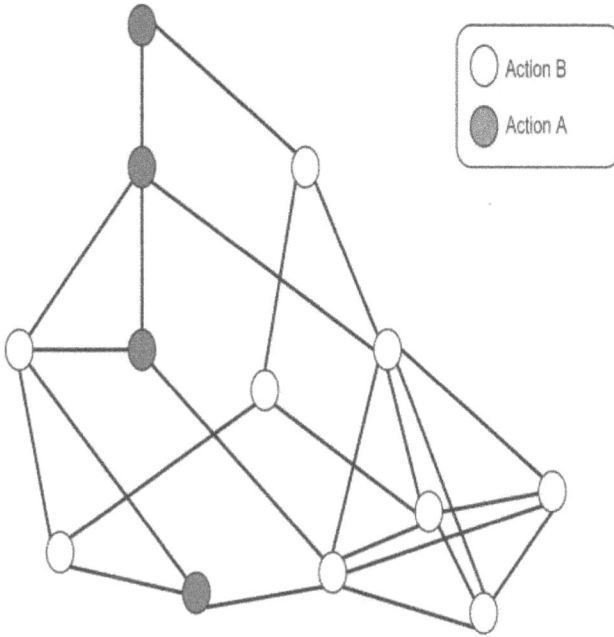

FIGURE 10.5   Initially four nodes have different behavior.

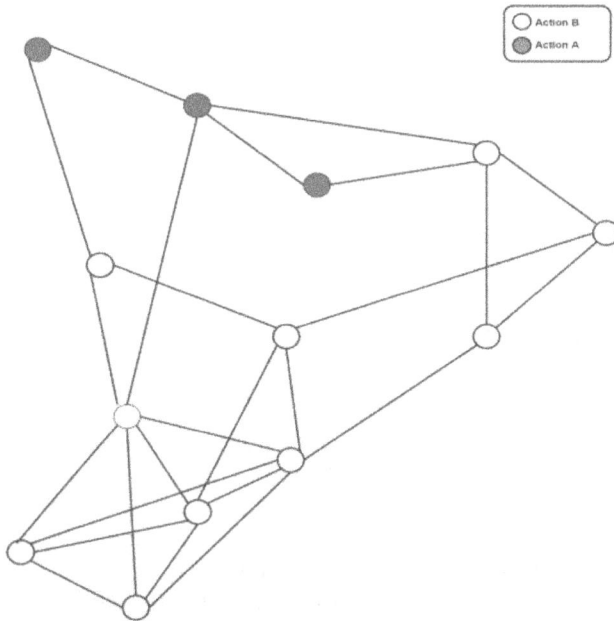

FIGURE 10.6   Partially cascaded structure.

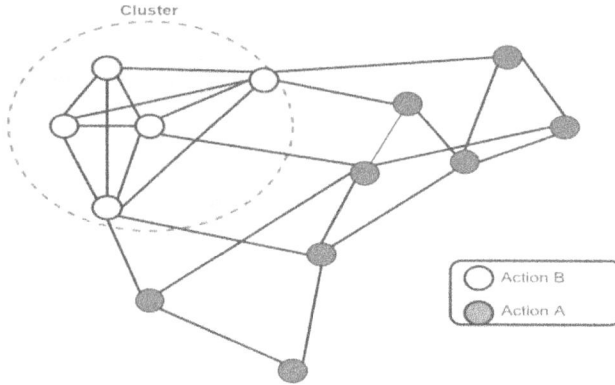

FIGURE 10.7   External nodes having different behavior.

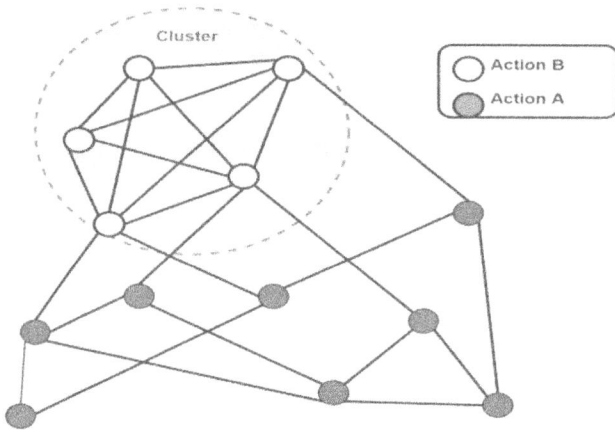

FIGURE 10.8   Cluster is not cascaded.

10.6   RESULT AND DISCUSSION

10.6.1   [M1] EXPLANATION OF INCREASING THE PAYOFF

[S1]: Initially two nodes adopt new behavior denoted by gray color nodes and the remaining nodes have their original behavior.

[S2]: When we are using the increased payoff method then every node of the community has adopted the new behavior. In this scenario, we can consider the example of a traffic controller at the airport where all the scenarios get reflected when a single plane lands at the airport. Here we can conclude that payoff is an important characteristic in determining whether it completes the cascade or not.

10.6.2   [M2] EXPLANATION OF KEY PEOPLE CONCEPT

[S3]: We have to select appropriate key people so that the entire community gets cascaded and achieves maximum benefits. This shows us that yes, there are certain key people or certain key nodes in this network, and if we start our cascade from there then it tends to become a complete cascade.

10.6.3   [C1] EXPLANATION OF IMPACT OF COMMUNITIES ON DIFFUSION

[S4]: Some of the nodes adopt behavior A and some of the nodes adopt behavior B. Hence, we can conclude here that even after cascading all the nodes in community 1 we can't cascade all the nodes in community 2 of the same behaviors.

10.6.4   [C2] EXPLANATION OF COMMUNITY AND CLUSTER

[S5]: If cluster density $>= (1 - q)$ then it is unable to cascade a particular cluster. Hence, the benefit of this challenge is that if we want to secure a particular cluster then this concept comes to light.

[S6]: Even after adopting the new behavior by entire external communicating nodes we are unable to cascade a particular cluster.

10.7   CONCLUSION

In this chapter, we have discussed a particular scenario. There is a network and on this network, every person has adopted some action or some behavior and living comfortably with it. When a new idea or new behavior comes into this network then people start changing their behavior. There are various algorithms proposed to cascade the behavior of the network. The first solution is to increase the payoff so that benefits associated with a gray idea could increase. As the benefit associated with the gray idea increases, more and more people start adopting this behavior. In the second idea, we have to select key people where if we start with a particular bunch of nodes, we are unable to create a complete cascade, but if we start from some other bunch of nodes which is all around associated with the remainder of the network then it will end up creating a complete cascade. Hence, choosing key people in a community is also an important criterion so that a complete cascade is obtained.

The third idea or challenge is the effect of cascades on the community, in which there are two communities and a new idea originates and then cascades to the first community, and the entire community adopts this idea, but it is difficult for a new idea to come and diffuse some other community completely. Hence, we can conclude that diffusing some other community completely is difficult but beneficial for security purposes. The fourth idea or challenge is that if there is a group of thickness $> (1 - q)$ in a particular network then, the proposed idea cannot diffuse the cluster or group within that network. Even if the proposed idea infects all these external nodes in this network, we have seen that it is unable to diffuse the cluster within the network. This idea is more useful when we want to secure a particular cluster in case of an epidemic or any terrorism.

The motivation behind this chapter prompts to illustrate cascading behavior in community structure with the goal that community gets handily recognized. Cascading behavior inside the community will happen according to the four mentioned cases. But there is some basic and representative scenario of cascading failures. In this chapter, an explanation has been given without touching any mathematical logic. The literature review mainly focuses on fundamental causes of cascading failure such as triggering, models characterizing, and actual components of cascading failure. Cascading failure primarily manages frameworks that generate power and IoT variants of such systems. We can also access some significant review articles on cascading failure of IoT applications on the ground of power generation, for instance [Guo 2017]. The research in the field of IoT applications that generate cascading failure is still underexplored. There is an appeal for important inventive thought and productive cascading failure model's solution so that it can catch the new intricate and dynamic nature of included IoT frameworks.

REFERENCES

Batagelj, A. F. V. (1992). Direct multi-criteria clustering algorithms. *Journal of Classification*, *9*, 43–61.

Blondel, V. D., Guillaume, J. L., Lambiotte, R., & Lefebvre, E. (2008). Fast unfolding of communities in large networks. *Journal of Statistical Mechanics: Theory and Experiment*, *2008*(10), P10008.

Chen, C., Wu, K., Srinivasan, V., & Zhang, X. (2013, August). Battling the internet water army: Detection of hidden paid posters. In *2013 IEEE/ACM International Conference on Advances in Social Networks Analysis and Mining (ASONAM 2013)* (pp. 116–120). IEEE

Cherifi, H., Palla, G., Szymanski, B. K., & Lu, X. (2019). On community structure in complex networks: challenges and opportunities. *Applied Network Science*, *4*(1), 1–35.

Easley, D., & Kleinberg, J. (2010). *Networks, crowds, and markets* (Vol. 8). Cambridge: Cambridge University Press.

Easley, D., & Kleinberg, J. (2012). *Networks, crowds, and markets*. Cambridge: Cambridge University Press.

Fionda, V., & Pirro, G. (2017). Community deception or: how to stop fearing community detection algorithms. *IEEE Transactions on Knowledge and Data Engineering*, *30*(4), 660–673.

Girvan, M., & Newman, M. E. (2002). Community structure in social and biological networks. *Proceedings of the National Academy of Sciences*, *99*(12), 7821–7826.

Guo, H., Zheng, C., IU, H. H. C., & Fernando, T. (2017). A critical review of cascading failure analysis and modeling of the power system. *Renewable and Sustainable Energy Reviews*, *80*, 9–22.

Khanfor, A., Ghazi, H., Yang, Y., & Massoud, Y. (2019, December). Application of community detection algorithms on social internet-of-things networks. In *2019 31st International Conference on Microelectronics (ICM)* (pp. 94–97). IEEE.

Wang, G., Wilson, C., Zhao, X., Zhu, Y., Mohanlal, M., Zheng, H., &Zhao, B. Y. (2012, April). Surf and turf: crowdsourcing for fun and profit. In *Proceedings of the 21st international conference on World Wide Web* (pp. 679–688).

Liu, B., Luo, J., Cao, J., Ni, X., Liu, B., & Fu, X. (2016, May). On crowd-retweeting spamming campaign in social networks. In *2016 IEEE International Conference on Communications (ICC)* (pp. 1–6). IEEE.

McAuley, J. J., & Leskovec, J. (2012, December). Learning to discover social circles in ego networks. In NIPS *2012*, 548–56. https://snap.stanford.edu/data/ego-Facebook.html

Minkel, J. R. (2008). The 2003 Northeast blackout – five years later. *Scientific American, 13*, 1–3.

Wu, J., Fang, B., Fang, J., Chen, X., & Chi, K. T. (2019). Sequential topology recovery of complex power systems based on reinforcement learning. *Physica A: Statistical Mechanics and Its Applications, 535*, 122487.

Xing, L. (2020). Cascading failures in the internet of things: review and perspectives on reliability and resilience. *IEEE Internet of Things Journal, 8*(1), 44–64.

Zhong, J., Zhang, F., Yang, S., & Li, D. (2019). Restoration of the interdependent network against cascading overload failure. *Physica A: Statistical Mechanics and Its Applications, 514*, 884–891.

11 Performance Evaluation of Machine Learning Classifiers for Memory Assessment Using EEG Signal

Shipra Swati and Mukesh Kumar
NIT Patna
Patna, India

CONTENTS

11.1 INTRODUCTION

Electroencephalography (EEG) records brain activity in the form of electro-physiological signals. Analysis of these signals helps in developing devices having the capacity to communicate with the human brain for providing better facilities in daily life. These devices require a brain–computer interface (BCI) to accomplish their tasks. EEG gives insights into the functioning of the brain because of different stimuli provided to it, which can be visual, audio, or both audio and visual i.e. audio-visual (in the form of multimedia). Thus, we get to know the status of the brain, attention performance of the brain, and health of the brain can be known by the memory power. Apart from being well established, EEG is inexpensive and non-invasive too, which makes it a promising tool for assessing different brain functions. Also because of these properties, it is more suitable for routine use as compared to other brain imaging modalities like magnetoencephalography (MEG) and functional magnetic resonance imaging (fMRI). This chapter discusses the importance of EEG for relating human memory with different cognitive tasks.

DOI: 10.1201/9781003145004-11

The rapid growth of the Internet of Things (IoT) in almost every aspect of human life exposed the scope of the Smart Healthcare system, which has strengthened remote doctor–patient communication for a better quality of life. In the recent past, the research community working on BCI has used EEG in IoT-enabled infrastructure for real-time cognitive interaction between patients with locked-in syndrome (LIS) and their assistive appliances, such as wheelchairs or speech aid (Zhang et al., 2018).

The brain is divided into different lobes, namely the Frontal, Parietal, Occipital, and Temporal lobes, each having specialized functions (Kumar & Bhuvaneswari, 2012). The frontal lobe has the control of memory, thinking ability, decision making, reasoning, impulse control, emotions, and speaking quality. In case this part gets injured, it may affect memory, emotions, and language. The Parietal Lobe is concerned with sensory information coming from different parts of the body. If there happens some damage to this part, the inability problem for recognizing and locating body parts may occur. The Occipital Lobe processes visual information and causes color blindness after getting an injury. The last one, Temporal Lobe, is responsible for sound and speech, precisely for hearing, recognizing language, and forming memories. It may create hearing loss and the problem of identifying languages after injury.

For recording EEG signals, electrodes are placed on brains following logical and ordered arrangements called montages. The 10–20 system has been the international standard for the past half a century. The significance of numbers "10" and "20" represents the distance measures between adjacent electrodes placed from right to left or front to back of the skull. The gap between positions of two consecutive electrodes is either 10% or 20% (Jurcak, Tsuzuki, & Dan, 2007). Figure 11.1 represents a diagrammatic representation of the placement of electrodes in connection with the brain lobes.

The nomenclature for each electrode is a combination of letters and numbers, which specifies a particular lobe and left or right hemisphere. Different electrode sights are denoted with certain letter codes, given as: frontal (F), parietal (P), temporal (T), occipital (O), central (C), pre-frontal (Fp), frontotemporal positions (FT), frontocentral (FC), temporal-posterior temporal (TP), centroparietal (CP) or parieto-occipital (PO) and anterior frontal (AF) (Acharya, Hani, Cheek, Thirumala, & Tsuchida, 2016). The "z" (zero) sites refer to the electrodes placed on the midline. Electrodes associated with even numbers are located on the right hemisphere of the brain, whereas those named with odd numbers are located on the left hemisphere.

The inclusion of IoT for addressing the problem in the proposed work requires the acquisition of brain wave through wireless headsets, which sends raw signals to a cloud server. The result is analyzed after the complete processing of raw data is passed to the actuator for the designated operation in the IoT framework (Konstantinidis et al., 2015).

The EEG data collected with these many sites are used in different clinical applications, including classification and recognition of emotional states (Alarcao & Fonseca, 2017), decoding mental loads (Edla, Mangalorekar, Dhavalikar, & Dodia, 2018), identifying neurological disorders like epilepsy (Vidyaratne & Iftekharuddin, 2017), Alzheimer's disease (Simpraga et al., 2017), Parkinson's disease (Xu et al., 2020), sleep disorder (Vimala, Ramar, & Ettappan, 2019), coma, and brain death

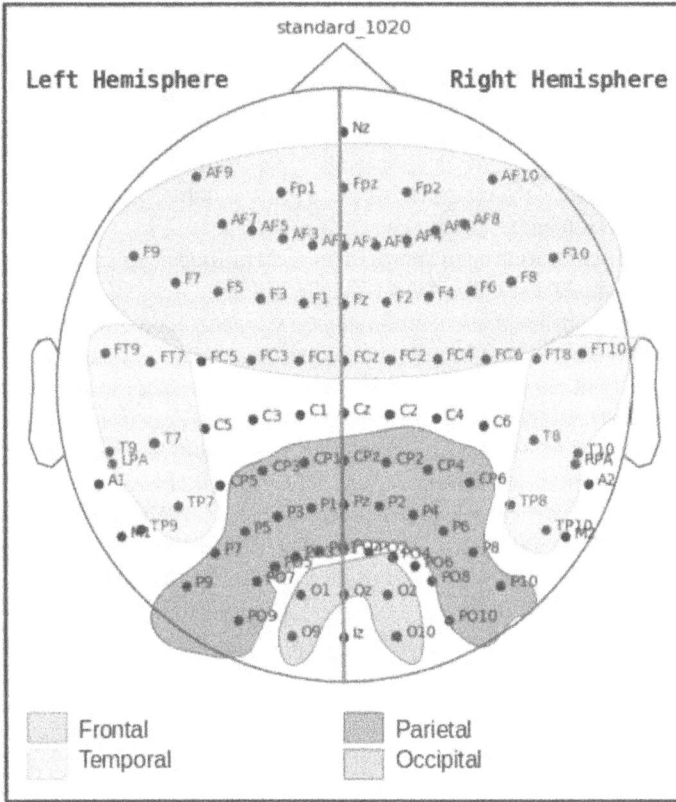

FIGURE 11.1 10–20 system for EEG recording.

(Zhu et al., 2019). For estimation of cognitive ability, California Verbal Learning Test (CVLT) is considered here (Elwood, 1995). All of these terms are briefly explained further.

There are various machine learning models, which can be used for classification tasks for the suggested methodology, including k-nearest neighbor (KNN), Naive Bayes (NB), Decision Tree (DT), Random Forest (RF), and Support Vector Machines (SVMs). This chapter proposes the use of EEG in estimating the memory strength of a human using different machine learning algorithms including KNN, RF classifier, and SVM only (Zheng, Santana, & Lu, 2015). Random Forest algorithms possess the power of several decision trees created randomly, so the decision tree classifier is eliminated. Also, because of the zero-frequency problem, the Naive Bayes algorithm is not considered for analyzing the issue at hand. The result obtained from the proposed methodology shows that SVM outperforms the other two classifiers KNN and RF in the memory ability classification task. The next section mentions some related research work carried out in this area. Then the logical representation of the proposed work is provided for better understanding with the required description.

11.2 RELATED WORK

Recent trends of researches in the field of recall strength of the human brain denote significant works carried out for the early detection of different diseases, which are related to memory loss. Priyanka et al. have emphasized the risk of increased damage to the human brain caused by delayed treatment of dementia as there are few effective treatments for this (Lodha, Talele, & Degaonkar, 2018). The World Health Organization (WHO) has reported that nearly around ten million new cases of dementia appear every year, resulting in approximately 50 million people suffering from this worldwide (*Dementia*, n.d.). The WHO also finds that depression puts a huge burden of diseases globally by its major contribution to their generation. Most of such diseases result in memory deficiency. Chien et al. have introduced an EEG-based major depressive disorder (MDD) detector by using a source memory task for the target population that involved emotionally neutral words (Wu et al., 2018). Zainab et al. have proposed an approach using EEG signals for predicting two cognitive skills, namely focused attention (FA) and working memory (WM). These factors have a strong impact on the outcomes of any particular learning process (Mohamed, El Halaby, Said, Shawky, & Badawi, 2018). A study related to the cognitive functioning of the brain is done by Poulami et al., which focuses on the cognitive state of the active part of the memory, i.e. working memory. It is mainly associated with the capability of the human brain to memorize and then recognize it later from a list of similar ones (Ghosh, Mazumder, Bhattacharyya, & Tibarewala, 2015). The virtue of learning and memorization are closely related to the ability of selective concentration depending on the varying nature of tasks. Li et al. have investigated the effect on the EEG spectrum during activities involving complex cognitive skills, including recognition, processing, and memorization of new information (Ko, Komarov, Hairston, Jung, & Lin, 2017). The case study considered by the author is a real classroom setting, which requires visual and listening attention from the participants. Introducing IoT for any EEG-based BCI system needs specific types of hardware and software framework too that agrees with the IoT framework. Francisco et al. have presented a prototype for eye-state classification using low-cost hardware for capturing EEG signals developed by their research group (Laport, Dapena, Castro, Vazquez-Araujo, & Iglesia, 2020). Such developments for inexpensive IoT-compliant EEG devices boost the research for the integration of the BCI domain with IoT, where results are explored by various machine learning techniques.

The study of the existing literature related to human brain functioning for knowing the memorizing ability leads to two very important factors for this presented work. The first one is the EEG signal and another one is a verbal learning task, both having the most significant role in the proposed system.

11.3 PROPOSED WORK

For assessing the memory capacity, we have used the most widely used neuropsychological tests called "California Verbal Learning Task (CVLT)," which was first proposed in 1987 and is currently available with its third edition (Farrer & Drozdick, 2020). This test has two lists L_A and L_B, each having elements X_{Ai} and X_{Bi}, where $1 \leq i \leq 16$. Each list has four semantic categories C_{Ai} and C_{Bi}, such that i ranges from 1 to 4.

$(\exists C_{Ax} \in C_{Ai}, \exists C_{Bx} \in C_{Bi} \rightarrow C_{Ax} \equiv C_{Bx})$ provided $\forall (C_{Ax} \in L_A) \wedge (C_{Bx} \in L_B) \rightarrow (C_{Ax} \neq C_{Bx}')$. The instructor reads $C_{Axi} \in L_A$ for the subjects involved in the test. The subject hears and recalls the list L_A five times, which is called free recall. Afterward, the subject hears another list L_B, also called interference list, and recalls this. Afterward, the subject should recall the first list L_A again (short delay recall). Another free recall takes place after 20 minutes, which is termed as long delay recall. In this way, short- and long-term memory performance can be measured using this test. It is being used for cognitive assessment for detecting diseases related to memory disorders (Graves et al., 2019) (Wolf & Rognstad, 2013).

For assessing memory ability, CVLT needs human assistance for executing different n tasks of recalling list items $T = \{t_1, t_2, ..., t_n\}$ and the observations $V = \{x | 1 \leq x \leq 16\}$ are also manually recorded. Let there be a function f_{CVLT} that maps results of tasks $t_i \in T$ to $x_i \in V$ as given in Figure 11.2.

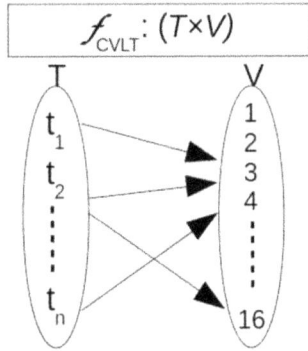

FIGURE 11.2   Mapping of CVLT tasks to recall values.

The purpose of this work is to provide a novel approach to relate EEG signal with this specific learning test (CVLT) for finding an insight into memory strength. Figure 11.3 represents a schematic diagram for this idea. It will widen the accuracy of diagnosing neurological health hazards.

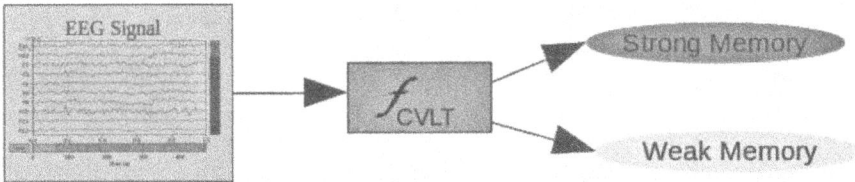

FIGURE 11.3   Use of EEG and CVLT for memory assessment.

We have introduced the use of machine learning models to learn the memory strength for different EEG signals using respective CVLT scores recorded for some volunteered subjects. Based on the performance of learning, these models can be used to estimate the memorizing capacity of humans on the basis of their EEG signals. Figure 11.4 shows the block structure of the proposed system. For this purpose,

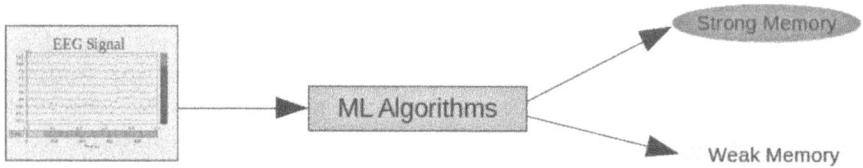

FIGURE 11.4 Block diagram of the proposed system.

the statistical features were extracted from the EEG signal using three effective classifiers known as kNN, RF, and SVM.

11.4   IMPLEMENTATION DETAILS

This chapter uses a recently released dataset named Leipzig Study for Mind-Body-Emotion Interactions, abbreviated as LEMON, which belongs to the larger dataset MPI Leipzig Mind-Brain-Body (MPILMBB) database (Babayan et al., 2019). This dataset has records of physiological, psychological, and neuroimaging measures for participants of young and elderly group. So, it can be helpful in the complete estimation of a subject's health, unlike other available EEG dataset DEAP (Koelstra et al., 2011) and SEED (*SEED Dataset*, n.d.). The number of young people is 153 (108 male, 45 female) with the age range of 20–35 years and the number of older people is 74 (37 male, 37 female) with an age range of 59–77 years. We are using the EEG data from the repository, which are recorded using 62-channel at rest called a resting-state electroencephalogram (RsEEG). One electrode VEOG is placed below the right eye, while the other 61 are scalp electrodes (mentioned in Table 11.1), which are placed according to the spatial distribution as mentioned in the introductory section. A generic architecture of the processing stages of EEG signal is given in Figure 11.5.

Gross et al. have used this EEG dataset to predict Internet addiction (IA), which has become a universal problem with the increasing existence of Internet-based

TABLE 11.1
Electrode Distribution over the Scalp

Frontal (F)	F5, F1, F2, F6, F7, F3, F4, F8, Fz
Temporal (T)	T7, T8
Parietal (P)	P7, P3, Pz, P4, P8, P5, P1, P2, P6
Occipital (O)	O1, Oz, O2
Central (C)	C3, Cz, C4, C5, C1, C2, C6
Pre-frontal (Fp)	Fp1, Fp2
Frontotemporal positions (FT)	FT7, FT8
Frontocentral (FC)	FC5, FC1, FC2, FC6, FC3, FC4
Temporal-posterior temporal (TP)	TP7, TP8
Centroparietal (CP)	CP5, CP1, CP2, CP6, CP3, CPz, CP4
Parieto-occipital (PO)	PO9, PO10, PO7, PO3, POz, PO4, PO8
Anterior frontal (AF)	AFz, AF7, AF3, AF4, AF8

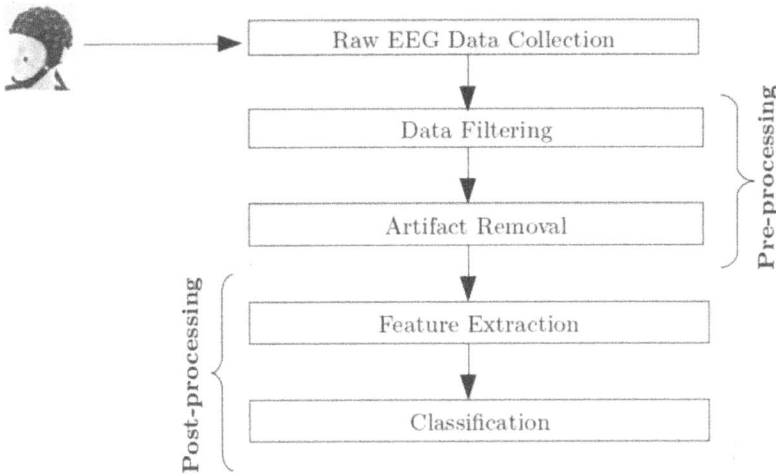

FIGURE 11.5   Stages of EEG processing.

applications in the personal and professional world (Gross, Baumgartl, & Buettner, 2020). Excessive use of the Internet leads to structural changes in the brain, which is similar to various disorders like depression, social isolation, attention deficit hyperactivity disorder (ADHD), and Internet gaming disorder (IGD). They emphasized that early prediction of IA can result in faster preventive interventions by increasing the speed and robustness of diagnosis.

The raw EEG data is recorded using 62-channel active ActiCAP electrodes for 16 minutes in a resting state with a sampling rate of 2500 Hz. An open-source library, MNE-python, is used to pre-process the raw data (Gramfort et al., 2013). First the downsampling of data is done to 250 Hz for further processing, and then the bandpass filter is applied on it having a low frequency range of 1 Hz (high-pass) and a high frequency range of 45 Hz (low-pass). The potential artifacts in EEG data components related to eye movement, eye blink, muscle activities, or electrode noises are removed using independent component analysis (ICA) (Jiang, Bian, & Tian, 2019). Figure 11.6 shows visibly distinguishable effects on the raw signal after these pre-processing steps.

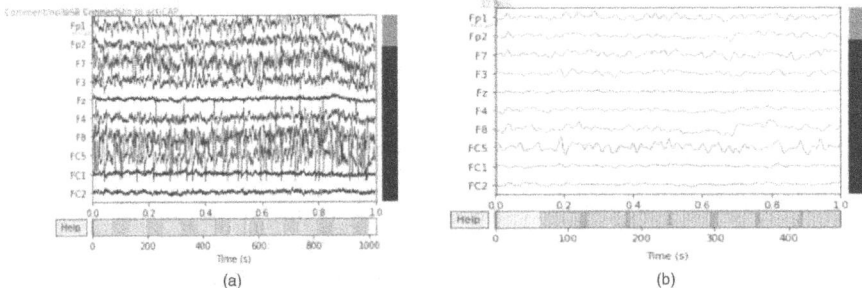

FIGURE 11.6   Effect of pre-processing on raw EEG signal. (a) Raw data. (b) Preprocessed data.

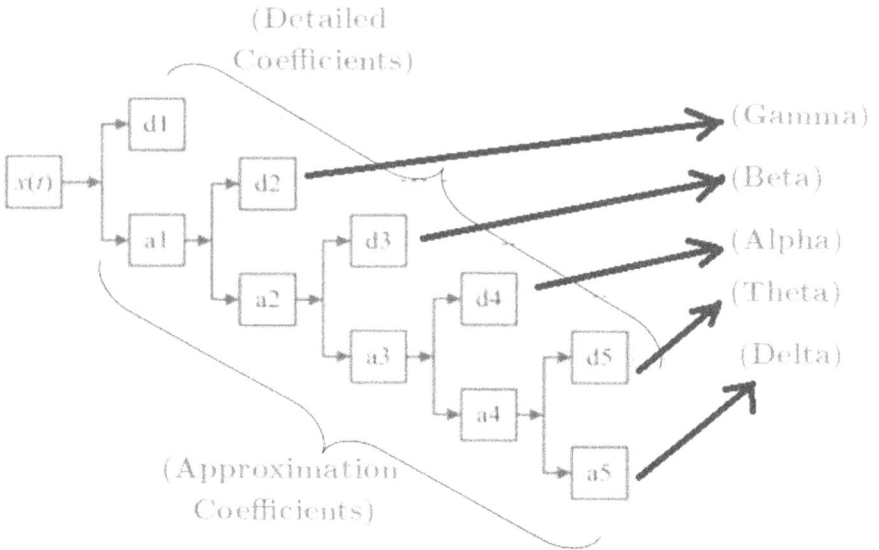

FIGURE 11.7   Wavelet decomposition of EEG signal.

There are five frequency-based features of EEG signal having delta band consisting of lowest frequency components and conversely gamma band has the highest one. In between there are three more frequency bands, theta, alpha, and beta. The Wavelet transform (WT) is applied to the signals captured through each EEG channel to decompose them into these five features (Al-Fahoum & Al-Fraihat, 2014). Figure 11.7 shows the sub-band division of a signal $x(t)$ as per WT for five levels in the frequency range 1 Hz–45 Hz, where $x(t)$ denotes EEG signal for a single subject. Table 11.2 mentions the frequency information and significance for human cognition for each band.

Let there be a matrix $X(m,n)$ represents the input data in the form of extracted EEG data, which are normalized and scaled, where m is the number of subjects selected for the classification task represented by row and n is the number of columns required for creating the feature vectors. $\forall X_i \in X, \exists X_{ij}$, such that $j = 1, 2, 3 ..., n$. Let the output data be represented as $Y \in Res_{CVLT}$, which is a column vector ($m \times 1$), such

TABLE 11.2

Frequency Domain Features of EEG Signal

Type	Frequency	Behavioral/Psychological State
Delta	<4 Hz	Sleep, dreaming
Theta	4–7 Hz	Creativity, insight, dreams, reduced consciousness
Alpha	8–15 Hz	Physically and mentally relaxed
Beta	16–31 Hz	Alert, consciousness, active thinking, panic attack
Gamma	>31 Hz	Hyper alertness and integration of sensory inputs

that $\forall y_i \in Y, 7 \le y_i \le 16$. The Res_{CVLT} represents the numeric measures as explained in Section 11.2. For the sake of binary classification, we divided these values in two groups according to this rule $\Rightarrow$: if $y_i < 10$, $y_i = 1$, i.e. strong memory, else $y_i = 0$, i.e. weak memory. The purpose of this research is to use different machine learning models to find a mapping between X and Y s.t. $Y = f(X)$, where $Y \in (0|1)$ and $f(X)$ represents features extracted from matrix X.

After pre-processing of the input data and appropriate labeling for target output, different machine learning methods are applied to build a model that will act as a CVLT test and predict the memory capacity of humans. The machine learning algorithms used for the proposed work are mentioned below.

- **k-nearest neighbor (KNN)** is a technique of supervised machine learning, which is most widely used for classifying EEG data for specific emotional states (Abu Alfeilat et al., 2019). It categorizes a new instance based on similarity measures with the nearest neighbors. It uses Euclidean distance as a measure to find out the nearest neighbors of a node.

  Let us say x is a new instance and x' is an existing one. Euclidean distance is the square root of the sum of the squared differences, which helps to know the similarity between these two points. More precisely, it gives an insight if x and x' belong to the same class or not. Following is the equation for the same:

$$d(x,x') = \sqrt{(x_j - x_{ij})^2} \tag{11.1}$$

  where j denotes the attribute of the data points. k in the kNN is the hyperparameter that represents the number of neighbors to use in the model. It also deals with the bias–variance tradeoff, which maintains a balance between underfitting and overfitting. As shown in Figure 11.8, choosing a large k results in a smooth decision boundary but have a high bias and low variance, whereas smaller k results in complex decision boundary but a high variance and low bias (Lantz, 2013).

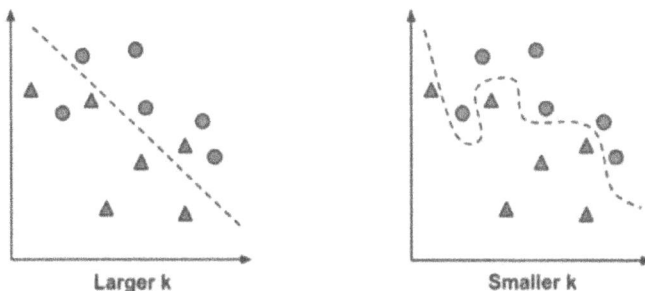

FIGURE 11.8   Impact of k in kNN.

- **Random Forest** is one of the most effective and incredible algorithms of machine learning. It is adjustable and needs less effort (Farooq & Kidmose, 2013). Instead of finding the most important feature from the given set, it tries to find the best feature from a defined subset of features generated randomly. So, it produces better and diverse results as compared to other algorithms.

During training, RFs construct many individual decision trees for predicting the particular class of a node. It is also referred to as ensemble techniques as it uses the collection of all results to reach the final conclusion, as shown in Figure 11.9. A node of the binary tree is processed using the following equation:

$$Imp_n = w_n I_n - w_{L(n)} I_{L(n)} - w_{R(n)} I_{R(n)} \qquad (11.2)$$

where
- Imp_n = importance of node n
- w_n = weighted number of samples reaching node n
- $L(n)$ = child node from left split on node n
- $R(n)$ = child node from right split on node n
- I_n = the Gini impurity value of node n, given by

$$\sum_{i=1}^{J} X_i (1 - X_i) \qquad (11.3)$$

X_i is the frequency of label i at a node and J represents the number of unique labels.

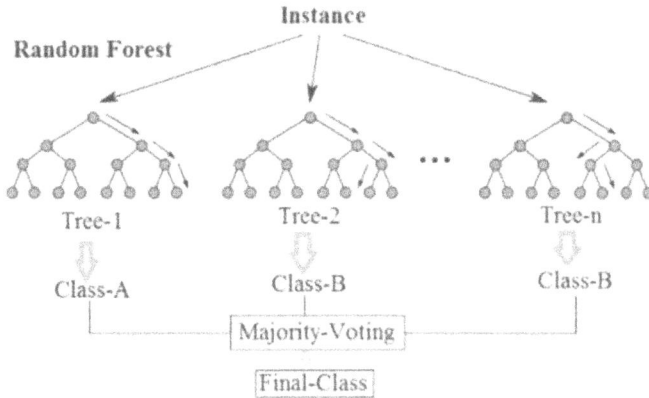

FIGURE 11.9   Random Forest classifier.

- **Support Vector Machine (SVM)** is a supervised machine learning algorithm that performs very well with a limited amount of data (Wei et al., 2018). This fast and dependable classifier creates hyperplanes in n-dimensional space according to the number of features for binary or multi-class classification. If $n = 2$, then the hyperplane takes the form of line; if $n = 3$, it is a plane, for $n > 3$, it is called a hyperplane.

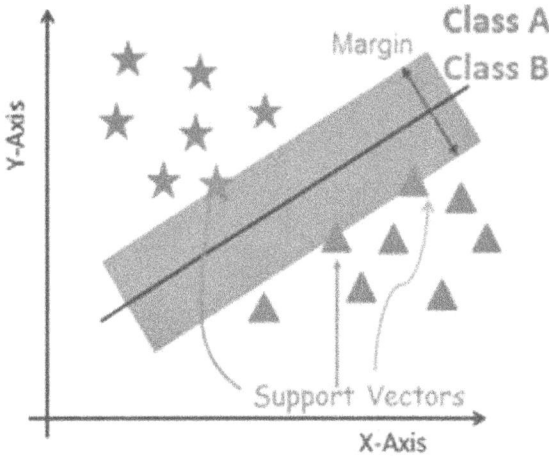

FIGURE 11.10 SVM classifier.

The support vectors are the data points closest to the hyperplane, as shown in Figure 11.10. The hyperplane helps to predict the class of a data point on the basis of hypothesis function as given below:

$$h(x_i) = \begin{cases} +1 \text{ if w.x} + b \geq 0 \\ -1 \text{ if w.x} + b < 0 \end{cases} \tag{11.4}$$

where w.x + b = 0 represents a hyperplane.

Different performance parameters such as *accuracy, sensitivity, precision,* and *F1-score* are investigated for KNN, RF, and SVM to evaluate the efficient implementation of the proposed model. The following sub-section further explains the mathematical interpretation of these evaluation metrics.

11.4.1 PERFORMANCE METRIC

Accuracy is defined as the number of correct predictions made divided by the total number of predictions. Mathematically accuracy can be stated as follows:

$$ACC = \frac{X + Y}{X + Y + X' + Y'} \times 100 \tag{11.5}$$

where X represents "True Positive," which gives a count for the number of subjects correctly identified as having strong memory. Y represents "True Negative," and it is the number of cases correctly identified to have a weak memory. X' represents "False Positive," which is the number of subjects incorrectly identified to possess strong memory. Y' represents "False Negative," which finds the number of subjects with weak memory to be incorrectly identified as having good memory strength.

Sensitivity (also called recall) is a measure to check the classifier's ability to correctly predict the true positives. For this system, sensitivity helps to determine the total number of subjects with a strong memory. Mathematically, this can be stated as,

$$Sensitivity = \frac{X}{X + Y'} \times 100 \tag{11.6}$$

Precision is a measure of the correctness of a positive prediction. So, it means that if a subject is predicted as having a good memory, how correct this prediction is. It is calculated using the following formula:

$$Precision = \frac{X}{X + X'} \times 100 \tag{11.7}$$

F1 Score is a performance measure, which considers both false positives and false negatives. It is the weighted average of Precision and Recall. For a model to be more accurate, the F1 Score should be higher. It can be stated mathematically as:

$$Precision = 2 \times \frac{Precision * Recall}{Precision + Recall} \tag{11.8}$$

11.5   RESULT AND DISCUSSION

In this section, the results achieved from the proposed system are analyzed. To train the Machine Learning models, we used 70% of the subjects as part of the training split, while for predicting the classification task, the remaining 30% of the subjects are used. The performance of the classification task is done by calculating the accuracy, sensitivity, precision, and F1-score of the three classifiers considered in this chapter. It helps to decide the effectiveness of these machine learning strategies.

The classification results for analyzing the performance of all the three machine learning techniques, namely KNN, SVM, and RF, are listed in Table 11.3. The comparative improvement for different measures is shown in Figure 11.11. It is quite clear that SVM performs better than the rest two classification algorithms, KNN and RF.

TABLE 11.3
Performance Analysis for Machine Learning Algorithms

Performance Analysis	Machine Learning Algorithms		
	KNN	RF	SVM
Accuracy	0.67	0.73	0.80
Sensitivity	0.91	1.0	1.0
Precision	0.71	0.73	0.80
F1 Score	0.80	0.85	0.89

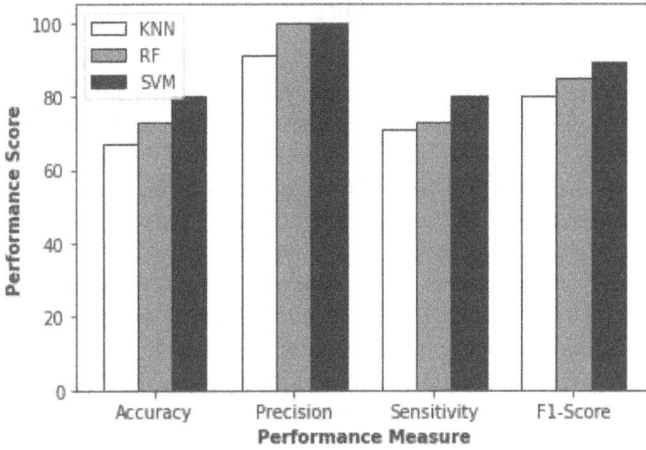

FIGURE 11.11   Performance measure of classification algorithms.

The small number of subjects results in the limitation of training and test set. The specific ratio of splitting the train and test data introduces a bias in training and hence that affects performance evaluation too. This can be handled following the k-fold cross-validation scheme, which also helps in increasing the reliability of the model. It divides the dataset into k disjoint folds and learns a classifier with (k-1) folds. Then with the remaining fold, an error value is calculated by testing the classifier. We have chosen the value of k to be 10, thus all the classifiers are trained 10 times, and the overall performance is computed as the average of all the results giving 0.80, 0.82, and 0.84 classification accuracy for KNN, RF, and SVM, respectively.

The research community of EEG has investigated the relation of human attention and working memory, but there is little known contribution for assessing the cognitive strength using EEG. Table 11.4 shows the comparative improvement of results achieved by the proposed approach with the classification accuracy obtained for other state-of-the-art for human concentration–related analyses using the same machine learning methods (Mohamed et al., 2018).

TABLE 11.4
Performance Comparison with Existing Work

Classifier Name	Cognitive Skills		
	Memory Strength (Our Method)	Focused Attention (Existing)	Working Memory (Existing)
KNN	0.80	0.69	0.46
RF	0.82	0.70	0.55
SVM	0.84	0.69	0.46

11.6   CONCLUSION AND FUTURE SCOPE

In the proposed system, we have tried to correlate the EEG signal with the verbal and learning ability test for assessing the memory strength of a healthy human with the help of different machine learning techniques, KNN, RF, and SVM. The comparison among results obtained through the proposed system finds SVM to be the most efficient among the three. We have also compared our work with other proposed methods for FA and WM. The classification tasks are related to memory functioning. However, they are not exactly similar to the memory assessment task considered here. So, we can say that the proposed work provides a new roadmap for assessing memory strength using EEG signal and machine learning models, which use CVLT for training purposes. This may guide the early detection of memory-related diseases like Alzheimer's. It can also help in mental load detection. The finding needs further research for extracting and minimizing EEG data for improving the existing result.

As the machine learning models may lack in identifying the significance of some specific pattern in EEG signals, it may result in loss of important predictions related to the health of the human brain. The proposed idea needs to be experimented with different deep learning techniques like recurrent neural network (RNN), long short term memory (LSTM), Transformers (uses convolutional neural network and attention models), etc. So, the application of deep learning models may have the capability to capture the hidden insights and phase-wise learning of human brains.

Implementation of the proposed idea in the IoT infrastructure will be a valuable addition to the medical adherence system, which will help in improving quality of life (QoL). Beneficiaries of the system may be health care providers and patients as well. One central system may receive EEG signals of a user captured through wireless EEG devices over a designated wireless sensor network (WSN). According to the discussed methodology, the classification task is performed using different machine learning methods. The early detection of brain-related disorders may prevent worsening the living span of a beneficiary. Proper security mechanisms imposed on the IoT-enabled system will have the added benefit of remote accessibility and privacy protection too.

REFERENCES

Abu Alfeilat, H. A., Hassanat, A. B., Lasassmeh, O., Tarawneh, A. S., Alhasanat, M. B., Eyal Salman, H. S., & Prasath, V. S. (2019). Effects of distance measure choice on k-nearest neighbor classifier performance: A review. *Big Data*, *7* (4), 221–248.

Acharya, J. N., Hani, A. J., Cheek, J., Thirumala, P., & Tsuchida, T. N. (2016). American Clinical Neurophysiology Society Guideline 2: Guidelines for standard electrode position nomenclature. *The Neurodiagnostic Journal*, *56* (4), 245–252.

Alarcao, S. M., & Fonseca, M. J. (2017). Emotions recognition using EEG signals: A survey. *IEEE Transactions on Affective Computing*, *10* (3), 374–393.

Al-Fahoum, A. S., & Al-Fraihat, A. A. (2014). Methods of EEG signal features extraction using linear analysis in frequency and time-frequency domains. *International Scholarly Research Notices*, *2014*, 1–7.

Babayan, A., Erbey, M., Kumral, D., Reinelt, J. D., Reiter, A. M., Röbbig, J., … others. (2019). A mind-brain-body dataset of MRI, EEG, cognition, emotion, and peripheral physiology in young and old adults. *Scientific Data*, *6*, 180308.

Dementia. (n.d.). https://www.who.int/news-room/fact-sheets/detail/dementia. (Accessed: 2020-09-30)

Edla, D. R., Mangalorekar, K., Dhavalikar, G., & Dodia, S. (2018). Classification of EEG data for human mental state analysis using random forest classifier. *Procedia Computer Science, 132,* 1523–1532.

Elwood, R. W. (1995). The California Verbal Learning Test: Psychometric characteristics and clinical application. *Neuropsychology Review, 5* (3), 173–201.

Farooq, F., & Kidmose, P. (2013). Random forest classification for p300 based brain computer interface applications. In *21st European signal processing conference (EUSIPCO 2013)* (pp. 1–5).

Farrer, T. J., & Drozdick, L. W. (2020). *Essentials of the California Verbal Learning Test: CVLT, CVLT-2, & CVLT3.* John Wiley & Sons.

Ghosh, P., Mazumder, A., Bhattacharyya, S., & Tibarewala, D. (2015). An EEG study on working memory and cognition. In *Proceedings of the 2nd international conference on perception and machine intelligence* (pp. 21–26).

Gramfort, A., Luessi, M., Larson, E., Engemann, D. A., Strohmeier, D., Brodbeck, C., ... others. (2013). MEG and EEG data analysis with MNE-Python. *Frontiers in Neuroscience, 7,* 267.

Graves, L. V., Holden, H. M., Van Etten, E. J., Delano-Wood, L., Bondi, M. W., Salmon, D. P., & Delis, D. C. (2019). New intrusion analyses on the CVLT-3: Utility in distinguishing the memory disorders of Alzheimer's versus Huntington's disease. *Journal of the International Neuropsychological Society, 25* (8), 878.

Gross, J., Baumgartl, H., & Buettner, R. (2020). A novel machine learning approach for high-performance diagnosis of premature Internet addiction using the unfolded EEG spectra. In *AMCIS'20 Proc.*

Jiang, X., Bian, G.-B., & Tian, Z. (2019). Removal of artifacts from EEG signals: a review. *Sensors, 19* (5), 987.

Jurcak, V., Tsuzuki, D., & Dan, I. (2007). 10/20, 10/10, and 10/5 systems revisited: their validity as relative head-surface-based positioning systems. *Neuroimage, 34* (4), 1600–1611.

Ko, L.-W., Komarov, O., Hairston, W. D., Jung, T.-P., & Lin, C.-T. (2017). Sustained attention in real classroom settings: an EEG study. *Frontiers in Human Neuroscience, 11,* 388.

Koelstra, S., Muhl, C., Soleymani, M., Lee, J.-S., Yazdani, A., Ebrahimi, T., Patras, I. (2011). DEAP: a database for emotion analysis; using physiological signals. *IEEE Transactions on Affective Computing, 3* (1), 18–31.

Konstantinidis, E., Conci, N., Bamparopoulos, G., Sidiropoulos, E., De Natale, F., & Bamidis, P. (2015). Introducing Neuroberry, a platform for pervasive EEG signaling in the IoT domain. In *Proceedings of the 5th EAI international conference on wireless mobile communication and healthcare* (pp. 166–169).

Kumar, J. S., & Bhuvaneswari, P. (2012). Analysis of electroencephalography (EEG) signals and its categorization–a study. *Procedia Engineering, 38,* 2525–2536.

Lantz, B. (2013). *Machine learning with R.* Packt Publishing Ltd, Birmingham, UK.

Laport, F., Dapena, A., Castro, P. M., Vazquez-Araujo, F. J., & Iglesia, D. (2020). A prototype of EEG system for IoT. *International Journal of Neural Systems, 30,* 2050018.

Lodha, P., Talele, A., & Degaonkar, K. (2018). Diagnosis of Alzheimer's disease using machine learning. In *2018 fourth international conference on computing communication control and automation (ICCUBEA)* (pp. 1–4).

Mohamed, Z., El Halaby, M., Said, T., Shawky, D., & Badawi, A. (2018). Characterizing focused attention and working memory using EEG. *Sensors, 18* (11), 3743.

SEED Dataset. (n.d.). http://bcmi.sjtu.edu.cn/ seed/. (Accessed: 2020-09-30)

Simpraga, S., Alvarez-Jimenez, R., Mansvelder, H. D., Van Gerven, J. M., Groeneveld, G. J., Poil, S.-S., & Linkenkaer-Hansen, K. (2017). EEG machine learning for accurate detection of cholinergic intervention and Alzheimer's disease. *Scientific Reports, 7* (1), 1–11.

Vidyaratne, L. S., & Iftekharuddin, K. M. (2017). Real-time epileptic seizure detection using EEG. *IEEE Transactions on Neural Systems and Rehabilitation Engineering*, *25* (11), 2146–2156.

Vimala, V., Ramar, K., & Ettappan, M. (2019). An intelligent sleep apnea classification system based on EEG signals. *Journal of Medical Systems*, *43* (2), 36.

Wei, Z., Wu, C., Wang, X., Supratak, A., Wang, P., & Guo, Y. (2018). Using support vector machine on EEG for advertisement impact assessment. *Frontiers in Neuroscience*, *12*, 76.

Wolf, T. J., & Rognstad, M. C. (2013). Changes in cognition following mild stroke. *Neuropsychological Rehabilitation*, *23* (2), 256–266.

Wu, C.-T., Dillon, D. G., Hsu, H.-C., Huang, S., Barrick, E., & Liu, Y.-H. (2018). Depression detection using relative EEG power induced by emotionally positive images and a conformal kernel support vector machine. *Applied Sciences*, *8* (8), 1244.

Xu, S., Wang, Z., Sun, J., Zhang, Z., Wu, Z., Yang, T., Cheng, C. (2020). Using a deep recurrent neural network with EEG signal to detect Parkinson's disease. *Annals of Translational Medicine*, *8* (14), 1–9.

Zhang, X., Yao, L., Zhang, S., Kanhere, S., Sheng, M., & Liu, Y. (2018). Internet of things meets brain–computer interface: a unified deep learning framework for enabling human-thing cognitive interactivity. *IEEE Internet of Things Journal*, *6* (2), 2084–2092.

Zheng, W.-L., Santana, R., & Lu, B.-L. (2015). Comparison of classification methods for EEG-based emotion recognition. In *World congress on medical physics and biomedical engineering, June 7–12, 2015* (pp. 1184–1187), Springer, Toronto, Canada.

Zhu, L., Cui, G., Cao, J., Cichocki, A., Zhang, J., & Zhou, C. (2019). A hybrid system for distinguishing between brain death and coma using diverse EEG features. *Sensors*, *19* (6), 1342.

12 Robotic Operating System and Human– Robot Interaction for Automated Guided Vehicles (AGVs)

An Application of Internet of Things in Industries

*Rahul Prakash, Mukesh Kumar,
and Dharmendra Kumar Dheer*
NIT Patna
Patna, India

Dilip Kumar Choubey
IIIT Bhagalpur
Bhagalpur, India

CONTENTS

12.1 INTRODUCTION

12.1.1 ABOUT ROBOTICS

Robotics is an interdisciplinary field of science and technology that started in the last century and is still growing. It has applications in areas such as engineering, transportation, medical science, manufacturing, civil areas, agriculture, and a lot more. In recent times robotics is growing in a new field known as human–robot interaction. It is a field in which robots are made to interact with humans to perform a specific task. This requires the robot to be intelligent to classify the actions or signals sent by humans and act accordingly.

12.1.2 ABOUT ROBOTIC OPERATING SYSTEM (ROS)

Robotic Operating System (ROS) is an open-source meta operating system (OS) [1], a software platform with tools and libraries providing hardware abstraction, low-level device control, message-passing between the processes. It is not a real OS like Windows, Linux, etc.; instead, it is a middleware framework supporting support client library for writing and running codes on multiple computers supporting developers to write code in C, C++, Python, JAVA, and other programming languages. The codes written can be used by different programmers in the form of packages. An ROS user does not need to write the code each time. Instead, it can be downloaded from the repository. It provides the feature of peer-to-peer communication for message passing via the publish/subscribe method. This message can be stored in an ROS bag and can be utilized later use.

12.1.3 DISTRIBUTIONS OF ROS

The first version of ROS 1.0 was developed in 2010 [1] and all the versions are listed below. Different versions of the ROS 1.0 release are shown in Table 12.1.

ROS 2.0 versions are mentioned below [2]. Table 12.2 shows the distributions of ROS 2.0.

12.1.4 NEED OF ROS

It is a software development platform for robots with many tools, libraries, and packages that makes development easier. It provides multilingual package management [3] worldwide by which software can work with multiple robots reducing the developing time and doesn't need to write the code each time. It also provides security and

TABLE 12.1
ROS 1.0 Distributions

Distributions	Release Date
ROS Neotic Ninjemys	May 23, 2020
ROS Melodic Morenia	May 23, 2018
ROS Lunar Loggerhead	May 23, 2017
ROS Kinetic Kane	May 23, 2016
ROS Jade Turtle	May 23, 2015
ROS Indigo Igloo	June 22, 2014
ROS Hydro Medusa	September 4, 2013
ROS Groovy Galapagos	December 31, 2012
ROS Fuerte Turtle	April 23, 2012
ROS Electric Emys	August 30, 2011
ROS Diamondback	March 2, 2011
ROS C Turtle	August 2, 2011
ROS Box Turtle	March 2, 2010

real-time processing [2]. In ROS, the code written, complied and debugged is reused by features like public repository, API, packages.

12.1.5 AUTOMATED GUIDED VEHICLES (AGVs)

Automated Guided Vehicles are industrial automatic vehicles developed in the 1970s [4] that are meant to perform a given specific task. The development of AGV is aimed to increase productivity and reduce production cost. The demand for a higher degree of automation in the industries leads to the use of AGV in industries such as manufacturing, logistic, transport, and others. The dimension and safety of AGV depend on the type of operation in which it is used. Market demand and the high user expectancy enhanced technological advancement.

12.1.6 BRIEF DESCRIPTION OF HUMAN–ROBOT INTERACTIONS (HRI)

Human–robot interaction is a multidisciplinary field that emerged in 1990 [5]. The complexity and the dynamics of the field make it challenging and, at the same time,

TABLE 12.2
ROS 2.0 Distributions

Distribution	Release Date
ROS 2 Foxy Fitzroy	June 5, 2020
ROS 2 Eloquent Elusor	November 22, 2019
ROS 2 Dashing Diademata	May 31, 2019
ROS 2 Crystal Clemmys	December 14, 2018
ROS 2 Bouncy Bolson	June 2, 2018

emerging rapidly in the current scenario of need in industries, service robots, medical, and many others. The robot capabilities in this field are enhanced as compared to other robotics fields. Interactions of the robot in the physical world can be any task obeying the laws of physics or can be a social behavior developed applying machine learning and neural networks. The robot interacts with the human in a social environment through perception, learning, decision making, and with high accuracy and precision during calculations. While designing the system, the safety of humans is of major concern as they work in the same workspace.

12.1.7 INDUSTRIAL INTERNET OF THINGS (IIoT)

Internet of things is being implemented in industries to fulfill the demands of automation of heavy machinery to remote monitor and manage faults in AGVs [6]. The operations in the large area of manufacturing industries using IoT-based AGVs are easily monitored rather than in classical industries. Material handling and logistics of raw materials are important tasks in the industries for productivity and time.

12.2 REQUIREMENTS OF ROS

The requirements for the running of ROS on a computer are an OS and programming languages that are described in the following sections [1, 7].

12.2.1 OPERATING SYSTEM

As discussed in Section 6.1.2, ROS is not a regular OS like Windows. It is developed to run on a pre-installed OS on a PC. For the proper running of ROS, the suggested OS to be used is Ubuntu. It is a Linux distribution with features that are used by ROS such as file system, process management system, compiler. ROS1.0 and ROS2.0 can be installed in Ubuntu and windows. In this chapter, the installation in Ubuntu is covered with ROS1.0.

12.2.2 PROGRAMMING LANGUAGES

ROS provides a feature of the client library to support various programming languages and thus supporting the developers to code in Python, C++, JAVA, Lua, Lisp, and Ruby. Users can import the program in this library using the preferred language.

12.3 INSTALLATION AND SETTINGS OF ROS

The installation of ROS requires the prerequisite Linux installation on the PC and the knowledge of its version as the computability issue with the ROS version. After the installation, environment setup is described that needs to be done.

12.3.1 INSTALLATION

There are many distributions of ROS that have been developed till 2020. Referring to the official page of ROS wiki, one of the three versions of ROS1.0 Kinetic, Melodic, and Noetic Ninjemys can be installed depending on the version of Ubuntu installed on the PC. This chapter covers the installation of ROS Noetic Ninjemys recommended for Ubuntu 20.4 on the official page of the ROS wiki.

The steps of installation are explained in the subsequent sections.

12.3.1.1 Configuring the Ubuntu Repositories

The programs that are stored in software archives are referred to as repositories which make it easy to install software and with a high level of security. In Ubuntu software center, three repositories are needed to be configured, namely universe, multiverse, and restricted.

12.3.1.2 Setting of Source List

Open a new terminal window and enter the following command to accept software form pakacge.ros.org:

```
sudo sh -c 'echo "deb http://packages.ros.org/ros/ubuntu
$(lsb_release -sc) main" >/etc/apt/sources.list.d/ros-latest.list
```

12.3.1.3 Setting of Key

To download packages from ROS, the repositories key is set up with the following command:

```
sudo apt-key adv –keyserver 'hkp://keyserver.ubuntu.com:80'
–recv-key C1CF6E31E6BADE8868B172B4F42ED6FBAB17C654
```

12.3.1.4 Updating of the Package Index

Updating of all Ubuntu packages is done with the below command:

```
sudo apt update
```

12.3.1.5 Installation

The full desktop version is recommended to install using the command given below. It contains tools for 2D/3D simulators, visualizing tools, etc.:

```
sudo apinstall ros-noetic-desktop-full
Enter the following command to install various tools:
sudo apt-get install python-rosinstall
```

12.3.2 Environment Setup

While installation of any of the above versions of ROS, a few settings are required to be done. Every time when a new terminal window is open following command needs to be executed:

```
source/opt/ros/noetic/setup.bash
```

12.4 IMPORTANT CONCEPTS

In this section, the terminologies, communication, levels of concept, launch files, and build system are defined for the understanding of ROS.

12.4.1 TERMINOLOGIES

In commonly used terminologies [1, 2, 3] node, master, message, topic, parameter, package, meta-package, service, publish/subscribe, action, bag, ROS build, catkin, graph, and name are defined.

12.4.1.1 Node

An ROS node is an executable program that utilizes the feature of the client library. For the functioning of the robot, each sensor or actuator needs to communicate to send or receive data. For this purpose, a node program is assigned for each functioning.

12.4.1.2 Master

Master is the central part of peer-to-peer communication for nodes. Each node is required to register its information to the master to publish and subscribe, without which they cannot communicate between them.

12.4.1.3 Message

The information shared during the publish/subscribe method between the nodes is called a message. Messages can be variables such as integer, Boolean, and floating type.

12.4.1.4 Topic

The message which is shared between nodes is first placed over the transmission bus called topic. The publishing node first registers the name of the topic to the master; then the subscriber node can directly access the message through the name of the topic.

12.4.1.5 Package

The package is the basic unit of ROS containing one or more nodes that are the smallest processors. It contains the file and folders necessary for running other dependencies, libraries, and nodes. There are lots of packages available in the repositories that are developed by users.

12.4.1.6 Meta Package

The packages with commonalities are called meta-package.

12.4.1.7 Service

It is a synchronous bidirectional communication used whenever there is a request and response occurs between the service client and service server, respectively, for a particular task. When the response to the request is completed, then the communication ends.

12.4.1.8 Publish and Subscribe

It is the method of transmitting and receiving messages over the same topic name registered with the master. The node which is transmitting the message is called the publisher node, and the node which is accessing the message is called the subscriber node. This way of communication is called asynchronous communication.

12.4.1.9 Action

Action is another message of communication. It is used when the response to a request takes a long time for some tasks. The action client sends the request to a specific goal to the action server and receives the response to it. The action server provides feedback with the result of the goal.

12.4.1.10 Parameter

ROS parameters are the parameters used in the nodes. The parameter name and value can be configured in the program. A hierarchy is maintained while naming the parameters to protect from colliding. A parameter server is a shared, multivariate dictionary that is accessible from the network. Nodes use these servers to store and retrieve the parameter in real-time.

12.4.1.11 ROS Build

It is the build system for ROS that was used before the catkin was developed. Up to ROS Fuerte, ROS was built using cmake wrapper script called rosmake, which is the part of rosbuild. It is used to create msg and srv files.

12.4.1.12 Catkin

Catkin is the build system for ROS after the fuerte versions. It combines the custom cmake macros and python script to provide top of cmake workflow. It was designed to be more conventional than rosbuild. Catkin can be easily implemented in a system that supports python and cmake.

12.4.1.13 Bag

A bag file in ROS is used to store the information of messages, topics, services that were shared between the nodes. These are stored to visualize the data later and to play, stop, and rewind whenever the environment is required to reproduce.

12.4.1.14 Graph

The graphical representation of node, topic, publisher, and subscriber are represented as a graph. It can be obtained by running rqt_graph.

12.4.1.15 Name

Node, topic, parameter, and services are collectively called graph resource name. These are identified by a short string as names. They are registered to the master and are searched using these names. Names can be modified during execution and have several forms such as global names, relative names, private names, and anonymous names.

12.4.2   COMMUNICATION

This section describes the message communication between nodes. Nodes are the smallest executable program that communicates via registering their information on ROS master. The message communication is of three types, namely topic, service, and action. The topic is an asynchronous unidirectional communication used for continuous data transfer. The topic uses *.msg extension for communication. Service uses request/respond by service client and server, respectively, with *.srv extension is used for service. Action provides a bidirectional message goal/result/feedback. Action uses *.action as extension.

12.4.3   LEVELS OF CONCEPT

The ROS architecture [7] has been divided into three levels of concept including file system level, computation level, and community level.

12.4.3.1   File System Level

At this level, the internal working of ROS is defined, folder structure and files are explained. It includes packages, meta-packages, manifests, repositories, service types, and message types.

The ROS file system is divided into the installation folder and workspace folder.

12.4.3.2   Computation Level

At this level, the communication process is defined between nodes and the systems required for ROS to handle processes to communicate between other computers. This level includes nodes, master, parameter server, topic, message, service, and bags.

12.4.3.3   Community Level

This level explains tools to share the algorithms, codes between the developers for the growth of ROS. This level includes distributions, repositories, ROS wiki, and mailing lists.

12.4.4   LAUNCH FILES

To start ROS master and many nodes at once launch files are used. The launch files are stored in the package directory. roslaunch will search in subdirectories of each package directory. The launch file names end with .launch.

12.4.5   BUILD SYSTEM

Catkin is the official build system for ROS and migrated from rosbuild build system. Catkin has dependencies such as Cmake, Python, GTest, GNU C++ compiler. ROS build system is responsible for the creation, compiling, and modification of ROS packages. The custom build system is catkin build system made of Cmake build system and python scripting. The build environment is described by CmakeLists.txt file in the package folder. The Cmake build system has multiple platforms that can be used to develop packages. The package folder contains two configuration files as CmakeList.txt and package.xml. The package.xml file contains the information of package name, author, license, and dependent packages.

TABLE 12.3
ROS Shell Commands

Commands	Description
roscd	Move directory to ROS package
rosls	Check file list of ROS package
rosed	Edit file of ROS package
roscp	Copy file of ROS package

12.5   ROS COMMANDS

While using ROS, the knowledge of commands is very important other than Linux commands. Following are the commands which are frequently used [1]:

12.5.1   ROS Shell Commands

These are called rosbash. These commands are used to use the bash shell in Ubuntu for ROS development environment. Generally, prefixes "ros" are added with suffixes such as "cd, ed, node, bag, core, run." Table 12.3 shows ROS shell commands that are used for packages.

12.5.2   Execution Commands

These commands are used for the execution of nodes. Execution can be done using "rosrun" for one node and "roslaunch" for multiple nodes. The command "roscore" is used as a name server for nodes. Table 12.4 describes different execution commands.

12.5.3   ROS Information Commands

These commands are used to get information on topics, nodes, and others. Among all information commands the essential command is "rosbag," which can record and play data. In Table 12.5, ROS information commands have been given.

TABLE 12.4
ROS Execution Commands

Commands	Description
roscore	Used communication between nodes
roslaunch	Launch multiple nodes at the same time
rosrun	Run the code
rosclean	Delete ROS log file

TABLE 12.5
ROS Information Commands

Commands	Description
rostopic	Check ROS topic information
rosnode	Check ROS node information
rosservice	Check ROS service information
rosparam	Edit and check ROS information of parameter server
rosmsg	Check ROS message information
rosbag	Play and record ROS messages

12.5.4 ROS Catkin Commands

These commands are used for building packages using catkin build system while "catkin_create_pkg" creates an empty package containing "CmakeLists.txt" and "package.xml" files. Table 12.6 represents different catkin commands.

12.5.5 ROS Package Commands

These commands are used for managing packages. ROS package commands are used for viewing information about packages and installing them. The command "rosinstall" will automatically install the package and update it if available. "rosdep" has options such as check, install, init, and update for packages. Different package commands are shown in Table 12.7.

TABLE 12.6
ROS Catkin Commands

Commands	Description
catkin_create_pkg	Creation of package
catkin_make	Build based on the catkin build system
catkin_prepare_release	Cleans log during the release
catkin_init_workspace	Initialize workspace of catkin

TABLE 12.7
ROS Package Commands

Commands	Description
rospack	View ROS package information
rosinstall	Installs additional ROS package
rosdep	Installs dependencies of ROS package
roslocate	Show information of ROS package
rosmake	Builds ROS package

12.6 ROS GUI TOOLS

In addition to command-line tools introduced in the above section, there are visualization and GUI development tools that help to visualize, plot, and analyze the data. These tools are rviz, rqt, rqt_image_view, rqt_graph, rqt_plot, rqt_bag. rviz is a 3D visualization tool for visualizing the messages in 3D. The rviz screen components are 3D view panel, Menu, display panel, tools, view panel, and time. GUI development tools are used for robot development. rqt is a GUI tool with additional plugins. The rqt initial screen contains files, plugins, perspectives, running. In this chapter only rqt_image_view, rqt_graph, rqt_plot, and rqt_bag are discussed. ROS GUI tools are discussed in Table 12.8.

12.7 ROS PROGRAMMING

So far different terminologies, levels of concept, message communication, commands, and GUI tools have been discussed. This section describes the creation of workspace, package, publisher, and subscriber program [2, 8].

12.7.1 CREATION OF WORKSPACE

While coding in ROS, the first step in ROS development is to create the workspace and package.

Workspace is created, and then the package is created, which is kept in the workspace itself.

The workspace folder is created and any name can be given to it and stored in the Ubuntu folder.

Now, to create catkin workspace in which src folder is contained enter the following command:

```
mkdir -p ~/catkin_ws/src
```

Now initialize the catkin workspace by entering the following command. CMakeLists.txt file is located inside src folder:

```
catkin_init_workspace
```

TABLE 12.8
GUI Tools with Description

Command	Description
rqt_image_view	Used to display the image taken by the camera
rqt_graph	It shows the relationship between nodes and messages transmitted on ROS network
rqt_plot	It plots data in 2D coordinates
rqt_bag	It allows visualizing the message

To build the catkin workspace, enter the following command:

```
catkn_make
```

Now inside catkin folder, there are three folders devel, build, and src. After the creation of the workspace, it is important to set the workspace path to access and visualize the packages inside it. Add the following line:

```
source ~/catkin_ws/devel/setup.bash
```

The src folder is the folder in which new packages from repositories are stored and are built. After the execution of catkin_make, src folder is checked and packages are built. The build folder contains build files and cache cmake files to prevent the rebuilding of all packages when catkin_make is running. Devel folder is used to store the executable when there is a successful build.

12.7.2   CREATING ROS PACKAGE

Packages are created with the help of the following command, which is run from the src directory of workspace:

```
catkin_create_pkg ros_package_name
```

This command is executed to form the src folder in the catkin workspace. Inside the package, there are src folder, package.xml, CMakeLists.txt, and include folder. The CMakeLists.txt folder contains the command for the source code and creates executables. The package.xml folder contains the information and package dependencies. The src folder contains the source code of ROS packages. The include folder contains the header files.

12.7.3   AN OVERVIEW OF A PROGRAM

ROS programming uses client libraries such as roscpp, rospy, and roslisp making the development of ROS nodes becomes easier. roscpp uses C++ for the development of the ROS node. For the use of python language in developing the node, rospy is used. roslisp uses lisp language for motion planning codes.

In developing a package using C++, the first step is to include the header file ros.h. This header file is used to create ROS nodes and is declared at the beginning of the code. Initialization of the node is needed for every node to start using ros::init(). After initialization of node, creation of a node handle is required to start it and operations like publishing and subscribing a topic. ROS_INFO_STREAM is used to send the message to different locations for display. To publish a topic in node, use the following command:

```
ros::Publisher publisher_object = node_handle.advertise<ROS
message type >("topic_name",1000)
```

after this add this line:

```
publisher_object.publish(message)
```

after publishing the message on the topic, use the following line to subscribe to the topic:

```
ros::Subscriber subscriber_obj = nodehandle.subscribe

("topic_name", 1000, callback function)
```

The callback function syntax is given as:

```
void callback_name(const ros_message_const_pointer &pointer)
{
//Access data
pointer->data
}
```

12.8 HUMAN–ROBOT INTERACTION FOR AUTOMATED GUIDED VEHICLE

This section describes the human–robot interaction, classification, and its application.

12.8.1 HUMAN–ROBOT INTERACTION (HRI)

Human–robot interaction is a new, emerging, and multidisciplinary field of study [5, 9] dedicated to designing, understanding, and evaluating robotic systems. HRI is related to human–computer interaction, artificial intelligence, robotics, mechanical design, electronics, kinematics, and dynamics. Interaction refers to the communication between the human and the robot. The robot should have abilities such as decision making, reasoning, learning, adapting, and more.

The interaction between humans and robots lies in the spatial dimension. It depends on the proximity of the two, which is divided into three sections, namely remote HRI, co-located HRI, and physical HRI.

12.8.2 CLASSIFICATION OF HRI

Human–robot interaction is broadly classified into three categories, namely remote HRI, co-located HRI, and physical human–robot interaction (pHRI) [10].

12.8.2.1 Remote HRI

Remote HRI refers to the location of the robot and human remotely. There are many application areas in which the robot has to be located in remote areas where the presence of humans can be a danger to them or inaccessible. The frame of reference for both is different in this category. The remote HRI deals with the problem of the interface to be used for communication. The problems can be the security of signals, noise, operator's knowledge, and others. The specified signals are required to send via the communication link to complete the task by the robot.

12.8.2.2   Co-Located HRI

When the human and the robot are located in the same frame of reference or the same sharing space but they don't come in physical contact with each other while completing the task it comes under co-located HRI. This category covers also social robotics behavior. The robot in this category has the ability of locomotion, perception, and learning. Social behavior refers to behaving and replying in an environment based on perception and learning. Industrial robots are also an example of this category. The operator is present in the same workspace for the task.

12.8.2.3   Physical Human–Robot Interaction (pHRI)

Physical human–robot interaction refers to the direct physical contact of the human and robot in the same frame of reference while performing the shared task. In this interaction, there is contact made between the robot and human by touch, motion, and pressure, known as haptic communication. This can be applied while guiding the person in the crowd or guiding a blind person while holding the hand. While in industries a lot of tasks are accomplished with the help of robotic arms side by side with humans. Tasks like cutting a log, moving a patient, feeding people require physical contact with people. It requires are proper modeling and calculation to complete the task accurately and without any harm to the personal. pHRI is further classified into three categories supportive, collective, and collaborative.

12.8.2.3.1   Supportive Interaction

In this type of interaction, the robot assists the operator with some tools or information required during the task. In the supportive interactions of robot with the human for some task, safety, and distance from the human are very much important.

12.8.2.3.2   Collaborative Interaction

In this type of interaction, the work is divided according to the ability and need for accuracy of work between the human and robot. Each one completes the divided task separately in a work.

The frequent interactions are done while changing the positions or exchanging some tools during task completion.

12.8.2.3.3   Cooperative Interaction

In this interaction, the robot independently participates in the task by self-decision-making and calculating the parameters of physics. This is a force interaction between the human and robot by making direct or indirect physical contact with some common object.

12.8.3   Application of Human–Robot Interaction (HRI)

In the application of human–robot interaction, the automated guided vehicle, its type, and safety toward humans are described. It has many applications areas such as service robots, robots for health care, self-driving cars, robots in society [9].

12.8.3.1 Automated Guided Vehicle (AGV)

Automated guided vehicles are self-driven vehicles [11] which are also referred to as mobile robots. They can be operated manually also, depending on the situation of the environment. They are meant for doing heavy-duty such as a forklift, skid tractors, tugger, unit load. The accuracy of these AGV should be high at the time of pick and placing the loads to avoid damage to the product and vehicle itself. As the functioning of AGV is in the same sharing space of the personal or operator it can be considered in pHRI. The safety of humans is the top priority as the sharing space or workspace is common. Designing and controlling AGV involves mathematical modeling and simulation of the vehicle. The motion planning of AGV is a crucial task and must be constrained to give high accuracy. However, this chapter focuses only on the theoretical part of AGV.

For ROS-based multiple AGVs working in industries for a specific task, there should not be a time lag between sending and receiving the data in between the AGV and jitters [12]. The presence of time lag can cause collisions, and also the information shared on the ROS network is of no use.

As the working space of AGV and humans are common, so there is a need for AGVs to learn the behavior of operators to interact. The AGV should be trained with the data of the human behavior in the surrounding space and predict future behavior as the AGV is centrally controlled, so the gestures of the operator should also be learned by the AGV, such that to pick up and drop the object with proper sensing and to calculate the amount pressure applied to hold.

This section discusses the current research on different fields of AGV [13]. While coordinating in the fleet of AGV, objectives include collision avoidance, localization, communication, path planning, guidance, and control is necessary.

In literature, many algorithms have been applied for localization and path planning, such as the D* Lite algorithm, for optimal control algorithm in trajectory tracking in industrial environment feedback control, fuzzy proportional–integral–derivative (PID), and the sliding model control are used, Dijkstra, artificial potential field and their combination for optimal path planning in warehouses. Classical algorithms were found not suitable as of large computational loads. So, the hybrid and improved algorithms are employed such as combining of Dijkstra and artificial potential field and applying improved Dijkstra algorithm. Improved genetic algorithm, deep learning and deep reinforcement learning have been applied for multi AGV path planning, to AGVs to select closet task among other multiple material handling tasks, respectively. Also the combination of genetic algorithm and Dijsktra algorithm is applied to solve the problem of AGV conflicts during transfer of payloads inside the factories with dynamic environment. In future high speed computation resources can be applied to overcome the computational loads.

The wireless communication for the controlling and managing the fleet of AGV is very much widespread for control and management. There is common use of artificial intelligence technologies and open source software such as ROS. RFID tags, QR codes, WIAFA, ZigBee networks are used for AGV management in industrial environments and still there is research gap with respect to 5G application for wireless communication.

For AGV navigation, control, and guidance, there are control algorithms such as fuzzy inference, PID, SMC, MPC, lyapunov are employed in the literature to solve these problems. Kalman Filters and simultaneous localization and mapping (SLAM)-based Extended Kalman Filters were used for the navigation. Laser-based vision systems are also being used for AGV industry. There is a research gap in AGV motion control and skidding during movement in factories and the exchange of control data using 5G communication. A charge-coupled device (CCD) camera is used for the vision system to improve the navigation of AGV.

Input power optimization, traffic management using 5G-based artificial intelligence solutions for human identification, docking waypoint selection in a dynamic environment for a high precision task is the area of research for AGV management.

12.8.3.2 Types of Automated Guided Vehicle (AGV)

Many AGVs exist in the industry [4] such as a forklift, piggyback, towing vehicle, underride, assembly, heavy load, people mover, and others. Few of them are mentioned in this chapter.

12.8.3.2.1 Forklift AGV

It is a floor-level load pickup and is adjustable at various heights. The logistic task is very simple. It can work independently or can work with other vehicles. Figure 12.1 represents forklift AGV.

12.8.3.2.2 Piggyback AGV

It works with traditional load aids. It does not lift from the floor but at some height which is maintained entirely in operation in the industry. The load handling can be done quickly and requires less space. A Piggyback AGV is shown in Figure 12.2.

12.8.3.2.3 Underride AGV

The underride AGV has small chassis for transporting loads on top of it. They are used in hospitals for clinical purposes. It underrides the roller container and lifts it for transport.

Figure 12.3 shows the Underride AGV.

12.8.3.2.4 People Mover AGV

They are used for safely transporting passengers in public areas, parks, parking areas, airports. An example of a people mover is shown in Figure 12.4.

12.8.3.2.5 Assembly Line AGV

Automatic guided vehicles are commonly used for assembly lines or can also replace traditional in-floor tow lines. These Automated Guided Vehicles are most commonly ordered as tape-guided configurations. An assembly line AGV has been shown in Figure 12.5.

12.8.3.3 Industrial Internet of Things for AGVs

In the literature [6], AGVs are used in manufacturing industries used for internal logistics are gaining popularity for their cooperative ability and incorporating the technology for production with a high demand of flexibility according to the customers. The navigation areas of AGVs in the industries are large as they have to travel in the internal and external areas to pick and drop the raw materials. This navigation in the industries demands automation with accuracy. The wired connection is not possible for these machines, so the communication system involved must be wireless technology. IoTs are used in industries to support navigation and also localization.

The amount of data generated becomes large, which gives rise to the problem of Big data. In industries IoT eases remote monitoring, managing, and detecting problems in AGVs using artificial intelligence. IoT-based AGVs in industries are used for material handling [14], which involves packaging, transporting, and movement. IIoT improves the performance of logistic service providers from labor-based to unmanned vehicles [15]. These IIoT-based vehicles are designed for efficient and accurate picking and dropping activities. The vehicles are to solve the problem of labor shortage and to increase the competitiveness of the logistic company. IoT is a global network of things that has different layers in its architecture, namely sensor layer, network layer, middleware layer, and application layer [16]. In the sensor layer, the different sensors that are involved in sensing the required parameters are present. The network layer comprises communication responsible for the data transmission, which includes a Wi-Fi router, Internet connectivity. Cloud computing, APIs, Big data analysis, web services are the parts of the middleware layer. In the application layer smart homes, smart industries, smart agriculture, etc., can be realized.

In industries, there are large inside and outside working spaces. The communication between the AGV performing a task while moving in whole space becomes limited because of the presence of large obstacles such as walls and containers and results in loss of signal, thus degrading the efficiency of industries. The primary task of AGV in the manufacturing industries is to move the objects precisely while avoiding obstacles. The implementation of IoT in industries gives rise to handling another task of handling the large amount of data generated during operation and security of the network responsible in the industry to protect from cyber-attack.

12.8.3.4 Human Safety and Reliability with AGVs

In this chapter, human interaction with AGV in terms of safety and reliability [17] is discussed. Interaction of AGV with a human can be defined as sensing and responding of AGV to humans as they come on the track or some area which is restricted only of AGV. Reliability is defined as the ability of a robot to tolerate failures. It becomes a crucial task to maintain a high tolerance level if the level of disturbances is high in the shared workspace. While working in the same workspace the vehicle should have high reliability so that the harm to humans or other vehicles should be minimized. Designing AGV depends on various factors such as purpose, cost, environment, and others. The type of industry in which AGV will be used and the amount of load it will carry play an important role while specifying the size and dimension of AGV. At the same time, the stability of AGV is also important; otherwise it will harm the personal working in the same shared area. While carrying heavy loads the AGV should not vibrate or should not fall aside from equilibrium position. While moving on the track AGV should recognize the potential threat around it and distinguish between an obstacle and a human. The safety of humans can be presented in different situations such as in restricted zones and tracks on AGV. If some vehicle is currently operating in a specific area, then areas around it are prohibited from entering the area. While carrying heavy loads and running on the defined path, it is a crucial task to respond as the operator comes into the path. The reaction time of AGV and the perception of personal in the working area should be designed properly. When an operator comes into the working area or the path, the AGV should stop immediately after recognizing and shift in the wait mode until the area or tracks get cleared. The design of the system and selection of parameters are such that there should be very little time lag between sensing and reacting to the input; otherwise some accident might occur, harming the vehicle or the personal.

12.9 CONCLUSION

The chapter focuses on the basic concept of ROS and the installation of ROS Noetic Ninjemys. The prerequisites have also been discussed required for the installation. As ROS is multilingual, the user can have the basic concept of programming language such as C++ and Python. The commands such as rosrun, rospy, roslaunch, etc., make the execution of the program easy to handle are used frequently have been discussed in brief. An overview of ROS programming is discussed to provide an insight into the code such as the creation of workspace using catkin and creating a package. The user need not require writing the code each time for a robot. Thousands of packages are available online that can be downloaded from a repository, which results in saving a lot of time during development. Application of human–robot interaction in AGV with challenges has been explained that needs to be overcome, and also classification of HRI has been discussed. Current research in AGV such as localization, path planning, and visualization has been explained. The future scope of research in AGV includes energy optimization, artificial intelligence, and localization in a dynamic environment. ROS-based AGV have the limitations such as time lag delay and jitters for wireless communications. There is a discussion given on industry 4.0 representing the implementation of IoT in AGV known as IIoT. The objective is to increase the efficiency, material handling, and quality production of goods. Different types of AGVs are also discussed based on the working space and type of work. The challenge in IIoT for AGV is to maintain the connectivity in the network and data loss while working in large industries due to obstacles. At last, the safety of humans and the reliability of AGV, which are of much concern, are explained.

REFERENCES

1. Pyo, YoonSeok, Cho, HanCheol, Woon, Ryu, Lim, TaeHoon. 2017. *ROS robot programming*. Robotis Co Ltd., GeumCheon-gu, Seoul, Republic of Korea.
2. Joseph, Lentin. 2018. *Robot operating system for absolute beginners*. Apress, New York.
3. Quigley, Morgon, Gerkey, Brian, Smart, William D. 2015. *Programming robots with ROS*. O'Reilly Media, Inc., Sebastopol, California.
4. Ullrich, Gunter. 2015. *Automated guided vehicle systems*. Springer, Heidelberg.
5. Goodrich, Michael A., Schultz, Alan C. 2007. *Human-robot interaction: A survey*. Now Publisher, Inc., Hanover, Massachusetts.
6. Cupek, Rafal, Ziebinski, Adam, Drewniak, Marek, Fojcik, Marcin. 2020. *Autonomous guided vehicles for smart industries – The state of the art and research challenges*. Springer, Switzerland. 330–343.

7. Bartneck, Christoph, Belparme, Tony, Eyssel, Friedeirke, Kanda, Takayuki, Keijsers, Merel, Sabanovic, Selma. 2020. *Human-robot interaction.* Cambridge University Press, Cambridge, UK.

8. Kane, Jason M. O'. 2018. *A gentle introduction to ROS.* http://www.cse.sc.edu/~jokane/agitr/

9. Martinez, Aaron, Fernandez, Enrique. 2013. *Learning ROS for robotics programming.* Packt Publishing Ltd., Birmingham, UK.

10. Siciliano, Bruno, Khatib, Oussama. 2016. *Handbook of robotics.* Springer.

11. Patil, Veranda, Bhatwadekar, S. G., 2018. Automated guided vehicle system. *International Research Journal of Engineering and Technology.* Volume 05, Issue 04 April. 4038–4043.

12. Tardioli, Danilo, Parasuraman, Ramvyas, Ogren, Petter. 2019. *Pound: A multi-master ROS node for reducing delay and jitter in wireless multi-robot networks. Robotics and Autonomous Systems.* Volume 111. 73–87.

13. Emmanuel, A. Oyekanlu, and Alexander, C. Smith. 2020. *A review of recent advances in automated guided vehicle technologies: Integration challenges and research areas for 5G-based smart manufacturing applications. IEEE Access.* Volume 8. 202312–202353.

14. Alhaddad, M. A. S. M., Kamarudin, K. 2019. *Development of IoT based mobile robot for automated guided vehicle application. Journal of Electronic and Information Systems.* Volume 01. Issue 01. 37–42.

15. Lee, CKM. 2018. *Development of an industrial internet of things (IIoT) based smart robotic warehouse management system.* CONF–IRM Proceeding 43. AIS Electronic Library (AISel).

16. Hassija, Vikas, Chamola, Vinay, Saxena, Vikas, Jain, Divyansh, Goyal, Pranav, Sikdar, Biplab. 2019. A survey on IoT security: Application areas, security threats, and solution architectures. *IEEE Access.* Volume 7. 82721–82743.

17. Rahimi, Mansoor, Karwowski, Waldemor. 2004. *Human-robot interaction.* Taylor & Francis, Washington DC, London.

13 A Review on IoT Architectures, Protocols, Security, and Applications

Dilip Kumar Choubey
Indian Institute of Information Technology
Bhagalpur, India

Vaibhav Shukla
Tech Mahindra
Mumbai, India

Vaibhav Soni
Maulana Azad National Institute of Technology
Bhopal, India

Jitendra Kumar
Vellore Institute of Technology
Vellore, India

Dharmendra Kumar Dheer
National Institute of Technology
Patna, India

CONTENTS

DOI: 10.1201/9781003145004-13

13.1 INTRODUCTION

To interact seamlessly in real world, devices/smart objects must work in tandem with speed, congestion-free capabilities far beyond what is required by people. Today Internet is basically human-oriented and the Internet of Things (IoT) in contrast mostly relies on automated communication networks and smart objects. The use of Internet Protocol (IP) technology in Wireless Sensor Network (WSN) is a key pre-requisite for the accomplishment of the IoT vision. Internet Protocol has widely been accepted and implemented by major researchers and industrial players. IP for Smart Objects (IPSO) Alliance is promoting the use of IP technology in embedded devices. There is a range of protocols that are promoted as the silver bullet of IoT communication for the higher-level M2M protocol in the protocol stack. Note that these IoT/M2M protocols focus on the application data transfer and processing serving the application context. Table 13.1 summarizes features of the protocols generally considered in IoT applications; here we focus on protocols suitable for constrained environments.

Even though older protocols like FTP, Telnet, and SSH are still present and working fine, but they have flaws such as being resource intensive and power intensive,

TABLE 13.1

List of IoT Application Layer Protocols

MQTT	• Aims to support two-way communications over unreliable networks.
	• Battery-powered devices with low power consumption.
	• Devices may sleep but not 95% of time; otherwise opt for MQTT-S or CoAP.
	• Network Address Translation (NAT) traversal to be addressed as an afterthought-important but not critical.
	• Traffic is expensive.
CoAP	• Pretty similar to the MQTT-S choice, but with several additions.
	• Web services–oriented architecture rather than messaging centric.
	• Suitable for building platforms for open development communities in Internet space
MQTT-SN	• Lightweight protocol for constrained devices.
	• Based on UDP mapping of MQTT.
	• Adds broker support for indexing topic names.
REST API	• One-way communication from device to the cloud.
	• NAT traversal around the globe is among the top priorities.
	• No specific limitations for the traffic + device themselves are not very resource constrained.
	• "Device Clouds" for the open development communities (e.g. Device Hive).

and they do not fit well with the IoT realm as they need to operate on low power, unreliable bandwidth, heterogeneous and constrained nature of operation and application domain.

This chapter will firstly take a view on related works associated with IoT architectures, protocols, applications in Section 13.2. In Section 13.3, we take a look at protocols selection assumptions and where protocols in consideration could be applied in various levels. Section 13.4 describes the protocols such as Message Queuing Telemetry Transport (MQTT), Constrained Application Protocol (CoAP), Message Queuing Telemetry Transport for Sensor Networks (MQTT-SN), and Representational State Transfer (REST) and compares their various aspects. In Section 13.5, we propose hybrid approaches that can be applied to make the existing system more coherence. In Section 13.6, we briefly look at other protocols and alliances in IoT.

13.2 RELATED WORK

Ray (2018) has summarized the current state of the art of Internet of Things (IoT) architectures in several domains systematically.

Datta and Sharma (2017) have discussed the survey on IoT architectures, protocols, security, and smart city–based applications.

Sethi and Sarangi (2017) have presented a state of the art methods, protocols, and applications. It proposed a novel taxonomy for IoT technologies.

Srinivasa (2019) has provided a detailed architecture of IoT in the form of layers. It covers the details of the hardware as well as software along with the challenges of IoT.

Atlam et al. (2018) have presented an overview of the IoT system. The state of the art and layered architecture of the IoT are discussed with highlighting its applications, challenges, and open issues.

Choubey et al. (2020) have used the cloud computing approach to enhance the productivity of crop production. The cloud computing approach has been applied to the image dataset.

Khanna and Kaur (2020) have presented a comprehensive literature survey on the concept of IoT. It includes various contributions of researchers in different areas of applications.

Alshohoumi et al. (2019) have presented a systematic review to study the existing IoT architectures. It includes architecture classification (the number of layers), limitations in each architecture, and considerations of different aspects or features in each layer such as storage, processing methods, security, and privacy.

Mrabet et al. (2020) have proposed an IoT five-layer architecture based on potential security threats and counter-measures.

Pathak and Tembhurne (2018) have presented an overview of the Internet of Things (IoT) and the standards used for it. The popular protocols are discussed along with their review.

There exists a range of communication protocols (wired/wireless family) such as IEEE 802.15.4 based protocols like ZigBee, Z-Wave, 6LoWPAN, ANT, BTLE, EnOcean, or even the IEEE 802.11 family of protocols (Wi-Fi), they are suitable for Internet of Things as they operate either in the sub-gigahertz frequency ranges or the crowded 2.4GHz radio spectrum. Given the heterogeneity of the IoT environment, the IEEE has pushed IPv6 and 6LoWPAN, a standard that allows running IP, as Internet Protocol at all levels.

The work deals (Dawson-Haggerty et al., 2010) with a RESTful architecture that allows devices and other producers of physical information to directly publish their data. (Kovatsch et al., 2010) proposed a REST/HTTP framework for Home Automation. (Mayer et al., 2010) proposed a toolkit/framework which permits the user to decide if there is data traffic to be exchanged over an IoT network. There is a need for new standard protocols that are based on standard Web technologies but that consider the limitations of constrained devices. Recently, Internet Engineering Task Force (IETF) Constrained RESTful environments (CoRE) Working Group has proposed a REST-based web transfer protocol called Constrained Application CoAP (Shelby et al., 2013), a standard over UDP that specifically addresses the prerequisites of the constrained nodes carrying sensors and actuators in many deployments. These nodes/smart objects usually run specialized operating systems such as (Dunkels et al., 2004). For the realization of RESTful environments, CoAP focuses on a binary HTTP-inspired open protocol, providing features such as resources discovery and subscription for a resource (resulting in push notifications). MQTT by IBM is a lightweight messaging protocol based on a unified publish/subscribe approach. It is designed for constrained devices and low bandwidth, high latency, or reliable networks requirements. In a similar fashion, MQTT-S is a cut-down version of the MQTT protocol, which is geared toward lossy radio protocols work in (Yasser, 2014) deals with various aspects of security challenges in context with cloud computing, issues mentioned in the document also require consideration and further investigation need to be articulated in context with IoT environment.

13.3 PROTOCOLS SELECTION: ASSUMPTIONS AND USE-CASE

13.3.1 ASSUMPTIONS

The fundamental assumptions for protocol selection associated with IoT application Use-Case are:

- Protocols need to serve a constrained environment; they have to address specific requirements present at various levels of Application Domain (Figure 13.1).
- Security is a prime concern.
- Various types of wireless connections can be used to form heterogeneous environments.
- Devices/smart objects might be ranging from tiny microcontroller units (MCUs) to high-performance systems supporting Small MCUs.
- Data might be stored in the cloud and processed in cloud-based platforms.
- Information regarding routing, resource discovery through wireless/wireline connections to the cloud storage is required.

Other assumptions proposed by the protocol developers require deeper insight and will strongly influence their choices. Before we look at this, we need to understand the protocols in question at various levels of the IoT realm.

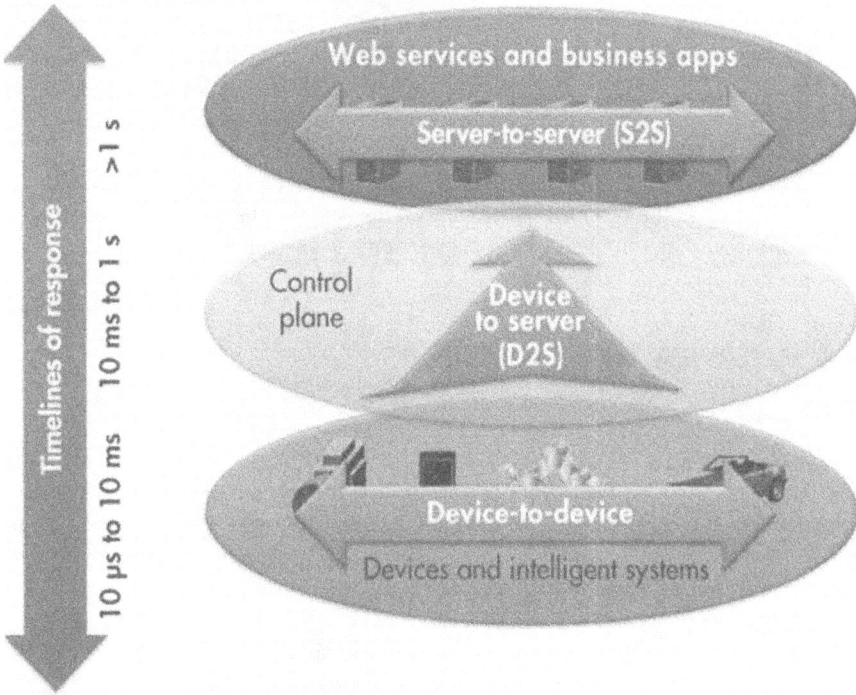

FIGURE 13.1 IoT protocols need to address response time at various levels.

13.3.2 Protocol Applicability Use-Case

The simple taxonomy in Figure 13.1 frames out the basic protocol use cases. There are basically three levels of application use case applicable in IoT scenario:

1. *Device-to-Device (D2D/M2M):* Devices must communicate with each other.
2. *Device-to-Server (D2S):* Data from devices then must be collected and sent to the server infrastructure.
3. *Server-to-Server (S2S):* That server infrastructure has to share device data. It is a case for cloud base realization, possibly providing it back to devices, to analysis programs, or to people.

The protocols can be described in this framework as:

- CoAP is a one-to-one protocol used for transferring state information between client and server over UDP. It is designed specifically for constrained resources, sensors, and devices/smart objects connected via lossy networks, which need to operate on low power, especially when there is a high number of sensors and devices within the network.

- MQTT is a many-to-many protocol for passing a message from multiple clients through a broker and communicating it to the servers (D2S). This protocol has been integrated with the IBM WebSphere application server.
- RESTful HTTP: Over TCP is particularly suitable for connecting consumer premise devices, given the near-universal availability of HTTP stacks for various platforms.
- Of course, it's not really that simple to figure out an appropriate protocol. For instance, the "control plane" represents some of the complexity in controlling and monitoring all IoT connections. Many other protocols cooperate in this region.

13.4   OVERVIEW OF IoT PROTOCOLS

13.4.1   MQTT

As its name suggests, i.e., Message Queue Telemetry Transport (Figure 13.2), its main task is telemetry, or remote monitoring. It was designed to collect device data from many devices/smart objects (data producers), and then it transports data to the IT infrastructure for further processing/storage. It relies on large networks of small devices that need to be monitored or controlled from the cloud infrastructure (*Website*, n.d.-a)

MQTT makes little approach toward device-to-device data transfer, nor does it "fan out" the data to many recipients. MQTT has simple, few controlling options which provide a clear, compelling single application. It is not meant to be particularly meteoric in nature. In this aspect, "real-time" access is typically measured in a few seconds.

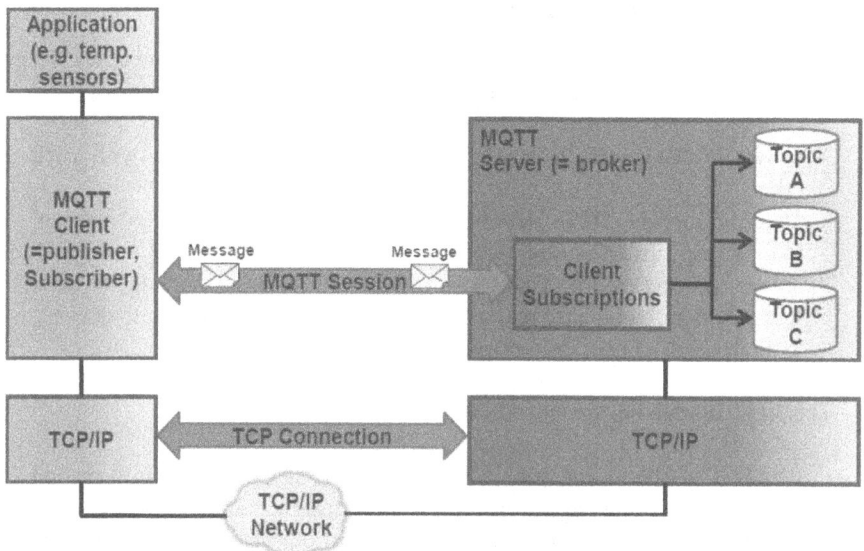

FIGURE 13.2   MQTT architecture.

MQTT is useful in domains like remote monitoring, controlling, sensing, etc. Where various nodes can combine together to take intelligent decisions, e.g. it can monitor leaks or vandalism in oil/gas pipeline with the help of detectors installed in that place. Here thousands of sensors are concentrated into a single location for analysis/data mining/processing/actuation. When the system finds a problem, it can take appropriate action to correct that problem. Other areas of MQTT application include power usage monitoring, lighting control, traffic-controlling, and even intelligent gardening.

13.4.2 MQTT-SN

MQTT-SN has its design inherited from MQTT, and it is adapted to constrained environments that comprise low bandwidth, high link failures rates, sleepy nature of objects, quick response time, short message length, etc. It is optimized for the implementation of low-cost, battery-operated devices with limited processing and storage resources capabilities. Compared to MQTT, MQTT-SN is characterized by the following differences:

- Topic string has been replaced by a topic ID (now fewer bytes required).
- Now predefined topic IDs is implemented which do not require a registration.
- Discovery procedure has been added for clients to find brokers (no need to configure broker addresses statically).
- In addition to persistent subscriptions persistent will message.
- Now there is support for sleeping clients by Off-line keep alive (it has a feature to receive buffered messages from the server once they wake up).

The architecture of MQTT-SN (Hunkeler et al., n.d.) is shown in Figure 13.3. It shows an example of MQTT-SN – MQTT network design for IoT. In a Zig Bee

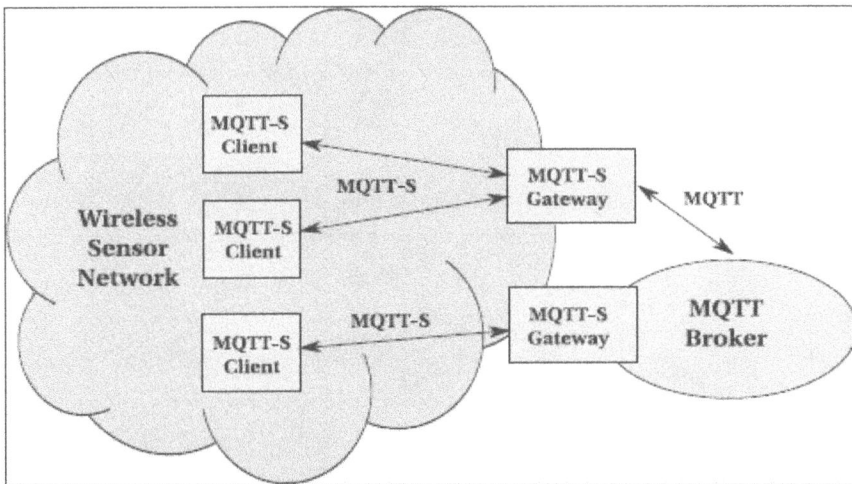

FIGURE 13.3   MQTT-SN architecture.

network, a gateway needs not to be hosted by a coordinator node but on an always-on router node to be able to receive client messages at any time. Due to the short payload length of the Zig Bee network/APS layer, the maximum length of an MQTT message gets restricted to 60 octets.

There are three kinds of MQTT-SN components:

- MQTT-SN clients: MQTT-SN clients connect themselves to an MQTT server via an MQTT-SN GW using the MQTT-SN protocol.
- MQTT-SN gateways (GW): Its main function is the translation between MQTT and MQTT-SN. An MQTT-SN GW may or may not be integrated with an MQTT server. In the case of a standalone GW the MQTT protocol is used between the MQTT server and the MQTT-SN GW. Depending on how a GW MQTT-SN, we can differentiate between two types of GWs, namely transparent and aggregating GWs (Figure 13.4).
- MQTT-SN forwarders: It simply encapsulates the MQTT-SN frames it receives on the wireless side and forwards them unchanged to the GW. In the opposite direction, it performs de-capsulation of frames it receives from the gateway and sends them to the clients, which is unchanged too. MQTT-SN clients can also access a GW via a forwarder in case the GW is not directly attached to their network.

13.4.3 CoAP

Due to the observed unsuitability of HTTP in resource-constrained environments, the IETF CoRE Working Group has defined CoAP, a web transfer protocol that is

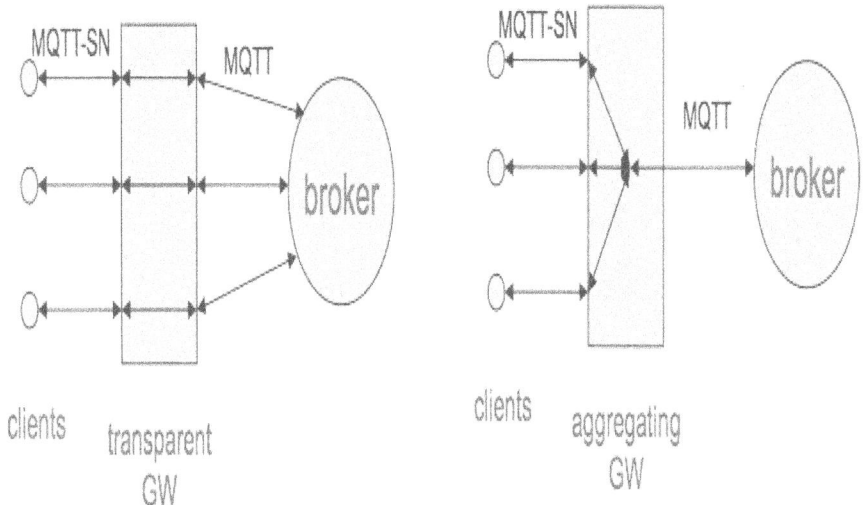

FIGURE 13.4 Transparent and aggregating gateways.

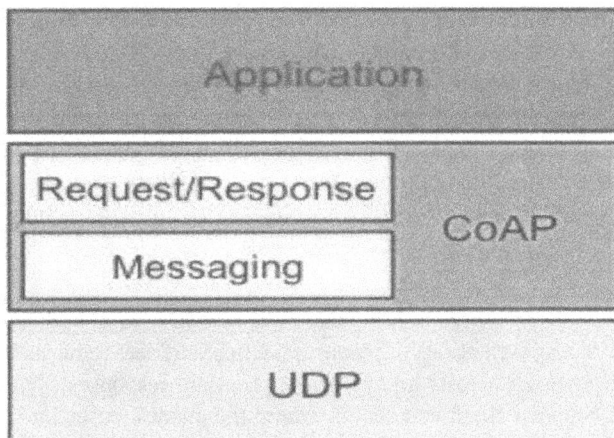

FIGURE 13.5   CoAP logical layers.

optimized for resource-constrained networks (*Online*, n.d.-a; *Website*, n.d.-b). It consists of the re-design of a subset of HTTP functions by taking into account the low processing power and energy consumption constraints of embedded devices, such as sensor motes. CoAP has been built on top of UDP, which has lower overhead and also provides multicast support (Figure 13.5).

One of the major design goals of CoAP was to minimize the message overhead and limit packet fragmentation. To this end, CoAP includes a short fixed-length compact binary header of 4 bytes followed by compact binary options. The length of the header of a typical request packet typically ranges between 10 and 20 bytes. The protocol stack of CoAP and its two-layer structure are illustrated in Figure 13.6, along with the stacks of HTTP and MQTT.

13.4.4   REST/HTTP

REST has emerged as the predominant Web API design model. RESTful style architectures conventionally consist of clients and servers. REST was initially

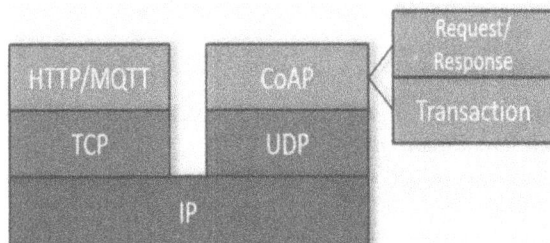

FIGURE 13.6   CoAP and its two-layer structure.

described in the context of HTTP, but it is not limited to that protocol. RESTful architectures may be based on other Application Layer protocols if they already provide a rich and uniform vocabulary for applications based on the transfer of meaningful representational state. A resource can be essentially any coherent and meaningful concept that may be addressed. A representation of a resource is typically a document that captures the current or intended state of a resource. The RESTful system tries to resolve the following core characteristics (Colitti et al., 2011).

- *Client-Server Interaction:* Requests and responses are built around the transfer of representations of resources. Clients initiate requests to servers; servers process requests and return appropriate responses. Processing is divided between client and server, where the client focuses on UI and the server on data storage, allowing separate evolution.
- *Stateless:* The server is stateless, meaning it doesn't have to know what state the client is in. The client is responsible for the session state, and each client request must contain all the information for the server to respond.
- *Layered System:* The service may be divided into many layers of other systems and each layer only has to know about the immediate component/layer it interacts with; these aspects are not visible to the client.
- *Caching:* The service decides which requests should be cached.
- *Uniform Interface:* The service enables clients to view and interact with representations of resources (with a URI) using a generic, finite set of requests, e.g. the HTTP verbs. It uses Hypermedia As the Engine of Application State (HATEOAS), e.g. a hyperlink that specifies which page a browser can load next.
- *Code on Demand (optional):* A client may be extended with executable code, e.g. JavaScript, but this requires a separate way to establish trust between the client and the server.

Based on architecture styles, semantics, and performance, the protocols described above can be compared (*Online*, n.d.-b). Table 13.2 summarizes the comparative study of IoT protocols.

13.4.5   BRIDIGING MQTT AND REST USING QEST BROKER

MQTT implements pub/sub, while the Hypertext Transfer Protocol (HTTP) – which is the basis of the REST pattern – is just a request/response protocol. Both kinds of architecture can be bridged using QEST (Figure 13.7) (Collina et al., 2012; Sanfilippo, 2012).

Thus, it realizes a new hybrid paradigm: QEST first modifies the broker semantic to retain and syndicate the last payload seen on a topic; secondly it exposes that payload as a REST resource.

TABLE 13.2
Comparative Study of IoT Protocols

Features	MQTT	REST/HTTP	CoAP
Abstraction	Publish/Subscribe	Request/Reply	Request/Reply
Architecture style	Brokered	P2P	P2P
Design orientation	Data centric	Document centric	Web centric
Data distribution	Supports 1 to zero, 1 to 1, 1 to n	1 to 1 only	1 to 1 typically
QoS	3	Provided by transport e.g. TCP	Confirmable or no confirmable messages
Interoperability	Partial	Yes	Yes
Message size	Small (2 bytes) Compact header	Lager because of text-based	Fixed compact 4-byte binary header and options
Performance	Typically 100s (persistent) to 1000 + (best effort) messages/second/ Subscriber. Scalability – broker implementation dependent	Typically 100s of requests per second	Typically 100s of requests per second
Hard real-time	No	No	No
Transports	TCP	TCP	UDP
Subscription control	Topics with hierarchical matching	N/A	Provides support for multicast addressing
Data serialization	Undefined	No	Configurable
Standards	Proposed OASIS MQTT standard	Is an architectural style rather than a standard	Proposed IETF CoAP standard
Encoding	Binary	Plain text	Binary
Extra libraries	C based (30 kb) Java baaed (100 kb)	Application dependent (e.g. JSON, XML)but typically not small	Based on C, Java, and Python
Licensing model	Open Source and commercially licensed	HTTP available for free on most platforms	Open Source and commercially licensed
Dynamic discovery	No	No	Yes
Mobile devices (Android, iOS)	Yes	Yes	Via HTTP proxy
6LoWPAN Devices	Yes	Yes	Yes
Multi-phase transactions	No	No	No
Security	Simple Username/Password Authentication, SSL for data encryption	Typically based on SSL or TLS	DTLS

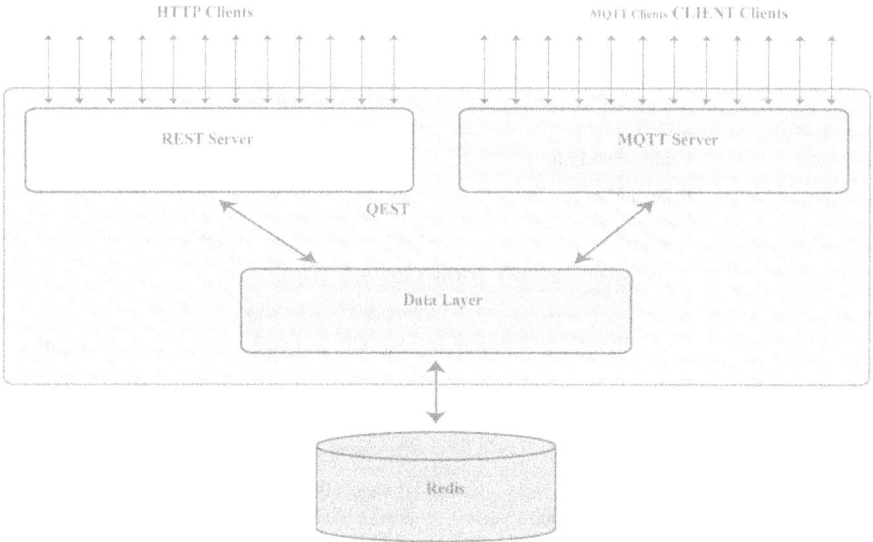

FIGURE 13.7   QEST broker architecture.

13.4.6   HTTP-CoAP Cross Protocol Proxy

Two protocols, CoAP and HTTP/REST, are both based on request/response without a publish/subscribe approach. In the case of CoAP, the use of 6LoWPAN and the automatic addressing of IPv6 is used to uniquely identify nodes. In the case of HTTP/REST the approach is different in that the request can be anything including a request to publish or a request to subscribe, so in fact it becomes the general case if designed in this way. Today, these protocols are being merged to provide a complete publish/subscribe request/response mode.

WebThings is an open-source software base for HTTP-CoAP proxy solution; it provides a number of different application layer components that can be pulled together to create a UNIX/Linux based HTTP-CoAP proxy, a CoAP-CoAP proxy, a Resource Directory, or any other client or server agent exposing a CoAP and/or HTTP interface (Figure 13.8).

By leveraging the multiplexing role of the cross proxy in a constrained network, and as a further step toward realizing efficient communication between web browsers

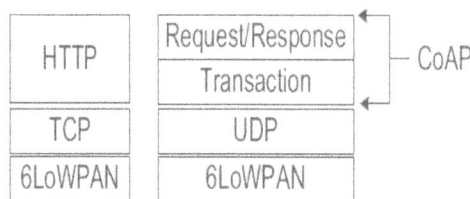

FIGURE 13.8   HTTP and CoAP protocol stack.

and smart objects, proxy functionalities may be extended to support more advanced cross-protocol interactions, such as:

1. concurrent tunneling of IPv4 in IPv6 combined with the HTTP-CoAP mapping,
2. the mapping of unicast HTTP requests to multicast CoAP messages, and the consequent aggregation of multiple responses in a single HTTP response payload and
3. the establishment of an observe relationship through a proxy using HTTP bidirectional technique (Figure 13.9)

Table 13.3 illustrates comparison of hybrid approaches. It contains three columns, namely Hybrid Approach (implementation), CoAP-REST (web things) and MQTT-REST (QEST broker) with their details.

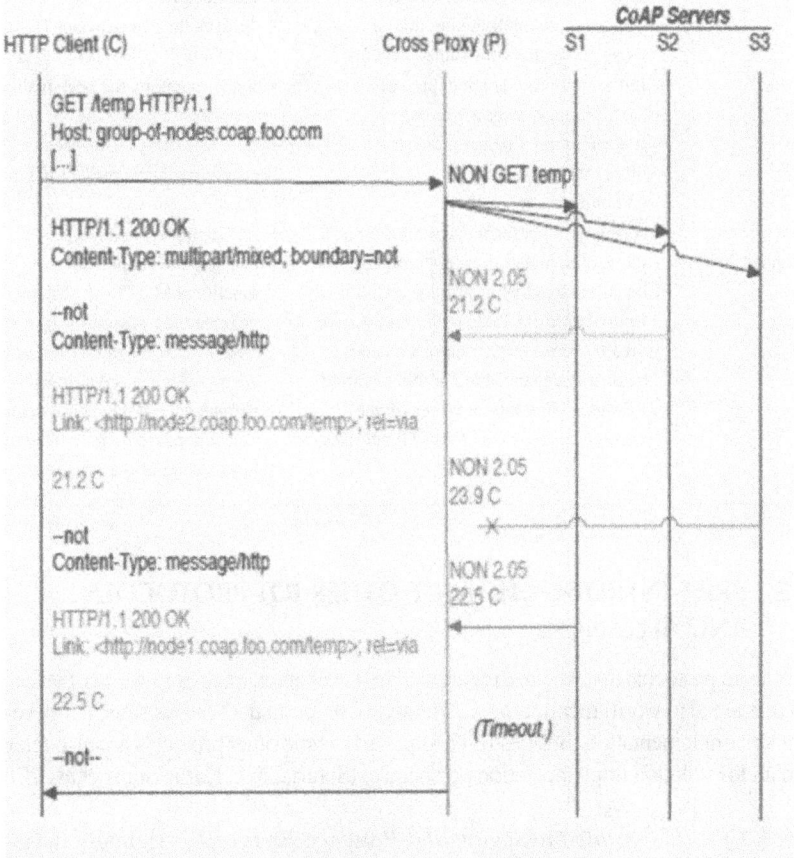

FIGURE 13.9 An example of mapping of unicast HTTP requests to multicast CoAP messages.

TABLE 13.3
IoT Hybrid Approaches Comparison

Hybrid Approach	CoAP-REST	MQTT-REST
Implementation	Web things	QEST broker
Features	• Based on HTTP CoAP process protocol proxy	• It modifies the broker semantics to retain the last payload on a topic.
	• Concurrent tunneling of IPv4 in IPv6 combined with the HTTP-CoAP mapping. The mapping of unicast HTTP requests to multicast CoAP messages and the consequent aggregation of multiple responses in a single HTTP response.	• It exposes the payload as a REST resource.
Advantage	• It can be implemented either using protocol-aware access or protocol agnostic-access (*Online*, n.d.-a).	• It can implement both "public-subscribe" and "request/response" architecture.
	• Can leverage multiplexing role of cross-proxy in a constrained network.	• Bridges the gap between Things and Web.
Drawbacks	• It does not provide a mapping of all the CoAP request/response code.	• It only supports the best quality of services
	• It does not provide support for observation and multicast communication.	• It lacks some peer-to-peer capability that is present in other implementations.
Future enhancements	• A seamless platform is required for a range of application layers protocols to be pulled together to provide a UNIX/Linux-based HTTP-CoAP, CoAP-CoAP proxy. A cross-proxy approach needs to be further enhanced to harness a unified Resource Directory or client-server agent exposing a CoAP-HTTP interface.	• QEST brokers need to be expanded to support fully functional MQTT. MQTT-SN can be integrated as well to support sensor networks (a realization of energy-efficient scenario using permanent TCP channel). Security aspect needs to be integrated to form a state-of-the-art system.

13.5 BRIEF INTRODUCTION OF OTHER IOT PROTOCOLS AND ALLIANCES

Apart from protocols that we have mentioned in this chapter, there are several other protocols that are also worth mentioning. Currently these are under various stages of development and implementation. Besides IETF standards, some other protocols are also relevant from an IoT solution implementation perspective (Sutaria, n.d.; Karagiannis et al., 2015):

a. *XMPP (Extensible Messaging and Presence Protocol)*– originally developed by the name "Jabber" it has roots in instant messaging and is a contender for mass scale management of consumer white goods, such as washers, dryers, refrigerators, and so on. It uses XML text format as its native type, making person-to-person communication a natural process. As it is based on a persistent TCP connection and lacks an efficient binary encoding, it's typically not been practical over LLNs (Low-power and Lossy Networks). But the recent work of XEP-0322, XEP-323, and XEP-324 aim to make XMPP suited for IoT. Finally it is considered to be excessively heavy.

b. *WebSocket* – WebSocket is a protocol that provides full-duplex communication over a single TCP connection between client and server. It is part of the HTML 5 specification. The WebSocket standard simplifies much of the complexity around bi-directional Web communication and connection management. WebSocket is neither a request/response nor a publish/subscribe protocol. In WebSocket a client initializes a handshake with a server to establish a WebSocket session. The handshake itself is similar to HTTP so that web servers can handle WebSocket sessions as well as HTTP connections through the same port.

c. *AMQP (Advanced Message Queuing Protocol)* – An open standard application layer protocol for message-oriented middleware. It provides asynchronous publish/subscribe communication with messaging. Its main advantage is its store-and-forward feature that ensures reliability even after network disruptions. The defining features of AMQP are message orientation, queuing, routing (including point-to-point and publish-and-subscribe), reliability and security.

d. *DDS (Data-Distribution Service for Real-Time Systems)* – The first open international middleware standard directly addressing publish-subscribe communications for real-time and embedded systems. It has problems in terms of scalability and dependence on various versions.

e. *Z-Wave* – The Z-Wave protocol is an interoperable, wireless, based on RF-communications technology designed specifically for control, monitoring, and status reading applications in residential and light commercial environments. Low Powered RF communications technology that supports full mesh networks without the need for a coordinator node Operates in the sub-1GHz band; impervious to interference from Wi-Fi and other wireless technologies in the 2.4-GHz range (Bluetooth, ZigBee, etc.).It is designed specifically for control and status apps, supports data rates of up to 100kbps, with AES128 encryption, IPV6, and multi-channel operation. The Z-Wave PHY and MAC layers are defined by ITU-T Recommendation G.9959.The frequencies used by Z-Wave are listed in Z-Wave Alliance Recommendation ZAD12837, "Z-Wave transceivers – Specification of Spectrum Related Components." It has full interoperability through layer 5 with backward compatibility to all versions. Shares the same position in the NIST/SGIP Catalog of Standards as the IEEE 802.11 and 802.15 and 802.16 families.

f. *ZigBee Alliance* – ZigBee (XBee) provides a set of application profiles for creating low-rate wireless mesh networks, which have been built upon the 802.15.4-2003 standard. While ZigBee is not directly comparable to IEFT standards like 6LoWPAN, it has been extensively implemented in small-scale ad-hoc networking smart-home and smart objects–related applications.

g. *DASH 7 Alliance* – It provides a unique solution as compared to standards it is a bidirectional, low-bandwidth, low-power, and long-range system targeted for objects sending sporadic and compact data. It is a tag-to-tag communication and operates at the 433 MHz frequency range. Typically it is useful in communication ranging up to 1 KM and this range reduces as the data transfer bit rate is increased and vice versa.

h. *BACnet* – BACnet, as per definition, is a "Data Communication Protocol for Building Automation and Control Networks." It was developed by the American Society of Heating, Refrigerating and Air-Conditioning Engineers (ASHRAE). It was standardized in 1995 and became an ISO standard in 2003. It is essentially used in HVAC systems (heating, ventilation, and air-conditioning), lighting control, access control, etc.

i. *LoRaWAN* – A group of computer and networking firms has formed an alliance to standardize the use of low-power wide-area networks, or LAPWANs, to drive the development of the Internet of Things (IoT), machine-to-machine, and smart city applications. The LoRa Alliance is dedicated to using protocols derived from LAPWAN, including LoRaWAN, to ensure interoperability of IoT applications between telecom operators and other firms who have joined the effort.

j. *LWM2M (Lightweight M2M)* – Lightweight M2M (LWM2M) is a system standard in the Open Mobile Alliance. It is based on technologies such as DTLS, CoAP, Block, Observe, SenML, and Resource Directory and combines them into a device-server interface along with an Object structure.

13.6 CONCLUSION AND FUTURE SCOPE

The Internet has brought a new era of computing; it has revolutionized how people interact with each other and work together. It has ushered in a pragmatic change of low-cost information retrieval methodology for everyone, which has the capacity to transform lives in ways that were hard to imagine in the early stages. The next wave of the Internet is not people centric; it's about intelligent and connected devices; these devices are known to be "smart objects." The IoT has proved its existence in our lives.

There is a range of IoT application layer protocols and more are emerging day by day that will support the next generation of IoT. The protocols/technologies that are in focus in this chapter include MQTT, MQTT-SN REST, and CoAP, each of which can be used to connect devices in a distributed network supporting the IoT realm. We have to also consider their suitability to support the different operational use-cases, which may include Inter- and Intra-Device communication, Device-to-Cloud communication, and Inter-Data Center communication. We need to take a measured approach while solving system functionalities such as performance, Quality-of-Service, reliability, interoperability, fault-tolerance, and security aspects.

We realize that there is a necessity for a common medium/framework to support the interaction between machines/smart objects and to allow new forms of applications between smart objects and end-users. We discussed the requirements for the two types of interactions, and we see that the state-of-art of communication protocols does not support all of them.

To solve these dilemmas we foresee that there is a requirement for common gateway architecture that can support the functionalities of CoAP, MQTT, REST, etc., so that it can bridge Things and Web in a seamless manner. Finally, we plan to integrate MQTT-SN with a QEST broker that can support sensor networks in IoT infrastructure.

REFERENCES

Alshohoumi, F., Sarrab, M., Alhamadani, A., & Al-abri, D. (2019). Systematic Review of Existing IoT Architectures Security and Privacy Issues and Concerns. *International Journal of Advanced Computer Science and Applications 10*(7), 232–251.

Atlam, H. F., Walters, R. J., & Wills, G. B. (2018). Internet of Things: State-of-the-art, Challenges, Applications, and Open Issues. *International Journal of Intelligent Computing Research 9*(3), 928–938.

Choubey, D. K., Kumar, A., Solutions, C. T., Srivastava, K., & Pahari, S. (2020). Notification and Image Analysis in Cloud. 1–5. https://doi.org/10.1109/ic-ETITE47903.2020.296

Colitti, W., Steenhaut, K., & Caro, N. De. (2011). REST Enabled Wireless Sensor Networks for Seamless Integration with Web Applications. https://doi.org/10.1109/MASS.2011.102

Collina, M., Corazza, G. E., & Vanelli-coralli, A. (2012). Introducing the QEST broker: Scaling the IoT by bridging MQTT and REST. *2012 IEEE 23rd International Symposium on Personal, Indoor and Mobile Radio Communications - (PIMRC)*, pp. 36–41, doi: 10.1109/PIMRC.2012.6362813.

Datta, P., & Sharma, B. (2017). A survey on IoT architectures, protocols, security and smart city based applications. *2017 8th International Conference on Computing, Communication and Networking Technologies (ICCCNT)*, 1–5.

Dawson-Haggerty, S., Jiang, X., Tolle, G., Ortiz, J., & Culler, D. (2010). sMAP – A Simple Measurement and Actuation Profile for Physical Information. 197–210. http://sensys.acm.org/2010/Papers/p197-Dawson-Haggerty.pdf

Dunkels, A., Gronvall, B., & Voigt, T. (2004). Contiki - a lightweight and flexible operating system for tiny networked sensors. *29th Annual IEEE International Conference on Local Computer Networks*, 455–462.

Hunkeler, U., Truong, H. L., & Stanford-clark, A. (n.d.). *MQTT-S – A Publish/Subscribe Protocol for Wireless Sensor Networks.*

Karagiannis, V., Chatzimisios, P., Vazquez-gallego, F., & Alonso-zarate, J. (2015). A Survey on Application Layer Protocols for the Internet of Things Research Motivation. 1–10. Transaction on IoT and Cloud computing, 3(1), 11–17.

Khanna, A., & Kaur, S. (2020). Internet of Things (IoT), Applications and Challenges. *Wireless Personal Communications 114*(2). https://doi.org/10.1007/s11277-020-07446-4

Kovatsch, M., Weiss, M., & Guinard, D. (2010). *Embedding Internet Technology for Home Automation.*

Mayer, S., Guinard, D., & Trifa, V. (2010). Facilitating the integration and interaction of real-world services for the web of things. *Proceedings of Urban Internet of Things-Towards Programmable Real-Time Cities (UrbanIoT).*

Mrabet, H., Belguith, S., Alhomoud, A., & Jemai, A. (2020). A Survey of IoT Security Based on a Layered Architecture of Sensing and Data Analysis. *Sensors 20*(13), 3625.

Online. (n.d.-a). http://datatracker.ietf.org/doc/rfc7252/

Online. (n.d.-b). www.prismtech.com

Pathak, A. D., & Tembhurne, J. V. (2018). Internet of Things: A Survey on IoT Protocols. 483–487. https://papers.ssrn.com/sol3/papers.cfm?abstract_id=3168575

Ray, P. P. (2018). A Survey on Internet of Things Architectures. *Journal of King Saud University – Computer and Information Sciences 30*(3), 291–319. https://doi.org/10.1016/j.jksuci.2016.10.003

Sanfilippo, S. (2012). *Redis.* http://redis.io

Sethi, P., & Sarangi, S. R. (2017). *Internet of Things : Architectures, Protocols, and Applications.*

Shelby, Z., Hartke, K., Bormann, C., & Frank, B. (2013). Constrained Application Protocol (CoAP). draft-ietfcore-coap-12.

Srinivasa, A. H. (2019). A Comprehensive Study of Architecture, Protocols and Enabling Applications in Internet of Things (IoT). *8*(11). http://www.ijstr.org/final-print/nov2019/A-Comprehensive-Study-Of-Architecture-Protocols-And-Enabling-Applications-In-Internet-Of-Things-iot-.pdf

Sutaria, R. (n.d.). *Making sense of interoperability : Protocols and Standardization initiatives in IoT.* 2–5.

Website. (n.d.-a). http://www.electronicdesign.com

Website. (n.d.-b). http://www.hinrg.cs.jhu.edu

Yasser, S. (2014). Security, Privacy and Trust Challenges in Cloud Computing and Solutions. *July*, 34–40. https://doi.org/10.5815/ijcnis.2014.08.05

14 Performance Analysis of Distributed Mobility Protocol for Multi-Hop IoT Networks

Shankar K. Ghosh
Indian Statistical Institute
Kolkata, India

Babul P. Tewari
Indian Institute of Information Technology
Bhagalpur, India

CONTENTS

14.1 INTRODUCTION

The next-generation wireless system is envisioned as a common platform where all devices will be IP enabled and independently connected to the Internet. Such devices include laptops, smartphones, television, refrigerator, sensor devices, radio frequency tags, and all other essential commodities. Such enhancements allow the end terminals (ETs) to remain inter-connected all the time through the Internet. The interconnected ETs may communicate with each other for the exchange of data of their relevant interest using wireless protocols. Such an all-IP network-based integrated architecture is referred to as the Internet of Things (IoT) [1].

Recently, IoT has got much attention because of its numerous applications in both indoor and outdoor environments. Applications of IoT include intelligent transport systems, security, and other related areas. Among different application areas, e-health remains one of the popular extended concerns of IoT. E-health addresses a health care system with electronic connectivity and processing. With the help of smart wearable devices, patients' health data like heartbeat, blood pressure, and glucose level can be collected using sensors and can be communicated to the e-health server over the Internet [2, 3]. A thorough experimental simulation of how respiration rate monitoring can be done with the help of an analog sensor has been represented in Ref. [4]. The influence of IoT and health applications in the context of 5G communication networks has been addressed in Ref. [5]. IoT applications toward smart health care have been addressed in Ref. [6]. Ref. [7] addresses the security and the privacy aspects of IoT applications in health care. A distributed approach for maintaining data confidentiality in IoT health care has been addressed in Ref. [8]. Security aspects of body area networks with clinical health care have been addressed in Ref. [9]. An architecture for smart health monitoring using the smartphone has been addressed in Ref. [10]. In this context, the role of the 5G network has been demonstrated as it causes higher bandwidth and lower delay. A wellness monitoring system has been proposed in Ref. [11] addressing human–computer interaction. The use of smartphones technology in e-health monitoring has been addressed in Ref. [12]. An environment-specific wellness state measurement of a person using his heart rate variability has been addressed in Ref. [13]. The authors have addressed the choice of a good neighborhood with the help of such a system. As housing and mental health both are strongly correlated [14]. In a similar context the wellness monitoring for the elderly people has been addressed in Ref. [15].

Therefore, it is evident that e-health service facilitates the patients in understanding their health status and in fixing their appointment for the next medical check-up. More specifically, IoT e-health has the following advantages: First, with the use of simple smart devices it can be applied to all classes of people. Second, IoT e-health allows a wide variety of technologies to work together seamlessly without much worrying about its complexity. Third, IoT e-health allows the processing of large-scale health data within a fairly reasonable time. Fourth, patients can monitor their health condition easily as they need not to bother with any complex integrated technology. Also the health monitoring becomes cost effective. Finally, IoT e-health allows a wide collaboration among medical professionals.

IoT-based e-health applications are usually delay-sensitive [16]. The end-to-end delay in an all-IP network in turn depends on the mobility management protocol used to provide ubiquitous connectivity [17]. To manage mobility in all-IP network, several protocols have been proposed in recent past [18]. Among the existing mobility management protocols, mobile IPv6 (MIPv6)-based protocols such as hierarchal mobile IPv6 (HMIPv6) and fast mobile IPv6 (FMIPv6) are widely known. It may be noted that the abovementioned MIPv6 protocols are host based, i.e. the mobile terminal need to perform all the mobility-related signaling. Such protocols require protocol stack modification at the MTs, which are quite unexpected for low-capacity IoT devices. To address this drawback, a proxy mobile IPv6 protocol was proposed. The operation of PMIPv6 is network based, i.e. the network performs all the mobility-related signaling and thereby reduces the overhead on ETs. However, the PMIPv6 protocol often suffers from a high end-to-end delay due to the redirection mechanism through the mobility anchor. Such constraint limits the performance of the delay-sensitive e-health services. In Ref. [19], a PMIPv6-based distributed mobility management (DMM) protocol has been proposed, which eliminates the necessity to redirect the packets through the mobility anchor, as the access routers (ARs) in DMM can perform mobility-related functionalities such as forwarding the ongoing sessions from one AR to another after handover. It may also be noted that apart from the terrestrial delay, the end-to-end delay also depends on the number of successive wireless hops that the packet needs to traverse to reach the ETs. Multi-hop wireless environments are particularly relevant in disastrous scenarios, where the cellular deployment is destroyed by natural catastrophe and the rescue team relies on low-power IoT devices to provide services in remote locations.

Investigating the performance of DMM protocol in multi-hop IoT scenarios has been quite limited in the preceding literature. In Ref. [20], an extension of onboard Transmission Control Protocol (obTCP) has been proposed for satellite-terrestrial hybrid networks, which consist of multiple wireless hops. In Ref. [21], handover performances of MIPv6 and seamless IP diversity-based generalized mobility architecture (SIGMA) have been analyzed. This analysis considers several system parameters including user mobility, handover decision metric, and traffic arrival rate. In Ref. [22], performance analyses of different IPv6-based mobility management protocols have been carried out in terms of handover latency and handover packet loss rate. In this analysis, layer 2 parameter details have not been considered. These analyses [20–22] mainly focus on mobility protocols having centralized architecture and do not explore the effectiveness of distributed characteristics. In Ref. [19], three distributed mobility solutions, namely PMIPv6 based, software-defined network (SDN) based, and routing-based solutions, have been presented and evaluated. In Ref. [23], the handover performance of DMM has been analyzed and compared with MIPv6 and SIGMA considering an ultra-dense network scenario. It may be noted that these analyses [19, 23] do not consider the challenges associated with the multi-hop wireless scenario. In this work, our objective is to analyze the effectiveness of the DMM protocol for a multi-hop IoT network in terms of end-to-end delay and throughput. We are particularly interested in investigating the performances when either number of wireless hops or packet loss probabilities in wireless hops are very high, which is very pertinent in disastrous scenarios. Our contributions can be summarized as follows.

We develop a performance model for both DMM and PMIPv6 considering a multi-hop wireless network scenario. Our proposed performance models explicitly consider packet loss probabilities at wireless hops, the number of wireless hops, and the path of packet routing incurred by the concerned mobility management protocol. In our performance model, we consider end-to-end delay and throughput as performance evaluation metrics.

Analyses based on our performance model reveal that the effectiveness of distributed characteristics can be exploited as long as the number of wireless hops and packet loss probabilities at wireless hops are low.

The rest of the manuscript is organized as follows. Section 14.2 describes a brief overview of PMIPv6 and DMM, followed by our proposed performance models in Section 14.3. Section 14.4 presents numerical results and discussions. Finally, Section 14.5 concludes the chapter.

14.2   A BRIEF DESCRIPTION OF PMIPV6 AND DMM

Increased mobile data traffic with their smart application requirements of high data rate is putting significant constraints on the performance of future generation networks. This is mainly due to the increased load on the existing core network (CN) [24]. Providing seamless network connectivity, particularly in heterogeneous networking scenarios such as high-speed 802.11n with Long Term Evolution (LTE) is also an important concern. To address the service continuity in such context centralized mobility support solutions for all-IP networks have been proposed by Internet Engineering Task Force (IETF) [25]. PMIPv6 is one such centralized mobility management protocol to manage the mobility that basically introduces a Local Mobile Anchor (LMA) [26]. The centralized approach, however, suffers from the issue of non-optimal routing, high handover latency, and scalability problems [25]. Furthermore, there is a considerable chance of single-point failure [27]. In order to resolve such an issue DMM aims to provide a valid architectural support that takes care of the increased traffic load of the CN [25, 28]. The major objective of DMM is to offload the CN by allowing different gateways for the mobile traffic depending on their connectivity. Thus, DMM becomes a suitable approach to mobility management for the fifth generation (5G) network even with dense networking scenarios of heterogeneous types [19]. The basic idea of DMM is to provide optimum routing support by distributing the mobility anchors close to the users [29].

14.3   PERFORMANCE ANALYSIS

In this section, we develop the performance model for PMIPv6 and DMM protocol. We consider a multi-hop wireless IoT scenario as shown in Figure 14.1. Here an IoT device communicates to the AR through other IoT devices. The AR in turn sends the packet to the correspondent node (CN) over the Internet. However, the path over which the packets are communicated from AR to CN explicitly depends on the concerned mobility management protocol. In the subsequent subsections, we derive the analytical expressions for end-to-end packet loss probability, end-to-end delay, and throughput. Here we employ the commonly used derivation techniques as presented

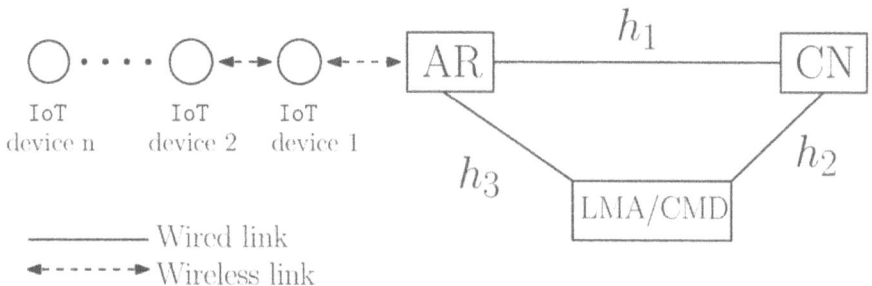

FIGURE 14.1   Considered system model.

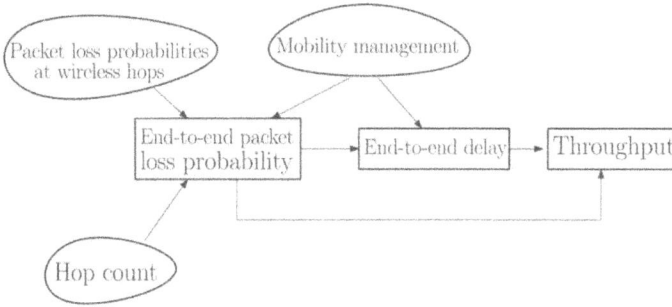

FIGURE 14.2   Block diagram of the proposed framework.

in Refs [20, 30]. The difference with the previous works lies in the underlying application scenario and analyzed mobility management protocols. The list of variables used in our analyses is depicted in Table 14.1.

14.3.1 END-TO-END PACKET LOSS PROBABILITY

End-to-end packet loss probability is defined as the probability that a packet is lost due to wireless transmission error while traveling from the IoT device to the CN. In the PMIPv6 protocol, packets are routed from AR to the CN through the local mobility anchor (LMA), i.e. the packet needs to travel a total of $h_2 + h_3$ hops. Hence, the probability that the packet will be successfully delivered to the CN while being sent from AR is $(1 - X)^{h_2 + h_3}$. On the other hand, considering the kth IoT device as the ET, the probability that a packet sent from the ET is successfully delivered to the AR

TABLE 14.1
List of Variables

Notation	Illustration
h_1	Number of hops between CN and AR.
h_2	Number of hops between CN and LMA/CMD.
h_3	Number of hops between LMA and AR.
X	Packet loss probability in wired links.
p_i	Packet loss probability in ith wireless link.
t_T	Transmission delay at each terrestrial hop.
t_p	Propagation delay at each terrestrial hop.
t_{ed}	Encapsulation-decapsulation delay for PMIPv6 protocol.
t_{pd}	Propagation delay at wireless links.
P_{end}^1	End to end packet loss probability for PMIPv6.
P_{end}^2	End-to-end packet loss probability for DMM.
D^1	End-to-end delay for PMIPv6.
D^2	End-to-end delay for DMM.

is $\prod_{i=1}^{k}(1-p_i)$. Hence, the expression for the end-to-end packet loss probability P_{end}^1 can be derived as:

$$P_{end}^1 = 1-(1-X)^{h_2+h_3} * \prod(1-p_i) \tag{14.1}$$

In DMM protocol, packets are sent from AR to CN through generalized IP routing. Here the control mobility database (CMD) just acts as a control entity to recover IP flows after handover. While traveling from AR to CN, the packet needs to travel just h_1 hops. Communication from the ET to the AR is similar to that of PMIPv6. Hence, the end-to-end packet loss probability P_{end}^2 for DMM protocol can be written as:

$$P_{end}^2 = 1-(1-X)^{h_1} * \prod(1-p_i) \tag{14.2}$$

14.3.2 END-TO-END DELAY

The end-to-end delay is defined as the expected delay required to successfully deliver a packet to the CN sent from the ET. The delay experienced in each terrestrial hop can be computed as $t_T + t_p + t_{pd}$. In PMIPv6 protocol, a packet needs to travel a total of $h_3 + h_2$ hops. Hence, the total delay experienced in terrestrial network, i.e. AR to CN, is $(t_T + t_p + t_{pd})*(h_3 + h_2)$.

Now, the expected number of transmissions in each wireless hop can be computed as $\frac{1}{1-p_i}$. Hence, the expected time spent in retransmission in each wireless hop is $\frac{t_p'}{1-p_i}$. Finally, successful transmission requires t_p' time. So, the expected transmission delay at each wireless hop is $t_p' + \frac{t_p'}{1-p_i} = t_p' * \left(\frac{2-p_i}{1-p_i}\right)$. In PMIPv6 protocol, a tunneling delay is associated between AR and LMA, i.e. $2t_{ed}$. Hence, the end-to-end delay D^1 for PMIPv6 protocol can be written as:

$$D^1 = (t_T + t_p + t_{pd})*(h_2 + h_3) + 2t_{ed} + \sum_{i=1}^{k}\left(\frac{2-p_i}{1-p_i}\right) \tag{14.3}$$

In DMM protocol, a packet needs to travel a total of h_1 hops. Communication in wireless hops is identical to that of PMIPv6. Hence, the end-to-end delay D_2 for DMM protocol can be written as:

$$D^2 = (t_T + t_p + t_{pd})*h_1 + \sum_{i=1}^{k}\left(\frac{2-p_i}{1-p_i}\right) \tag{14.4}$$

14.3.3 THROUGHPUT

Throughput is defined as the mean data rate achieved by the CN in our considered scenario. Following the well-known formulation as presented in Refs [20, 30] and [21], the expression for throughput for PMIPv6 protocol g^1 can be written as:

$$g^1 = \frac{s}{RTT} * \frac{C}{(P_{end}^1)^{\frac{1}{2}}} = \frac{s}{2*D^1} * \frac{C}{(P_{end}^1)^{\frac{1}{2}}} \tag{14.5}$$

where the round trip time (RTT) can be computed as twice the end-to-end delay D_1. Here C is a predefined constant. Similarly, the expression for throughput for DMM protocol g^2 can be written as:

$$g^2 = \frac{s}{RTT} * \frac{C}{\left(P_{end}^2\right)^{\frac{1}{2}}} = \frac{s}{2 * D^1} * \frac{C}{\left(P_{end}^2\right)^{\frac{1}{2}}} \tag{14.6}$$

14.4 RESULTS AND DISCUSSIONS

To carry out the numerical analysis, we adopted the parameter values used in Refs [20, 30] and [21]. The considered parameter values are as follows: $h_1 = h_2 = 12$, $h_3 = 5$, $X = 0.001$, terrestrial link bandwidth has been set to 100 Mbps, propagation delay p at both terrestrial and wireless links has been set to 5 ms, $t_{ed} = 5$ ms, $s = 2$ KB, and $C = (1.5)^{\frac{1}{2}}$.

Figure 14.3 depicts the effect of wireless link conditions on end-to-end packet loss probability. The result has been depicted for different hop counts. Here the packet loss probability in wireless links varies from 0.1 to 0.3 with a step of 0.15. For the sake of simplicity and tractability, we assume that packet loss probabilities in all of the wireless links are equal, i.e. all p_is are equal ($1 \le i \le 5$). The result shows that the end-to-end packet loss probability monotonically increases with increasing packet loss probability at the wireless links up to a threshold (0. 25 in Figure 14.3). After that threshold, the packet loss probabilities at wireless links do not have much effect

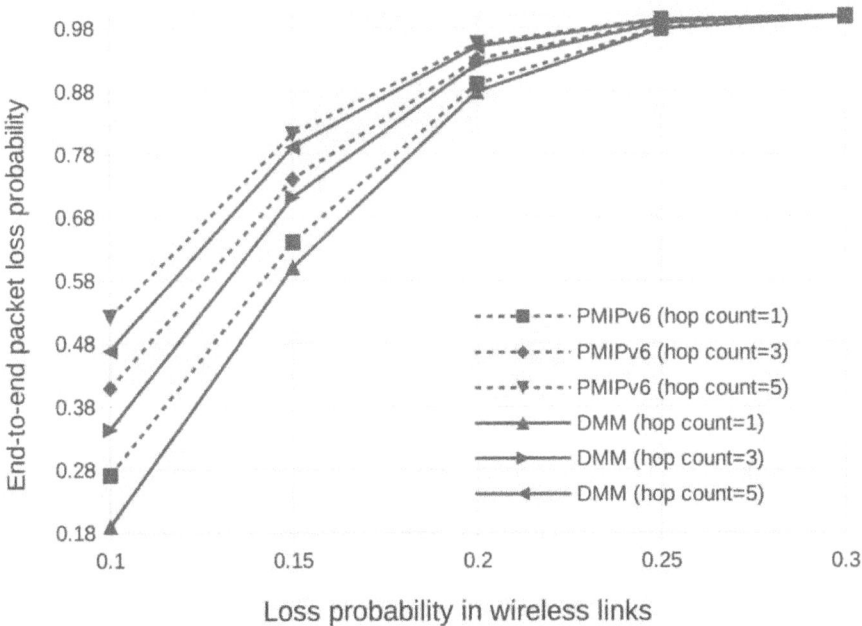

FIGURE 14.3 Effect of wireless link condition on the end-to-end packet loss rate.

on end-to-end packet loss probability. This is quite expected as the rise in packet loss probability in successive wireless links results in higher losses when the packets are traveling between the MT and the CN. As the packet loss probability at wireless links increases beyond a threshold, communication at wireless hops become almost ineffective. As a result, the end-to-end packet loss probability shows a steady trend. It may be noted that the DMM protocol outperforms the PMIPv6 protocol in terms of end-to-end packet loss probability. This is because in PMIPv6 the packets need to traverse a higher number of terrestrial hops compared to DMM. This results in higher packet losses. The performance gain of DMM over PMIPv6 decreases as the packet loss probability at wireless links increase. This is because the end-to-end loss probability is dominated by the wireless links, as the packet loss probability in wireless hops is comparatively higher than the terrestrial links.

Figure 14.4 depicts the effect of wireless link condition on end-to-end delay. Here the packet loss probabilities at wireless links vary from 0.1 to 0.5 with a step of 0.1. Here the hop count has been kept fixed to 1. Results show that the end-to-end delay monotonically increases with increasing packet loss probabilities at wireless links for DMM and PMIPv6. It may also be noted that the DMM protocol significantly outperforms the PMIPv6 protocol. The reasons behind this are as follows. In the PMIPv6 protocol, packets communicated by the ET to the CN need to traverse through the LMA. Due to such redirection, the end-to-end delay in PMIPv6 is high. On the other hand, in DMM protocol, packets sent by the ET can reach the CN through generalized IP routing without any redirection. As a result, the end-to-end delay in DMM is lower computer to that of PMIPv6 protocol. Since losses at wireless links trigger retransmission of packets at corresponding links resulting in higher

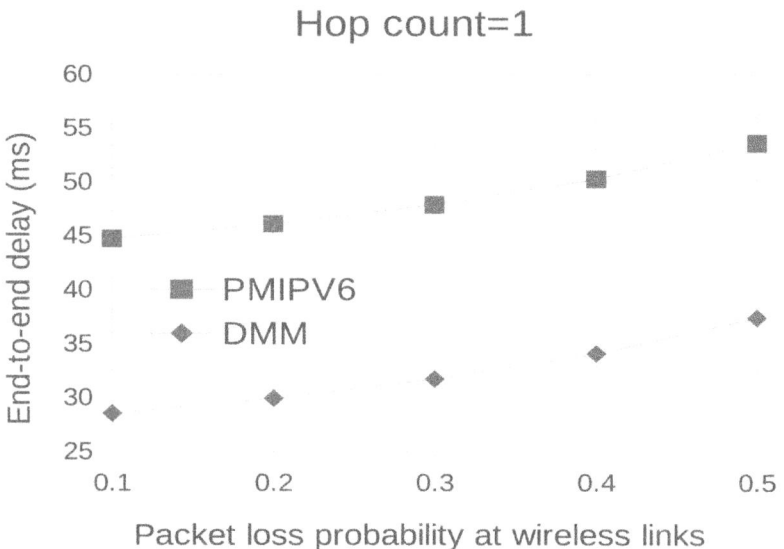

FIGURE 14.4 Effect of wireless link condition on end-to-end delay.

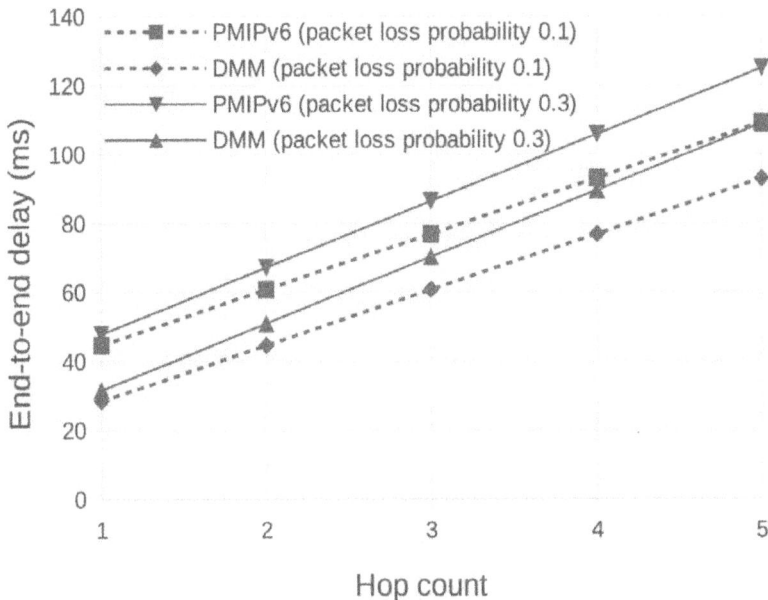

FIGURE 14.5 Effect of wireless hop count on end-to-end delay.

delay at the concerned wireless hops, the end-to-end delay increases with increasing packet loss probabilities.

Figure 14.5 depicts the effect of wireless hop count on end-to-end delay. Here the number of hop count varies from 1 to 5 with a step of *unity*, while packet loss probabilities at each wireless link have been kept fixed to 0.1. The results show that the end-to-end delay monotonically increases with an increasing hop count. Furthermore, the DMM protocol outperforms the PMIPv6 protocol. The reasons are similar to those described previously.

Figure 14.6 depicts the effect of wireless hop counts on throughput. The results show that the DMM protocol significantly outperforms the PMIPv6 protocol in terms of throughput. However, the performance gain of DMM reduces with an increasing hop count. The reasons behind this are as follows. With increasing hop count both the end-to-end delay and end-to-end packet loss probability increase, as explained in Figures 14.3 and 14.5. As the user throughput is *inversely proportional* to the end-to-end delay and square root of end-to-end packet loss probability, the user throughput decreases with an increasing hop count. Since the performance gain of DMM reduces with increasing hop count in terms of delay and packet loss, a similar effect is seen for throughput as well.

From these results, we conclude that the distributed characteristics of DMM can be exploited in multi-hop IoT network as long as the number of wireless hops is low. The effectiveness of distributed characteristics reduces as the number of hop count increases.

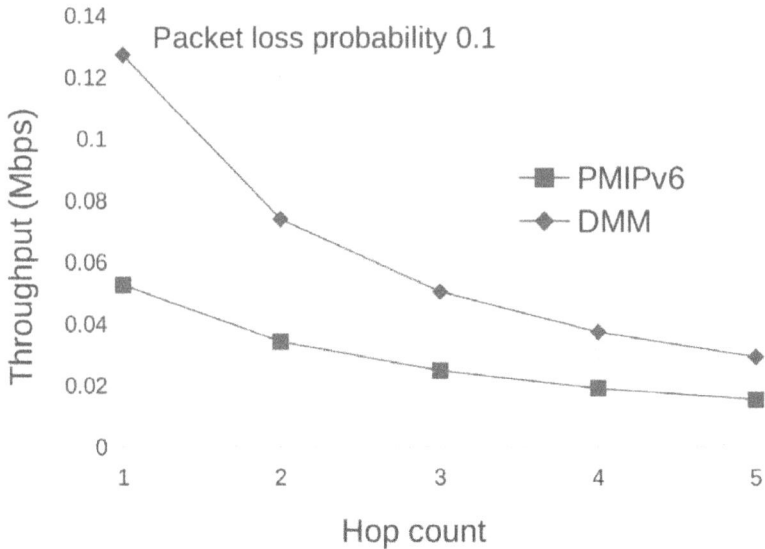

FIGURE 14.6   Effect of wireless hop count on throughput.

14.5   CONCLUSIONS AND FUTURE WORKS

In this work, we have analyzed the performance of the DMM protocol in a multi-hop wireless IoT network in terms of end-to-end delay and throughput. Such study is particularly relevant in the context of e-health services. From our study we conclude that deployment of multiple IoT devices may not always be beneficial for e-health services as long as DMM is used for mobility management. In fact, the effectiveness of distributed characteristics reduces as the number of hop count increases. Our future research scope includes the following:

- Analyzing the performances of routing-based and software-defined radio-based distributed protocols in the considered disastrous scenario. The proposed model will be validated against extensive system-level simulations. The results obtained from such analysis can be used to develop a new mobility management protocol for e-health applications.
- The situation becomes particularly challenging if the IoT devices are moving randomly. We aim to develop a decision algorithm to decide the number of hops that will be optimal for communication. To develop such an algorithm, first we will analyze the end-to-end throughput considering the mobility pattern of the IoT devices, traffic pattern, and data rate requirement of e-health applications. To capture the effect of mobility, we plan to use the properties of the Erdos–Renyi graph.
- The massive IoT network is expected to generate a huge amount of traffic. To deal with the traffic demand, our aim is to develop reinforcement learning–based resource allocation algorithms for optimal usage of bandwidth and

resources. We are particularly interested in developing value approximation methods such as deep Q learning to develop resource allocation algorithms. The performance of the developed algorithms will be validated against testbed experiments.

REFERENCES

1. Atzori, L., Iera, A., Morabito, G. "The internet of things: A survey," Computer Networks, Vol. 54, pp. 2787–2805, 2010.
2. Farahani, B., Firouzi, F., Chang, V., Badaroglu, M., Constant, N., Mankodiya, K. "Towards fog-driven IoT eHealth: Promises and challenges of IoT in medicine and healthcare," Future Generation Computer Systems, Vol. 78, pp. 659–676, 2018.
3. Chiuchisan, I., Chiuchisan, I., Dimian, M. Internet of Things for e-Health: An approach to medical applications. In Proceedings of 2015 International Workshop on Computational Intelligence for Multimedia Understanding (IWCIM), pp. 1–5, 2015.
4. Aileni, R. M., Pasca, S., Strungaru, R., Valderrama, C. Biomedical signal acquisition for respiration monitoring by flexible analog wearable sensors. E-Health and Bioengineering Conference (EHB) 2017, pp. 81–84, 2017.
5. Brito, J. M. C. Trends in wireless communications towards 5G networks? The influence of e-health and IoT applications. Computer and Energy Science (SpliTech) International Multidisciplinary Conference, pp. 1–7, 2016.
6. Firouzi, F. Farahani, B., Ibrahim, M., Chakrabarty, K. "Keynote paper: From EDA to IoT eHealth: Promises challenges and solutions," IEEE Transactions on Computer-Aided Design of Integrated Circuits and Systems, Vol. 37, no. 12, pp. 2965–2978, 2018.
7. Soufiene, B. O., Bahattab, A. A., Trad, A., Youssef, H. RESDA: Robust and Efficient Secure Data Aggregation Scheme in Healthcare Using the IoT. Internet of Things Embedded Systems and Communications (IINTEC) 2019 International Conference, pp. 209–213, 2019.
8. Yang, Y., Zheng, X., Tang, C. "Lightweight distributed secure data management system for health internet of things," Journal of Network and Computer Applications, Vol. 89, p. 26, 2017.
9. Esmaeili, S., Tabbakh, S. R. K, Shakeri, H. "A priority-aware lightweight secure sensing model for body area networks with clinical healthcare applications in internet of things," Pervasive and Mobile Computing, Vol. 69, p. 101265, 2020.
10. Lioret, J., Parra, L., Taha, M., Tomas, J. "An architecture and protocol for smart continuous eHealth monitoring using 5G," Computer Networks, Vol. 129, pp. 340–351, 2017.
11. García, L., Parra, L., Romero, O. et al. "System for monitoring the wellness state of people in domestic environments employing emoticon-based HCI," The Journal of Supercomputing, Vol. 75, pp. 1869–1893, 2019.
12. Parra, L., Sendra, S., Jiménez, J. M., Lloret, J. "Multimedia sensors embedded in smartphones for ambient assisted living and e-health," Multimedia Tools and Applications, Vol. 75, no. 21, pp. 13271–13297, 2016.
13. Laquesta, R., Garcia, L., Garcia-Magarino, I., Lloret, J. "System to recommend the best place to life based on wellness state of the user employing the heart rate variability," IEEE Access, Vol. 5, pp. 10594–10604, 2017.
14. Martin, C. J., Platt, S. D., Hunt, S. M. "Housing conditions and ill health," BMJ Clinical Research, Vol. 294, no. 6580, pp. 1125–1127, 1987.
15. Arshad, A., Khan, S., Alam, A. H. M. Z., Tasnim, R., Boby, R. I. Health and wellness monitoring of elderly people using intelligent sensing technique. International Conference on Computer and Communications Engineering, Kuala Lumpur, Malaysia, pp. 231–235, 2016.

16. Lv, Y., Dias, M. P. I., Ruan, L., Wong, E., Feng, Y., Jiang, N., & Qiu, K. "Request-based polling access: Investigation of novel wireless LAN MAC scheme for low-latency e-health applications," IEEE Communications Letters, Vol. 23, no. 5, pp. 896–899, 2019.
17. Fei, A., Pei, G., Liu, R., Zhang, L. Measurements on delay and Hop-count of the internet. IEEE Global Communications Conference (GLOBECOM), 1998.
18. Lee, J. H., Bonnin, J. M., You, I., Chung, T. M. "IEEE comparative handover performance analysis of IPv6 mobility management protocols," IEEE Transactions on Industrial Electronics, Vol. 60, no. 3, pp. 1077–1087, 2013.
19. Giust, F., Cominardi, L., Bernardos, C. J. "Distributed mobility management for future 5G networks: Overview and analysis of existing approaches," IEEE Communications Magazine, Vol. 53, pp. 142–149, 2015.
20. Ghosh, S. K., Kundu, P., Sardar, B., Saha, D. An extension of on-board TCP (obTCP) for satellite-terrestrial hybrid networks. Fourth International Conference of Emerging Applications of Information Technology, pp. 146–151, 2014.
21. Ghosh, S. K., Ghosh, S. C. An analytical framework for throughput analysis of real time applications in All-IP networks. IEEE 31st International Conference on Advanced Information Networking and Applications (AINA), pp. 508–515, 2017.
22. Lee, J. H., Bonnin, J. M., You, I., Chung, T. M. "Comparative handover performance analysis of Ipv6 mobility management protocols," IEEE Transactions on Industrial Electronics, Vol. 60, no. 3, pp. 1077–1088, 2013.
23. Ghosh, S. K., Ghosh, S. C. "Analyzing Handover Performances of Mobility Management Protocols in Ultra-dense Networks," Journal of Network and Systems Management, Vol. 28, pp. 1427–1452, 2020.
24. Panwar, N., Sharma, S., Singh, A. K. "A survey on 5G: The next generation of mobile communication," Physical Communications, Vol. 18, no. 2, pp. 64–84, 2016.
25. Carmona-Murillo, J., Soto, I., Rodriguez-Perez, F. J., Cortes-Polo, D., Gonzalez-Sanchez, J. L. "Performance evaluation of distributed mobility management protocols: Limitations and solutions for future mobile networks," Mobile Information Systems, Vol. 2017, pp. 1–15, 2017.
26. Gundavelli, S., Leung, K., Devarapalli, V., Chowdhury, K., B. Patil, Proxy mobile IPv6, RFC 5213, IETF, 2008.
27. Chan, H., Liu, D., Seite, P., Yokota, H., Korhonen, J. "Requirements for Distributed Mobility Management," IETF RFC 7333, August 2014.
28. Zuniga, J. C., Bernardos, C. J., De La Oliva, A., Melia, T., Costa, R., Reznik, A. "Distributed mobility management: A standards landscape," IEEE Communications Magazine, Vol. 51, no. 3, p. 8087, 2013.
29. Liu, D., Zuniga, J. C., Seite, P., Chan, H., Bernardos, C. J. "Distributed mobility management: Current practices and gap analysis," IETF RFC 7429, January 2015.
30. Chowdhury, P. K., Reaz, A. S., Atiquzzaman, M., Ivancic, W. Performance analysis of SINEMO: Seamless IP-diversity based network mobility. IEEE International Conference on Communications (ICC), pp. 6032–6037, 2007.

15 Comparative Analysis of Emotional State Classification Using Different Machine Learning Techniques

Chhotelal Kumar and Mukesh Kumar
NIT Patna
Patna, India

CONTENTS

15.1 INTRODUCTION

Emotion plays a very significant role in everyday communication. Previously, emotion was recognized through a variety of methods, including emotional words, text, vocal tone (i.e. speech), gestures, and facial expressions. Emotion can also be elicited by observation of vocabulary such as well-known words and non-verbally such as voice tone, but all of these can be easily hidden because of social fear. Even facial expression is useless, if a person is suffering from facial paralysis. Electroencephalogram (EEG)-based technology has historically been used in medical science applications and healthcare equipment. Nowadays, physiological signal (EEG) is used to detect emotion by researchers because it is more accurate and the individual has almost no control over it (Bos, 2006; Horlings, 2008; Murugappan, 2008; Li, 2009; Lin, 2009).

EEG is a recording of the electrical activity of the human brain impulses obtained by putting electrodes and some Leitmotiv devices on the brain's scalp surface (Teplan, 2002). Emotion recognition using EEG signals produces more precise emotions that can be used in a variety of fields such as healthcare, learning through the Internet, gaming, virtual worlds, marketing, monitoring, law, entertainment, cyber worlds, counseling, and human social interactions (Yoon and Park, 2007; Liu, 2011; Hondrou, 2012; Rached, 2013; Ali, 2016). The actual emotional state of patients in the assessment of diseases, especially those with expression issues, will allow the doctor to provide more adequate medical care. The patient's EEG signals are transmitted to the cloud via smart IoT devices where the processing of EEG signals is done and then will be sent to a cognitive module. By observing sensor readings such as EEG, the device determines the patient condition.

The cortex is the most important part of the human brain and is divided into four lobes (Ribas, 2010): (i) frontal lobe, (ii) temporal lobe, (iii) parietal lobe, and (iv) occipital lobe. The frontal lobe is a bigger and more developed part of the complex human brain that regulates essential cognitive abilities such as problem solving, judgment, emotional expressions, and memory. It is a portion of the cerebral region of the human brain. Both conscious and long-term memory are mostly created and preserved by the temporal lobe. It is essential for both object as well as language recognition and plays a role in interpreting and processing sensory information. The purpose of the parietal lobe is to process sensory information from various senses such as touch, cold, heat, and pain and it also manipulates the objects. The purpose of the occipital lobe is color determination, object recognition, face recognition, movement, memory information, etc. Figure 15.1 shows an image of a cortex.

Brain waves are generated by synchronized electrical pulses. Hertz is the unit of measurement for brain waves is defined as the number of cycles per second. These types of brain waves are constantly evolving. These are the foundations of our actions and emotions. The brainwave's frequencies can be divided into three categories: slow, moderate, and fast. Brainwave bands can be divided into five categories (Niedermeyer, 2005):

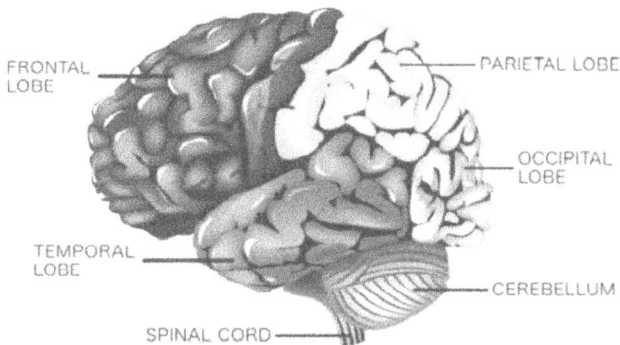

FIGURE 15.1   Cortex of the cerebrum.

(i) alpha, (ii) beta, (iii) delta, (iv) gamma, and (v) theta. Alpha has greater amplitude and having a frequency that varies between 8 and 13 Hz. The amplitude is greater. It is mainly found in the posterior regions of the right and left hemispheres. These kinds of brain waves are prominent during comparatively owing thoughts. Beta has a frequency range of 13 to 30 Hz. This may be seen in order to be transmitted, mainly in the frontal region on both sides of the human brain. It may be at the lower level or absent in regions of cortical injury. The frequency of delta lies between 1 and 4 Hz. Like beating a drum, this seems to have the greatest amplitude at the frequency of the slowest wave. Delta wave is most common in one-year-old infants, although it seems to be more common in adolescents. Gamma has a greater frequency of more than 30 Hz. There are a few examples of rapid brainwaves. It is simultaneously collecting data from different parts of the brain and moves data around invisibly. Theta frequency has been measured in the range of 4 to 7 Hz. It is well-known for having the slowest speed. It occurs when one gradually departs from the outside world as well as concentrates on the sensations produced within. Figure 15.2 shows a representation of brain waves.

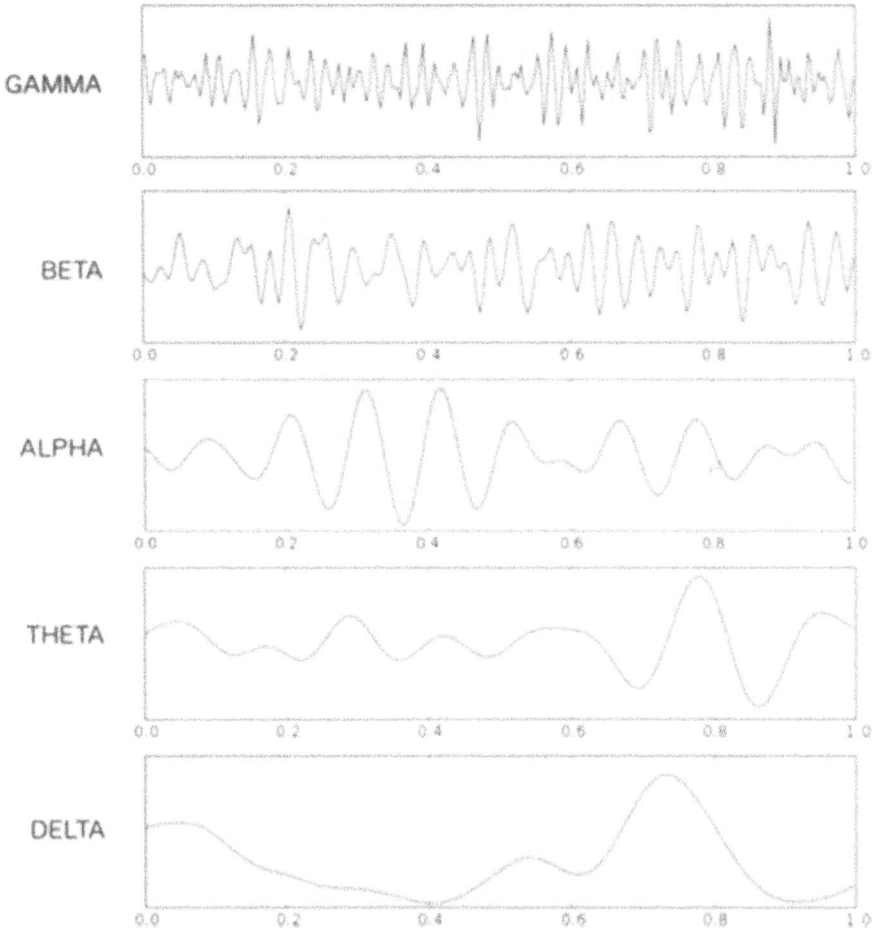

FIGURE 15.2 Categorization of brain waves based on the frequency range.

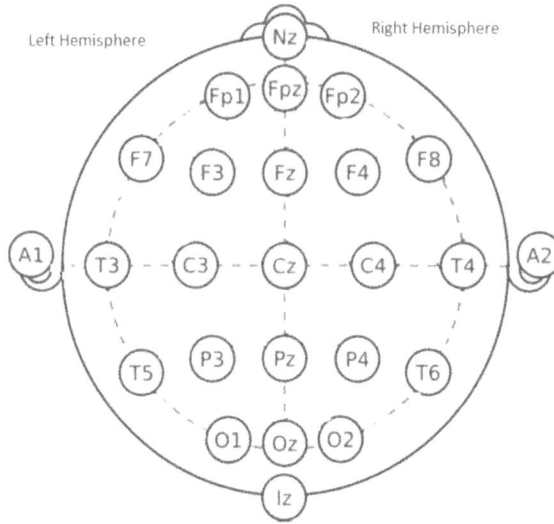

FIGURE 15.3 The 10/20 International System.

For collecting the EEG signals, there is a standard rule for placement of electrodes onto the human scalp known as the 10/20 International System (IS) shown in Figure 15.3 (Trans Cranial Technologies, 2012). This method focused on capturing the relationship between the placement of electrodes and the underlying area of the cerebral cortex. The numbers 10 and 20 represent the proportion of the entire front-back region or the skull's right-left distance. As a result, each region is represented by a unique alphabet that identifies the lobe uniquely, as well as a numerical value that indicates the hemisphere position. The frontal lobe is represented by the letter F, while the temporal lobe is represented by the letter T, the middle lobe is represented by the letter C, the Parietal lobe is represented by the letter P, and the occipital lobe is represented by the letter O. The middle line electrode is designated by the letter Z (zero). Even numbers indicate that electrodes are located on the right hemisphere, while odd numbers indicate that electrodes are located on the left hemisphere.

In recent years, emotion recognition using EEG has received a lot of interest. The procedure below must be followed in order to understand emotions recognition using EEG signals. Figure 15.4 shows a typical process for identifying emotions using EEG signals with the following steps.

1. Familiarize the subject with the stimuli to be tested.
2. Measure the voltage fluctuations in voltage generated within the human brain.
3. Noise and artifacts from the recorded raw EEG signals must be removed in order to get a better result.
4. From the resultant EEG data, a different set of features are collected.
5. A classifier is trained on the extracted feature set on the basis of the training dataset.

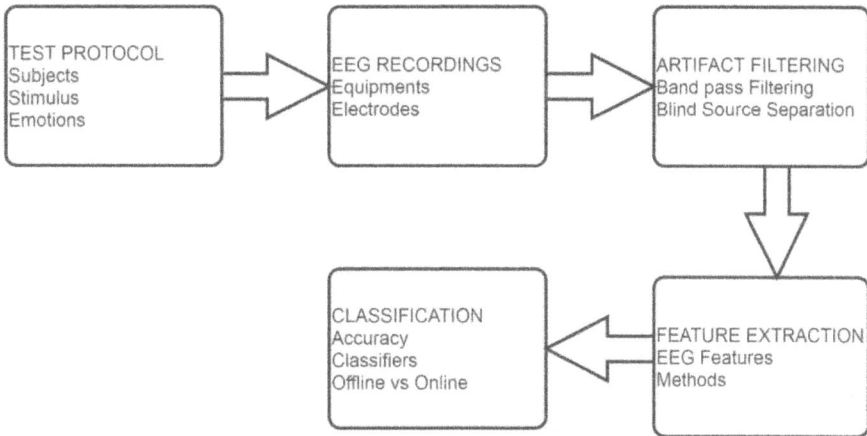

FIGURE 15.4 Emotion detection process using EEG signals.

Different Machine learning techniques are used to classify emotions into valence, arousal, dominance, and liking in this chapter, and a comparative analysis is performed to know which machine learning classifier classifies better emotion from EEG signals. Our major contributions in this chapter are summarized as follows.

The rest of the chapter is organized as follows: in Section 15.2, we review the related works. In Section 15.3, we introduced the proposed emotion recognition system using EEG signals. The design of the proposed solution and its analysis are presented in Section 15.4. Furthermore, Section 15.5 discusses the implementation of the proposed work and evaluates its performance based on accuracy. Finally, Section 15.6 summarizes the chapter.

15.2 RELATED WORKS

In recent years, from the EEG signals different types of features in various domains such as time domain, frequency, and time-frequency are extracted. On the extracted features different machine learning classification methods are applied for classifying the emotions into various classes, such as valence, arousal, dominance, and liking. This segment discusses the domain's background, state-of-the-art literature survey, and recent methods and trends.

Horlings et al. (2008) proposed an emotion recognition technique using human brain activity in which they proposed a system that analyzes the EEG signals into five classes on the two dimensions of emotions i.e. valence and arousal. They have collected their own EEG signals dataset from the people that were emotionally stimulated by pictures. For training and testing, a threefold cross-validation method is used. They achieved classification accuracies of 32% and 37% using Support Vector Machine (SVM) for recognizing the valence and arousal dimension, respectively, but when the only extreme values from both dimensions were used then classification accuracy was much better and that was 71% and 81%, respectively. Naser and Saha (2013) presented an emotion detection method from EEG signals using publicly available DEAP (Koelstra, 2011) dataset in which Dual-tree complex wavelet packet transform method is used for feature extraction from EEG signals. On the basis of singular value decomposition (SVD), QR Factorization with column pivoting (QRcp), and F-Ratio, the most discriminating features from input features are selected. Selected features are fed to the SVM to classify emotions into four different binary classes of low/high valence, low/high arousal, low/high dominance, and low/high liking. Atkinson and Campos (2016) proposed a method for emotion recognition based on Brain–Computer Interface (BCI) by combining an efficient feature selection algorithm and kernel-based classifiers on the publicly available DEAP (Koelstra, 2011) dataset. In this approach, a significant number of useful features were extracted, which includes statistical features, frequency bands, Hjorth parameters, and fractal dimension for each of the EEG channels. Statistical features such as median, kurtosis coefficient, and standard deviation have been extracted. A set of relevant features are extracted by using minimum Redundancy Maximum Relevance (mRMR) from the extracted features and then selected features are fed to the SVM classifier, which gives an accuracy of 60.72% for arousal and 62.4% for valence. Zhuang et al. (2017) introduced a method for emotion recognition using multidimensional information from EEG signals. On the basis of Empirical mode decomposition (EMD) feature extraction and emotion detection is done where EEG signals were decomposed into Intrinsic Mode Functions (IMFs) by using EMD and the Proposed method is verified on the publicly available DEAP dataset (Koelstra, 2011). Three features of IMFs, namely the first difference of phase, the first difference of time series, and the normalized energy, are fed as a feature vector to the SVM for classification. The accuracy for valence and arousal given by this method are 69.10% and 71.99%, respectively.

Petrantonakis and Hadjileontiadis (2010) recognized user-independent emotions from brain signals in which a novel feature extraction technique called Hybrid Adaptive Filtering (HAF) with Genetic Algorithms (GAs) is used to extract emotion-related EEG features to the EMD-based representation of EEG signals. Additionally, from the HAF-filtered EEG signals EEG features were extracted using Higher-Order Crossings (HOCs) method. Four different classification methods i.e. Quadratic Discriminant Analysis (QDA), K-NN, Mahalanobis Distance (MD), and SVM are used. Through a sequence of facial expressions 6 emotions, namely happiness, fear, anger, surprise, sadness, and disgust, were obtained from 16 healthy subjects by placing three EEG channels. By applying the HAF-HOC to the EEG data provides classification accuracy of up to 85.17%.

Yoon and Chung (2013) proposed an emotion recognition technique from EEG signals using Bayesian weighted-log-posterior function and perceptron convergence algorithm in which emotions are represented on two dimensions i.e. valence and arousal. For extracting the features Fast Fourier transform method was used, and for feature selection Pearson correlation coefficient method was applied. They proposed a probabilistic classifier based on Bayes' theorem of probability and a perceptron convergence algorithm. They have used a public dataset to verify the proposed method. For both valence and arousal dimensions, an emotion is divided into two-level classes as well as three-level classes. The average accuracy achieved for the valence and arousal class was 70.9% and 70.1%, respectively, in the case of two-level classes. The average accuracy achieved for the three-level classes was 55.4% and 55.2%, respectively.

Liu et al. (2016) proposed an emotion recognition technique from EEG recordings using publicly accessible DEAP (Koelstra, 2011) dataset. Total 12 different features have been used, in which, from time domain statistical features, HOCs, Fractal dimension, Hjorth feature, and non-stationarity index have been used. From the frequency domain Power Spectral Density has been used. From the time-frequency domain Discrete Wavelet Transform has been used and from Multi-Electrode Features Differential Asymmetry (DA) and Rational Asymmetry (RA), Magnitude Squared Coherence Estimate (MSCE) have been used. For feature selection Maximum Relevance Minimum Redundancy (mRMR) method is used. Selected features are fed to the K-Nearest Neighbor (KNN) and RF classifiers. The accuracy for arousal and valence was 71.2% and 69.9% using RF. Jadhav et al. (2017) proposed a method for emotion recognition in which Grey Level Co-occurrence Matrix features were extracted in terms of texture image for each sample of EEG signals. KNN classifier is evaluated on publicly available DEAP (Koelstra, 2011) dataset to classify the emotions into four different classes happy, sad, angry, and relaxed. The classification accuracy for statistical, Power Spectral Density and HOC are 66.25%, 70.1%, and 69.59%, respectively. Li et al. (2018) introduced an emotion recognition technique from EEG signals, in which signals were captured from different channels in different frequency bands. They classified the emotions into valence and arousal dimensions. To evaluate the classifiers publicly available pre-processed DEAP (Koelstra, 2011) dataset was used. By using DWT EEG signals were divided into four frequency bands, and to the K-NN classifier, energy and entropy were fed as features. The valence dimension classification accuracies of the 10, 14, 18, and 32 EEG channels based on the Gamma frequency band were 89.54%, 92.28%, 93.72%, and 95.70%, respectively, and the arousal dimension classification accuracies were 89.81%, 92.24%, 93.69%, and 95.69%, respectively. From the above results it was evident that when the number of channels increases, classification accuracy improves. The gamma frequency band has a higher classification accuracy than the beta frequency band, which is accompanied by the alpha and theta frequency bands.

Tripathi et al. (2017) used two different Neural Networks Models, a Simple Deep Neural Network and a Convolutional Neural Network, for emotion recognition using EEG signals on publicly available DEAP (Koelstra, 2011) dataset. The simple Deep Neural Networks model gives an accuracy of 75.78% for valence and 73.28% for arousal class, respectively. The Convolutional Neural Network model gives an accuracy of 81.41% for valence and 73.35% for arousal class, respectively. Chen et al. (2019) proposed EEG-based emotion recognition on combined features using Deep CNN. From the publicly available DEAP (Koelstra, 2011) dataset, they have extracted temporal features, frequential features, and a combination of both the features. To make the binary classification on valence and arousal dimensions, many machine learning classifiers are evaluated such as SVM, Linear Discriminant Analysis (LDA), Bagging Tree (BT), Bayesian linear discriminant analysis (BLDA), and deep convolution neural network (CNN). It was observed that the deep CNN model gives the best recognition accuracy for both valence and arousal dimensions without any feature engineering.

15.3   PROBLEM DEFINITION

Many researchers have suggested various emotion recognition techniques based on facial expression or audio stimuli. However, the main issue with these techniques is that they are all dependent on the human's outer state, or even if they have been used EEG signals for emotion detection they have mostly classified emotion into two classes or three classes. Even different researchers have taken signal data from a different number of electrodes. We found a number of flaws in the existing EEG-based emotion recognition method. These issues are listed as follows.

1. Emotion can be triggered by the observation of vocabulary such as well-known phrases, as well as non-verbally such as voice tone, but due to social anxiety, these can all be easily obscured.
2. If an individual has facial paralysis, then the facial expression for emotion detection is useless.

Problem: Emotion recognition from EEG signals using various machine learning techniques such as SVM, KNN, Random Forest (RF), Gaussian Naive Bayes (GNB), and Decision Tree (DT). A comparative analysis is performed with the proposed classifiers and is used to decide which classifier gives better accuracy.

15.4   METHODOLOGY

This section discusses about the various machine learning classifiers used for emotion recognition using EEG signals. These classifiers use EEG data to predict various types of emotions. The machine learning models are trained on a pre-processed DEAP dataset, which can be used for classification. The training amount of data is varied from model to model. Entire EEG data is divided into the training set and testing set. Prediction is made in terms of the accuracy of the model on the testing set by using a confusion matrix. Figure 15.5 shows a block diagram of emotion recognition from EEG signals using different classifiers.

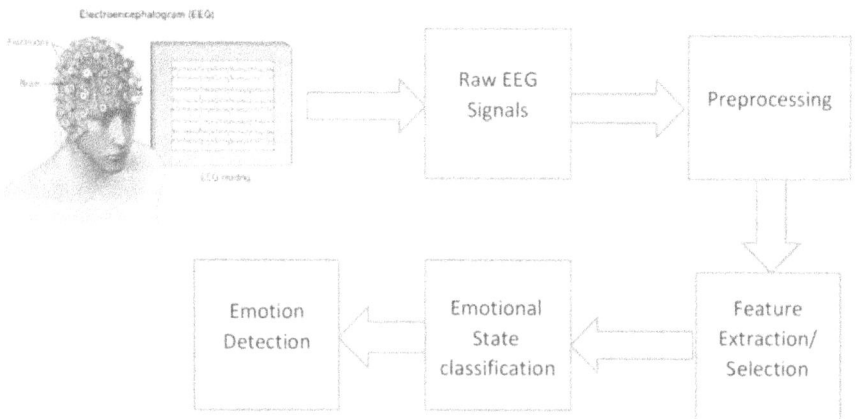

FIGURE 15.5   Block diagram for emotion recognition from EEG signals using various classifiers.

15.4.1 DEAP DATASET

DEAP (Koelstra, 2011) is a free publicly available dataset for recognition of emotions in which peripheral physiological signals of 32 participants (subjects) are collected at 512 Hz of the sampling rate. All of the participants were between the ages of 19 and 37. While EEG recordings, Every person had watched 40 induced music videos, each music video clip of duration 60 seconds. Each video clip represents a single emotion and is graded on a scale of 1 to 9. Twenty of the 40 music video samples are high valence visual stimuli, while the other 20 are low valence visual stimuli. The situation is the same in terms of arousal. Participants rated their valence, arousal, liking, dominance, and familiarity on a scale of 1 to 9 after watching the music video. The EEG signals were recorded using 32 electrodes that were placed according to the international 10-20 model. Every electrode recorded a 63-seconds EEG signal, with the first 3 seconds serving as the baseline signal before the trial. Out of 32 participants, 22 had their frontal faces video captured. The sampling frequency for EEG signals was initially set to 512 Hz. To reduce service time, the EEG signals were down-sampled at the frequency of 256 Hz. Detailed statistics of the dataset is given in the Table 15.1.

15.4.2 RAW EEG SIGNAL

Data is obtained from the human brain by placing the electrodes on the human scalp. The total number of electrodes used in this study is 32 and electrode placement follows the international 10/20 scheme (Trans Cranial Technologies, 2012). Raw EEG signals are processed, amplified, digitized, and placed on a computing system for data processing.

15.4.3 PREPROCESSING

In the pre-processing phase, the actual data was pre-processed after the EEG data was obtained i.e. the downsampling, elimination of EOG, removal of noise, and electromyogram (EMG) artifacts, to decrease the computational complexity of extraction of features. The default pre-processing method was applied to this DEAP dataset:

1. The data was down-sampled to 128 Hz.
2. The electrooculogram (EOG) artifacts have been eliminated, as accomplished in Koelstra (2011).
3. A bandpass filter of frequency ranging between 4.0 and 45.0 Hz was used throughput.
4. The data was averaged to the common reference.
5. The data was decomposed into 60-second trials and a 3-second baseline pre-trial (well before-trial).

TABLE 15.1
DEAP Dataset

Array Name	Array Shape	Array Contents
Data	40*40*8064	Number of trial/video*channel*data
Labels	40*4	Number of trial or video*label

15.4.4 FEATURE EXTRACTION

We evaluate a large variety of EEG emotion detection features that are being suggested in the past. The Time domain, frequency domain, and time-frequency domain characteristics are commonly distinguished. Features are usually determined from a single electrode's reported signal, but literature has also found a few features integrating signals of more than one electrode. Power spectral density (PSD) in the frequency domain from the EEG signals of DEAP (Koelstra, 2011) dataset were extracted. The logarithms of the power spectral density features from theta, slow alpha, alpha, beta, and gamma power bands were extracted for all 32 electrodes. The ranges for theta, slow alpha, alpha, beta, and gamma are defined as 4–8 Hz, 8–10 Hz, 8–12 Hz, 12–30 Hz, and 30+ Hz, respectively. Aside from spectral features, the difference in spectral power of all symmetric pairs of electrodes on the left and right hemispheres was calculated to assess the possibility of asymmetry of human brain function caused by emotional stimulation. The number of features taken into consideration from all the 32 electrodes is 216. For each electrode all the five frequency bands, namely theta, slow alpha, alpha, beta, and gamma, have been taken (i.e. 32*5 = 160). Asymmetry of spectral power between 14 pairs of electrodes in four bands of alpha, beta, theta, and gamma, i.e. 14*4 = 56.

15.4.5 EMOTIONAL STATE CLASSIFICATION

For identifying emotions, there are a variety of algorithms available but the biggest flaw in such algorithms is the lack of accuracy. There need to be carried out more research in order to evaluate different algorithms and proposed algorithms with improved accuracy. Only a few types of emotions could be recognized until now. Research should be performed for more types of recognition of emotions. In this proposed solution, Emotion recognition is done from EEG signals using various machine learning techniques such as SVM (Noble, 2006) with Principal Component Analysis (PCA), SVM (Noble, 2006) without PCA, KNN with PCA, KNN without PCA, RF, GNB, and DT (Swain, 1977).

15.4.5.1 Support Vector Machine

SVM (Noble, 2006) is a supervised machine learning model that is used for classification and regression problems. For linear classification SVM performs well but for non-linear classification, it works well by using the kernel method. The main objective of SVM is to find a hyperplane in an N-dimensional feature space that differentiates the two-class better. The maximum distance between data points of both the classes is known as margin. Selection of hyperplane with low margin will have a very high chance of miss classification. So, selecting a hyperplane with a high margin always leads to a better classification. If the number of features is two, then the hyperplane will be a line. If the number of features is three then the hyperplane will be a 2D plane. It becomes very difficult to imagine when the number of features will be more than 3.

15.4.5.2 K-Nearest Neighbor

KNN (Peterson, 2009) is a supervised machine learning algorithm that is used for both classification and regression problems. It works on the principle that similar things exist in close proximity. It is a non-parametric algorithm. In the training phase

KNN only stores the data, and when it gets unseen data it classifies this unseen data point to its very much similar category.

15.4.5.3 Decision Tree

DT (Swain, 1977) is a tree-structured supervised machine learning model that is used for both classification and regression problems. It falls under the traditional divide-and-conquer learning strategy. In the DT, internal nodes denote the features, branches denote the decision rules, and the leaf nodes denote the outcome of the classifier. It starts from the root node, which explores the further branches, which leads to the construction of the DT. If the tree is large then it will raise the problem of overfitting. If the tree is small then it may not capture all the important features. Therefore there is a technique called pruning which decreases the size of the DT without reducing the accuracy.

15.4.5.4 Random Forest

RF (Chaudhary, 2016) is a supervised machine learning model that is used for both classification and regression problems. RF contains many DTs on different subsets of the given dataset and takes mean value to improve the predicted accuracy of that dataset. When the number of trees is more, then the accuracy will be better, and it prevents overfitting.

15.4.5.5 Gaussian Naive Bayes

GNB (Murphy, 2006) follows Gaussian normal distribution that supports continuous data. It is a variant of the Naive Bayes classifier. Naive Bayes is a supervised machine learning model that is used for classification, which is based on Bayes' theorem of probability. There is a strong assumption that the value of a particular feature should be independent of any other feature in the dataset.

15.4.5.6 Principal Component Analysis

PCA (Karamizadeh, 2013) is an unsupervised method that is basically used to reduce the number of features in a dataset that is known as dimensionality reduction. PCA is a statistical method that converts correlated features into linear uncorrelated features by using orthogonal transformation. These transformed linear uncorrelated features are called principal components. It is frequently used in predictive modeling and exploratory data analysis. It is a feature extraction technique that keeps the most important features and drops the least important features. The number of principal components may be less than or equal to the number of original features.

15.5 EXPERIMENTAL ANALYSIS

In this section, the performance of the proposed Emotion Recognition technique from EEG signals using a variety of classifiers is analyzed. We used the DEAP (Koelstra, 2011) dataset to interpret the results. Signals from 32 subjects are obtained in this study. Subjects include both males and females of age between 19 years and 37 years. Valence, arousal, dominance, and liking are the four types of emotions. Accuracy for each participant is calculated separately. Accuracy for each participant is evaluated for four classes, which are low/high valence, low/high arousal, low/high dominance, and low/high liking.

TABLE 15.2

Results for Classifiers

	Accuracy Score for Each Class			
Classifier	Valence	Arousal	Dominance	Liking
KNN	56.64	56.25	67.57	65.23
SVM	63.43	64.06	67.65	67.18
KNN-PCA	53.51	58.59	68.75	69.53
SVM-PCA	64.06	64.30	69.50	64.06
RF	63.59	63.59	67.03	67.18
DT	53.90	55.89	54.68	57.42
Gaussian NB	57.42	58.20	49.60	65.23

The performance is measured using accuracy, which is computed as follows:

$$Accuracy = \frac{TP + TN}{TP + FP + TN + FN} \quad (15.1)$$

where TP denotes true positive, TN denotes true negative, FP denotes false positive, and FN denotes false negative (Table 15.2).

By using the KNN classifier, we got an accuracy of 56.64% for valence, 56.25% for arousal, 67.57% for dominance, and 65.23% for liking. By using the SVM classifier, we got an accuracy of 63.43% for valence, 64.06% for arousal, 67.65% for dominance, and 67.18% for liking. By using the KNN-PCA classifier, we got an accuracy of 53.51% for valence, 58.59% for arousal, 68.75% for dominance, and 65.53% for liking. By using the SVM-PCA classifier, we got an accuracy of 64.06% for valence, 64.30% for arousal, 69.50% for dominance, and 64.06% for liking. By using the RF classifier, we got an accuracy of 63.59% for valence, 63.59% for arousal, 67.03% for dominance, and 67.18% for liking. By using the DT classifier, we got an accuracy of 53.90% for valence, 55.89% for arousal, 54.68% for dominance, and 57.42% for liking. By using the GNB classifier, we got an accuracy of 57.42% for valence, 58.20% for arousal, 49.60% for dominance, and 65.23% for liking. It is found out that SVM with PCA gives better accuracy in the case of valence, arousal, and dominance but in the case of liking, KNN with PCA gives better results in terms of accuracy. We found that SVM with PCA classifies better, with a classification accuracy of 64.06% for valence, 64.30% for arousal, 69.50% for domination, and 64.06% for liking. Figure 15.6 shows the accuracy results for the valence class using different machine learning classifiers. Figure 15.7 shows accuracy results for arousal class using different machine learning classifiers.

Figure 15.8 shows accuracy for dominance class using different machine learning classifiers. Figure 15.9 shows accuracy for liking class using different machine learning classifiers.

Figure 15.10 shows accuracy for all classes using different machine learning classifiers.

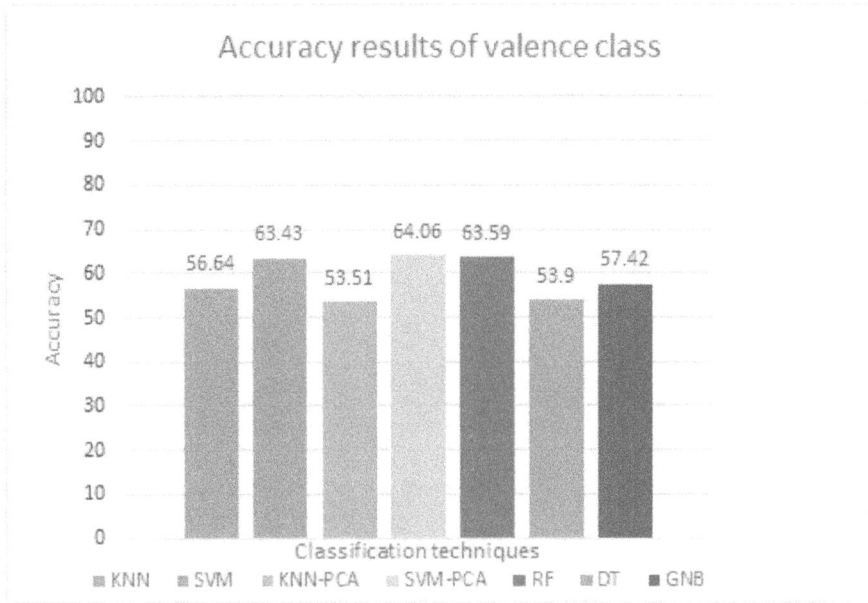

FIGURE 15.6   Accuracy for valence class.

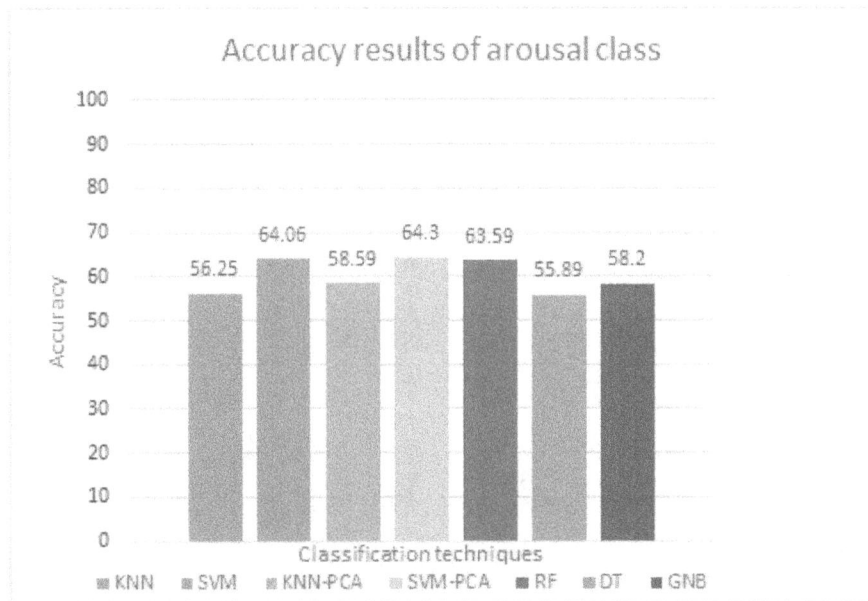

FIGURE 15.7   Accuracy for arousal class.

Accuracy results of dominance class

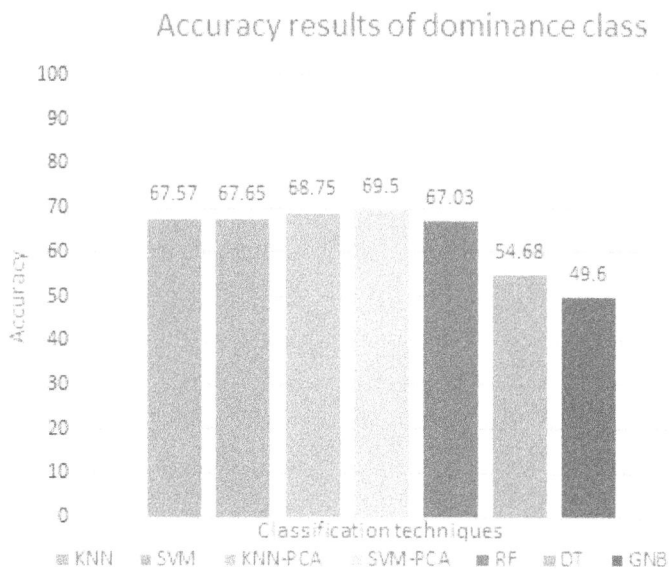

FIGURE 15.8   Accuracy for dominance class.

Accuracy results of liking class

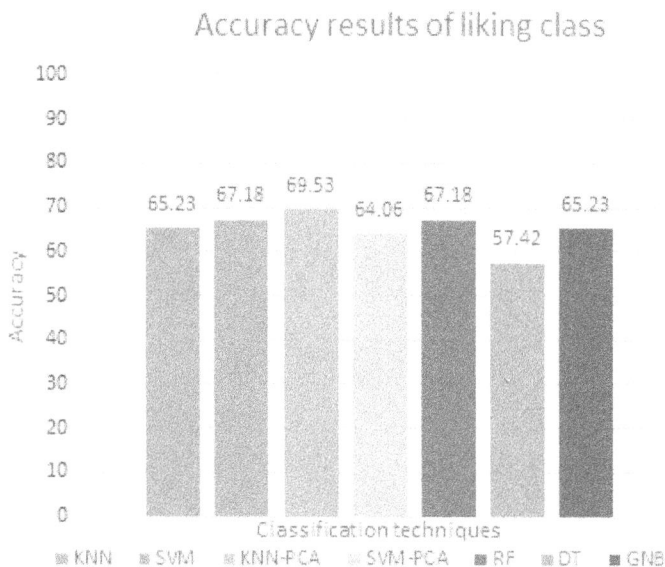

FIGURE 15.9   Accuracy for liking class.

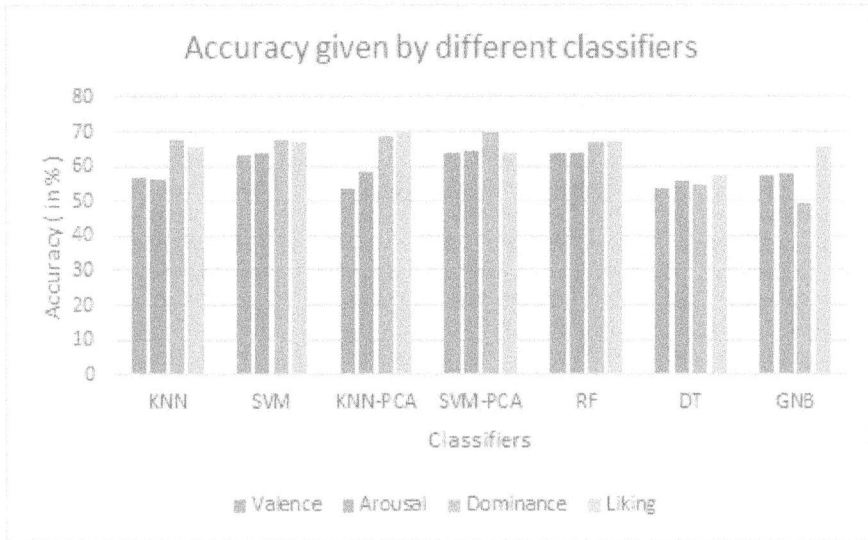

FIGURE 15.10 Accuracy for all classes.

15.6 CONCLUSION

The detection of emotions from EEG signals when showing music videos is investigated in detail. Multiple classifying videos belong to low/high arousal, low/high valence, low/high dominance, and low/high liking but some samples belong to multiple classes at the same time. Hence, observing the nature of the dataset, we can conclude that this problem falls under the category of a multi-label and binary class but not a multi-class problem. We have used various machine learning classifiers to classify emotions. Finally, by observing the accuracy from the above results for valence, arousal, dominance, and liking class it is concluded that the performance of SVM with PCA (SVM-PCA) is better with respect to others. A comparative analysis of emotional state using various machine learning techniques has been accomplished on the DEAP dataset. We can use multiple datasets to recognize the emotions from multiple sources to explore human emotions. The patient's state tracking and the EEG processing results are shared with healthcare professionals, who can then determine the condition of the patient and offer medical assistance.

Further, human emotions can be investigated by applying various deep learning models.

REFERENCES

Ali, Mouhannad, Ahmad Haj Mosa, Fadi Al Machot, and Kyandoghere Kyamakya. "EEG-based emotion recognition approach for e-healthcare applications." In 2016 Eighth International Conference on Ubiquitous and Future Networks (ICUFN), pp. 946–950. IEEE, 2016.

Atkinson, John, and Daniel Campos. "Improving BCI-based emotion recognition by combining EEG feature selection and kernel classifiers." Expert systems with applications 47 (2016): 35–41.

Bos, Danny Oude. "EEG-based emotion recognition." The influence of visual and auditory stimuli 56, no. 3 (2006): 1–17.

Chaudhary, Archana, Savita Kolhe, and Raj Kamal. "An improved random forest classifier for multi-class classification." Information processing in agriculture 3, no. 4 (2016): 215–222.

Chen, J. X., P. W. Zhang, Z. J. Mao, Y. F. Huang, D. M. Jiang, and Y. N. Zhang. "Accurate EEG-based emotion recognition on combined features using deep convolutional neural networks." IEEE Access 7 (2019): 44317–44328.

Hondrou, Charline, and George Caridakis. "Affective, natural interaction using EEG: sensors, application and future directions." In Hellenic Conference on Artificial Intelligence, pp. 331–338. Springer, Berlin, Heidelberg, 2012.

Horlings, Robert, Dragos Datcu, and Leon J. M. Rothkrantz. "Emotion recognition using brain activity." In Proceedings of the 9th International Conference on Computer Systems and Technologies and Workshop for PhD Students in Computing, pp. II–1. 2008.

Jadhav, Narendra, Ramchandra Manthalkar, and Yashwant Joshi. "Electroencephalography-based emotion recognition using gray-level co-occurrence matrix features." In Proceedings of International Conference on Computer Vision and Image Processing, pp. 335–343. Springer, Singapore, 2017.

Karamizadeh, Sasan, Shahidan M. Abdullah, Azizah A. Manaf, Mazdak Zamani, and Alireza Hooman. "An overview of principal component analysis." Journal of signal and information processing 4, no. 3B (2013): 173.

Koelstra, Sander, Christian Muhl, Mohammad Soleymani, Jong-Seok Lee, Ashkan Yazdani, Touradj Ebrahimi, Thierry Pun, Anton Nijholt, and Ioannis Patras. "DEAP: A database for emotion analysis; using physiological signals." IEEE transactions on affective computing 3, no. 1 (2011): 18–31.

Li, Ma, Quek Chai, Teo Kaixiang, Abdul Wahab, and Hüseyin Abut. "EEG emotion recognition system." In In-vehicle Corpus and Signal Processing for Driver Behavior, pp. 125–135. Springer, Boston, MA, 2009.

Li, Mi, Hongpei Xu, Xingwang Liu, and Shengfu Lu. "Emotion recognition from multichannel EEG signals using K-nearest neighbor classification." Technology and health care 26, no. S1 (2018): 509–519.

Lin, Yuan-Pin, Chi-Hong Wang, Tien-Lin Wu, Shyh-Kang Jeng, and Jyh-Horng Chen. "EEG-based emotion recognition in music listening: A comparison of schemes for multiclass support vector machine." In 2009 IEEE International Conference on Acoustics, Speech and Signal Processing, pp. 489–492. IEEE, 2009.

Liu, Jingxin, Hongying Meng, Asoke Nandi, and Maozhen Li. "Emotion detection from EEG recordings." In 2016 12th International Conference on Natural Computation, Fuzzy Systems and Knowledge Discovery (ICNC-FSKD), pp. 1722–1727. IEEE, 2016.

Liu, Yisi, Olga Sourina, and Minh Khoa Nguyen. "Real-time EEG-based emotion recognition and its applications." In Transactions on Computational Science XII, pp. 256–277. Springer, Berlin, Heidelberg, 2011.

Murphy, Kevin P. "Naive Bayes classifiers." University of British Columbia 18, no. 60 (2006): 1–8.

Murugappan, M., Mohd Rizon, Ramachandran Nagarajan, S. Yaacob, I. Zunaidi, and Desa Hazry. "Lifting scheme for human emotion recognition using EEG." In 2008 International Symposium on Information Technology, vol. 2, pp. 1–7. IEEE, 2008.

Naser, Daimi Syed, and Goutam Saha. "Recognition of emotions induced by music videos using DT-CWPT." In 2013 Indian Conference on Medical Informatics and Telemedicine (ICMIT), pp. 53–57. IEEE, 2013.

Niedermeyer, Ernst, and FH Lopes da Silva, eds. Electroencephalography: basic principles, clinical applications, and related fields. Lippincott Williams & Wilkins, 2005.

Noble, William S. "What is a support vector machine?." Nature biotechnology 24, no. 12 (2006): 1565–1567.

Peterson, Leif E. "K-nearest neighbor." Scholarpedia 4, no. 2 (2009): 1883.

Petrantonakis, Panagiotis C., and Leontios J. Hadjileontiadis. "Emotion recognition from brain signals using hybrid adaptive filtering and higher order crossings analysis." IEEE Transactions on affective computing 1, no. 2 (2010): 81–97.

Rached, Taciana Saad, and Angelo Perkusich. "Emotion recognition based on brain-computer interface systems." In Brain-Computer Interface Systems: Recent Progress and Future Prospects, pp. 253–270. 2013.

Ribas, Guilherme Carvalhal. "The cerebral sulci and gyri." Neurosurgical focus 28, no. 2 (2010): E2.

Swain, Philip H., and Hans Hauska. "The decision tree classifier: Design and potential." IEEE transactions on geoscience electronics 15, no. 3 (1977): 142–147.

Teplan, Michal. "Fundamentals of EEG measurement." Measurement science review 2, no. 2 (2002): 1–11.

Trans Cranial Technologies. "10/20 System Positioning Manual." 2012.

Tripathi, Samarth, Shrinivas Acharya, Ranti Dev Sharma, Sudhanshi Mittal, and Samit Bhattacharya. "Using deep and convolutional neural networks for accurate emotion classification on DEAP dataset." In Proceedings of the Thirty-First AAAI Conference on Artificial Intelligence, pp. 4746–4752. 2017.

Yoon, Hyun Joong, and Seong Youb Chung. "EEG-based emotion estimation using Bayesian weighted-log-posterior function and perceptron convergence algorithm." Computers in biology and medicine 43, no. 12 (2013): 2230–2237.

Yoon, Won-Joong, and Kyu-Sik Park. "A study of emotion recognition and its applications." In International Conference on Modeling Decisions for Artificial Intelligence, pp. 455–462. Springer, Berlin, Heidelberg, 2007.

Zhuang, Ning, Ying Zeng, Li Tong, Chi Zhang, Hanming Zhang, and Bin Yan. "Emotion recognition from EEG signals using multidimensional information in EMD domain." BioMed research international 2017 (2017): 9 pages.

16 A Survey on Antennas for IIoT Application

Prashant Kumar Singh and Shashank Kumar Singh
UCET VBU
Hazaribag, India

Sandipan Mallik
NIST
Berhampur, India

Dilip Kumar Choudhary
G. H. Raisoni College of Engineering
Nagpur, India

Anjini Kumar Tiwary
Birla Institute of Technology
Mesra, India

CONTENTS

16.1 INTRODUCTION

Nowadays, Internet of Things (IoT) technology became very popular among various research groups due to its utilization in numerous sectors like industry, healthcare, irrigation, identification, biomedical, home appliances, and so on. The advent of IoT revolutionized the communication and information world drastically. It is developed in a manner to become the part of everyday life. The communication through this technology is not only limited to conventional human-human (H-H) interaction; however, this incorporates the human-machine (H-M) and the extension of this provides machine-machine (M-M) interaction. In IoT technology, all devices/machines/

DOI: 10.1201/9781003145004-16

objects have their own identities and works as live member for communication and interaction between themselves using internet services without any intervention of human. The feasibility of M-M connectivity developed the field of Industrial Internet of Things (IIoT), which can be considered as a subset of IoT.

In industries, IIoT technology used in real time automation and helps to supervise and inspect the industrial process without the involvement of human. For the above said purpose, there is a need to disseminate and collect the data. The collection of data is done through the sensors; however, wireless technology is used for the data dissemination. The wireless data dissemination needs a means of transmitting and receiving wireless data in the form of electromagnetic (EM) waves, which can be done by antenna. This makes the antenna as one of the vital components of IoT/IIoT technology. A triple band CP microstrip patch antenna[1] is designed, which radiates at 5.8 GHz, 6.76 GHz, and 8.4 GHz frequencies. The presented antenna designed for satellite-based IoT application. In Reference[2], an array antenna is proposed for IoT/ISM band application. The antenna operates at 5.8 GHz and comprises an array of two patches with I-beam slots and defected ground surface (DGS). An antenna-on-package[3] for IoT application using 3-D printing technology is demonstrated, which is designed for 2.45 GHz. The sensors used for IoT/IIoT applications come in different size and shapes depending upon the applications. Hence, there is a need to design application specific miniaturized antenna with low profile, low cost, easier integration, and low power characteristics; which makes microstrip technology as a better candidate. Various IIoT technologies work on different bands, which demands multiband/wideband antennas too.

This chapter presents a brief literature survey on the antennas for IoT/IIoT applications, which details about the various antennas designed for RFID tag and readers, smart construction, wearable IoT, and RF energy harvesting. Further, a novel dual band microstrip patch antenna with partial ground is proposed and presented. The center frequencies of the proposed antenna are 2.5 GHz and 5.6 GHz having the bandwidth of 0.52 GHz and 2.67 GHz, respectively, which covers IoT bands. The return losses at 2.5 GHz and 5.6 GHz are −41.6 dB and −39.28 dB, respectively. The simulation results of the antenna show its suitability for IoT/IIoT devices.

16.2 ANTENNAS FOR IIoT/IoT APPLICATIONS

16.2.1 RADIO FREQUENCY IDENTIFICATION (RFID)

The key purpose of RFID technology is to identify and track the object using exchange of information amid RFID tags and readers wirelessly through EM waves. Nowadays, this technology is employed in many applications like industrial robotics, road tall collection, logistics transportation, cattle management, attendance system, wireless sensors (like gas, humidity, temperature), smart cards, library management, healthcare and so on. The antennas associated with RFID tags are linearly polarized (LP) and in practical situations the orientation of tags are random. The solution to cope with the problem of random orientation angle between RFID tags and readers is the use of circularly polarized (CP) RFID reader antenna, which gives the better flexibility and reliability[4]. The available methods used to develop microstrip antennas with CP are single feed[5], dual feed[6], and quad feed[7]. Multiple frequency bands like 125 KHz, 13.56 MHz, 0.92 GHz, 2.45 GHz, and 5.8 GHz are allotted for RFID system. Accordingly, various

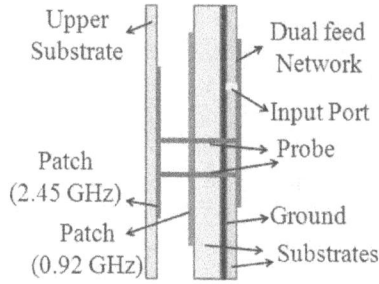

FIGURE 16.1   Side view of multilayered RFID reader antenna[4].

reader and tag antennas with mono, dual and multiple band performance are proposed by the researchers. In Reference 8, a dual band RFID reader antenna is presented for 0.92/2.45 GHz with LP. The dual band CP antennas[9–10] are also investigated for RFID reader. Some of common antenna design requirements for the reader and tag are high gain, compact size, directional, and symmetrical pattern.

A compact multilayered directive antenna structure is also proposed for handheld RFID reader[4]. This antenna provides dual band performance at 0.92 GHz and 2.45 GHz frequencies with CP radiation. There are three substrate layers; however, top substrate and middle one are separated by air as shown in Figure 16.1. The CP radiation is achieved by using the dual feed system. The top circular patch is excited through this feeding network and middle rectangular patch is excited through top patch using coupling mechanism. The connection between the feeding network and top patch is done by via and copper wires (probes). The two radiators provide radiation at two different frequencies and shows CP for both. All the layers are supported by Teflon sets in the form of small hollow pillars, nuts and screws. The presented work gives compact antenna (due to stacked patches for two resonating frequencies), symmetrical and stable radiation pattern, and return loss above 10 dB. The range test for reading is also presented, which demonstrates the readable range above 3.3 m for 0.92 GHz and 5.6 m for 2.45 GHz.

Many works are available in literature for RFID tags in chip-based as well as chip-less configurations. The chip-less configurations have advantage of comparatively low cost. The chip-less tag comprises two antennas[11], one for receiving the interrogation signal from RFID reader antenna and other for retransmitting the encoded signal. The two antennas should be arranged in a manner to provide less interference, which may be done by choosing the LP as vertical for one and horizontal for other. These tag antennas are generally ultra-wide band monopole antennas.

16.2.2   SMART CONSTRUCTION

The IIoT as well as Industry 4.0 concepts are closely linked with the concept of IoT, which bring the real-time inspection, monitoring, automation and operation to manufacturing machineries. The IIoT is the part of IoT and specifically used in industry applications to improve the productivity and capability of making decisions smartly[12]. The increased use of IoT/IIoT technology demands the higher data rate with low latency and the upcoming 5G technology is especially designed in a way to incorporate these issues also[13]. An example of the incorporation of IIoT technology is in construction industry. For smart construction, the IIoT may help by providing regular inspection of sites in any weather/construction conditions, tracing and tracking the vehicles, materials, construction equipment and other resources to reduce the wastage and to improve the scheduling, operation and team coordination[14]. A MIMO antenna[15] with dual band performance is presented in literature for smart construction purpose. Even the directivity can be improved by using MIMO antenna; however, the large coupling amid elements of antennas degrades the efficiency and data rate in conventional MIMO system. The mutual coupling can be reduced by various techniques like use of metamaterial structure, shorting pins in substrate, dielectric groove, use of extra dielectric layer to cover patch. These methods are good for single band. In Reference 15, the reduction in coupling at both frequencies had done by using novel, different sized adjacent metallic arcs (with small separation) on substrate between the patches. The presented MIMO antenna illustrates reflection coefficient of −27 dB at 1.8 GHz and −35 dB at 2.6 GHz, which shows its suitability for IIoT-based smart construction. The antenna size in terms of area as well as volume is appreciably reduced by using Planar Inverted F Antenna like structure on FR4 substrate using shorting pins and creating defect in ground.

16.2.3 Wearable Antennas

The wearable antenna technology has the potential to solve the challenges associated with invasive human tracking and surveillance system in healthcare industry[16]. The invasion of electronic devices (for security purpose, tracking health and living condition) in human body can be eliminated or partially eliminated by using the industrial wearable IoT devices. Future garments may be able to transmit/receive/process the wireless body area network (WBAN) data in noninvasive manner. The health related data updates through nearby devices will help the doctor and patient for diagnosis. Wearable technology may encompass in many more applications like message and call alert, social media notification, app alert, and so on. For proper wireless dissemination of data various works had been done to design wearable antennas. Wearable antennas are generally using fabric as the substrate material and showing its significance in various applications like smart healthcare, telemedicine, and navigation. The antennas on garments may suffer from change in radiation pattern, polarization mismatch, and change in resonating frequency, etc. due to stretching of fabric, alteration of shape, variation in orientation. For practical purpose, wearable antennas should be flexible, compact, inexpensive, comfortable, and light weight. The microstrip technology for wearable antenna design is famous among the researchers due to its light weight, low power, low profile, easy fabrication, and easier integration with flexible fabrics. There are a variety of technologies available to enhance the gain, improve efficiency, lower side and back lobes. The electromagnetic band-gap (EBG) structure[17–19] (periodic structures comprise conductor/dielectric elements) is one of those techniques, which can be incorporated in wearable microstrip antennas to improve the antenna performance and overcome the constraints like low gain, narrow bandwidth, high backward radiation and low radiation efficiency. It also helps in suppressing the surface waves. A hexagonal shaped patch antenna[17] on flexible EBG substrate is presented for wearable application using felt textile as substrate material. The study shows the enhanced gain and return loss due to the incorporation of EBG

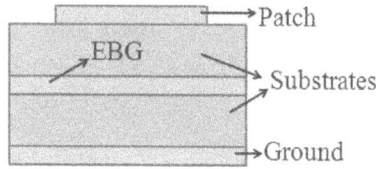

FIGURE 16.2   Layered view of patch antenna with EBG structure between two substrates [19].

structure. In Reference 17, EBG structure is introduced between two substrates. The dielectric materials chosen for this antenna is jeans with 1.6 mm thickness and 1.6 as dielectric constant. The material used for patch, ground, and EBG designs is copper. The layered view of the presented antenna in Reference 17 is shown in Figure 16.2. Here, the impacts of different EBG structures are well presented by analyzing the antenna performance without EBG and with different EBGs. The paper illustrates the enhancement in radiation efficiency and gain due to addition of EBG.

In Reference 20, a reconfigurable textile antenna for 2.4 GHz off-body communication is presented to compensate the power loss due to polarization mismatch between transmitter and receiver. The reconfiguration in polarization (LP, Left-Hand CP and Right-Hand CP) is done by using four PIN diodes as switch at truncation ends of rectangular patch antenna on top of felt textile. The ground conductor is integrated at the back side of felt. An on-body wearable antenna is designed using Jeans substrate for 5.8 GHz employing coaxial feed technique[16]. The antenna composed of semicircular conductor patch at top and ground at bottom of jeans fabric. The curved part of patch is stitched through conductive thread (works as via post) and other part left open. The simulated result shows return loss and gain of −18 dB and 6.02 dBi, respectively. A triple band wearable dipole antenna[21] is also suggested for IoT application using denim textile. The antenna structure comprises E-shaped two symmetrical patches and rectangular patch at back side with L-shaped slot, which is passively coupled. The observation of antenna characteristics is done on different body part as well as with different deformations. The wearable antennas for breast cancer detection[22], breast tumor detection[23], respiration rate measurement and monitoring[24] are also presented in literature.

16.2.4   RF ENERGY HARVESTING

The wireless sensors are the backbone of IoT/IIoT devices. Due to vast application of IoT devices in everyday life, there is a requirement of huge number of sensors. The challenge with installation of these sensors is "How to provide the regular power to these huge numbers of sensors." The wired energy transfer or battery is not the proper solution as wired connection is not possible in many of the situations like fine pipe, airtight space, water pipe, narrow space[25], and battery needs regular monitoring and change. Hence, the better solution is to provide the power using wireless technology. This can be done by converting the RF energy into electrical energy, which is known as RF energy harvesting. Nowadays, the antennas for RF energy harvesting became very hot topic among the academic and industry research. The basic block diagram for RF energy transmission is shown in Figure 16.3. The transmitting antenna should have to transmit max energy in a particular direction. So, the horn antenna or antenna array[25] may be the better choice for transmission purpose. However, Omni-directional

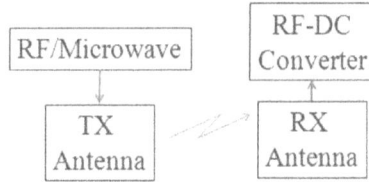

FIGURE 16.3   Block diagram of RF energy harvesting system.

microstrip patch antennas can be used for receiving the RF energy for further rectification purpose. A dual band microstrip antenna[25] with two rectangular slots in ground for RF energy harvesting application is presented for 2.45 GHz and 5.8 GHz resonating frequencies. A complete rectenna (antenna + RF/DC converter) circuit is proposed for 2.4 GHz[26] and 2.45 GHz[27] frequency using rectangular patch antenna as RF receiving terminal for powering IoT wireless sensors.

16.3   PROPOSED MICROSTRIP ANTENNA DESIGN

Here, a novel microstrip antenna with dual band performance is proposed for IoT/IIoT application. The presented antenna comprises mirror image of "6" shaped patch with microstrip feed and partial ground. Partial ground is used for getting Omni-directional radiation. This antenna is designed for center frequency of 2.5 GHz and 5.6 GHz. High Frequency Structure Simulator (HFSS 2020) is used for the electromagnetic simulation of all the designs presented here. The substrate used for the antenna is Taconic TLX having relative permittivity of 2.55 with 0.0019 loss tangent and 0.5 mm thickness.

To reach the design goal, initially an antenna with rectangular patch and partial ground is designed for wide band characteristic of 3 GHz to 7 GHz frequency band. This configuration is illustrated in Figure 16.4(a) with all the dimensional parameters. Here, the optimized value of the illustrated dimensions Wp, Lp, W50, L50, Wg, Lg, Wa are 15 mm, 11 mm, 1.41 mm, 21.5 mm, 20 mm, 19.8 mm, 6.795 mm, respectively. Figure 16.4(b) illustrates the simulated result of return loss for this configuration. The return loss obtained is below −15 dB in the range of 3.15 GHz to 6.73 GHz.

Further, an "L" shaped patch is added at top corner of rectangular patch, as shown in Figure 16.5(a), to obtain dual band performance. Here, the rectangular patch is responsible for radiation in the frequency band centered at 5.6 GHz and the patch with added "L" shaped structure radiates amid 2.5 GHz band. Keeping same dimensions of antenna (in Figure 16.4(a)), the obtained dimensional parameters for the added patch are Ll = 9.5 mm, Wl = 14.5 mm, Lb = 6 mm, and Wb = 10.5 mm. The return loss obtained through simulation is −31 dB at 2.5 GHz and −28 dB at 5.6 GHz, as depicted in Figure 16.5(b). The −10 dB bandwidth of the antenna are 0.54 GHz (2.32 GHz–2.86 GHz) and 2.54 GHz (4.54 GHz–7.08 GHz), while −20 dB bandwidth is 0.15 GHz and 1.3 GHz around 2.5GHz and 5.6 GHz, respectively.

For further improvement in return loss, a rectangular slot is etched in rectangular patch at center as depicted in Figure 16.6(a). This is the proposed antenna, where the complete patch looks like mirror image of number "6." The optimized dimensions of the slot are Sl = 9.1 mm and Sw = 5 mm. The dimension used for the substrate for

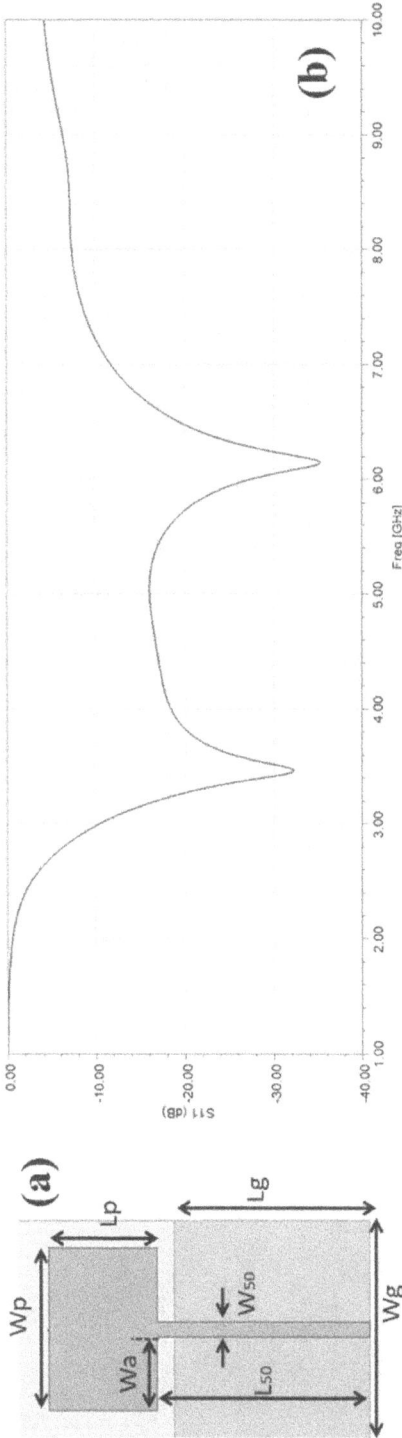

FIGURE 16.4 (a) Antenna with partial ground and 50 ohm microstrip fed rectangular patch; and (b) Its return loss characteristic plot.

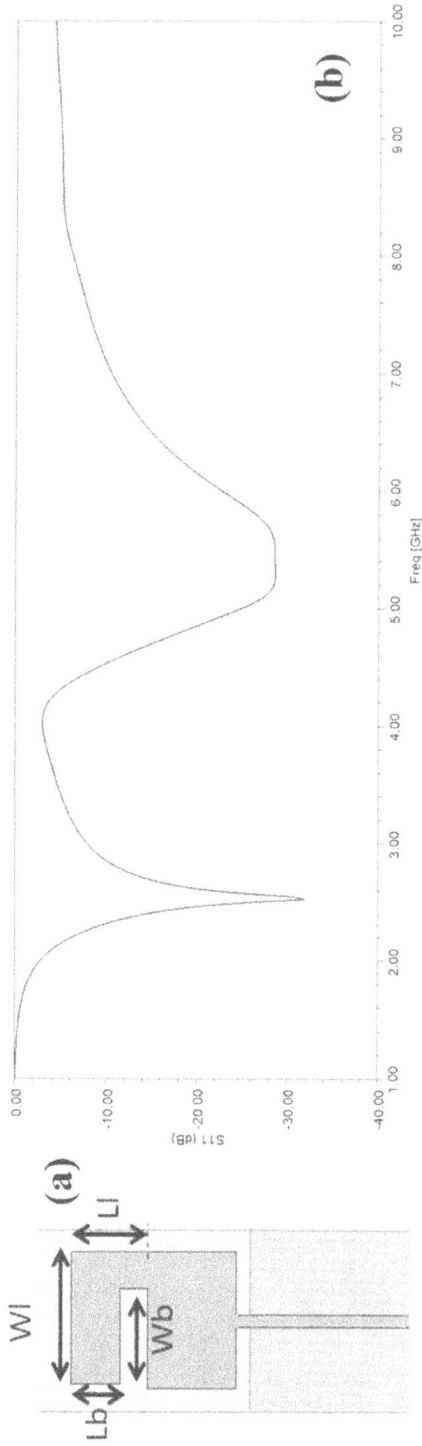

FIGURE 16.5   (a) Antenna using the combination of rectangular and "L" shaped patch; and (b) The simulated return loss for the antenna.

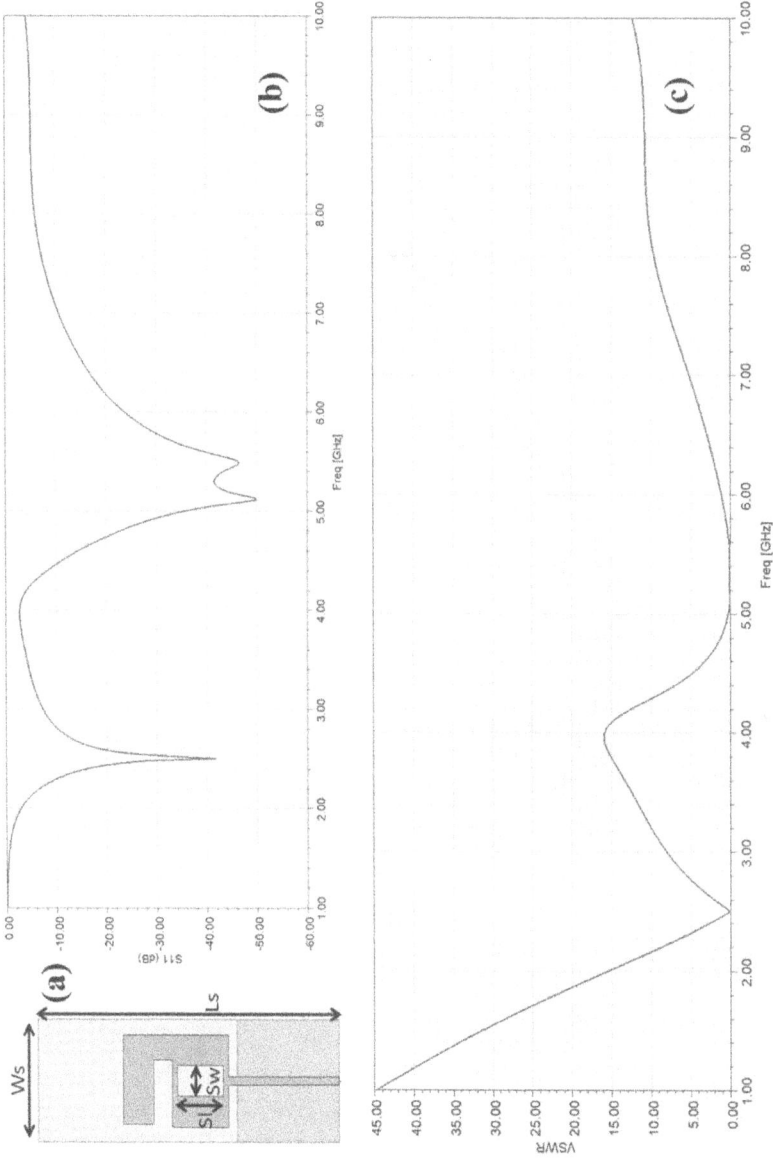

FIGURE 16.6 (a) Proposed microstrip antenna configuration; (b) The return loss plot for proposed antenna showing dual band characteristics; and (c) the VSWR plot of proposed antenna.

all the designs are Ls = 58.5 mm and Ws = 20 mm. Figure 16.6(b) and (c) illustrates the simulated return loss and VSWR plot; however, Figure 16.7(a)–(d) depicts the 2D gain plot for 2.5 GHz and 5.59 GHz, and 3D radiation pattern for 2.5 GHz and 5.59 GHz, respectively. The return loss at 2.5 GHz is −41.6 dB and at 5.6 GHz is −39.28 dB. The −10 dB bandwidth is 0.52 GHz (2.3 GHz–2.82 GHz) and 2.67 GHz (4.46 GHz–7.13 GHz); however, the −20 dB bandwidth is 0.15 GHz and 1.44 GHz. The VSWR of the proposed antenna for the bands 2.43 GHz to 2.6 GHz and 4.71 GHz to 6.25 GHz is well below 2, that depicts the matched condition.

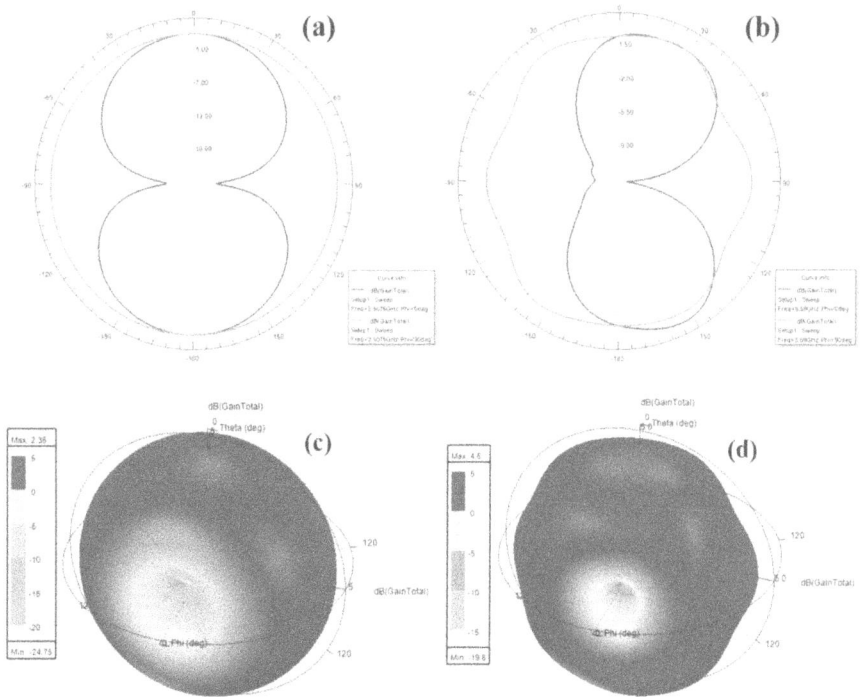

FIGURE 16.7 2D gain plot for (a) 2.5 GHz and (b) 5.59 GHz; and the 3D radiation pattern for (c) 2.5 GHz and (d) 5.59 GHz.

16.4 CONCLUSION

The modern society will use a variety of IoT/IIoT enabled devices in industry, home appliances, healthcare industry, and telemedicine, etc. This requires various application specific and multiband antennas for wireless dissemination of the data. However, all antennas for IoT/IIoT application have common requirement of miniaturization, light weight, cheap, easy fabrication, and easy integration as they may be used in smart watch, finger ring, water pipe, etc. The microstrip technology is one of the best technologies for antenna design to fulfill above mentioned requirements.

In this work, a short review of various IoT/IIoT antennas is presented, which are designed for RFID technology, smart construction, wearable technology and RF energy harvesting. Further, a novel microstrip antenna configuration with dual band performance for IoT/IIoT application is proposed and detailed. The proposed antenna comprises partial ground at the back of dielectric substrate and a 50 ohm microstrip fed patch (shape like mirror image of number "6") at top of the substrate. The proposed antennas shows appreciable return loss of −41.6 dB at 2.5 GHz and −39.28 dB at 5.6 GHz frequencies. The VSWR of the proposed antenna at these frequencies is very close to 0 dB. The simulation result depict −10 dB bandwidth of 0.52 GHz (2.3 GHz–2.82 GHz) and 2.67 GHz (4.46 GHz–7.13 GHz), which covers the IoT/IIoT bands of 2.4 GHz to 2.4835 GHz, 5.150 GHz to 5.350 GHz, and 5.470 GHz −5.725 GHz. Considering the simulation results, the proposed antennas may be a better candidate for future IoT/IIoT, Wi-Fi, and other wireless devices.

REFERENCES

1. Sanil N, Venkat PAN, Ahmed MR. 2018. Design and Performance Analysis of Multiband Microstrip Antennas for IoT applications via Satellite Communication. *In: 2018 Second International Conference on Green Computing and Internet of Things (ICGCIoT), Bangalore, India*; 60–63.
2. Olan-Nuñez KN, Murphy-Arteaga RS, Colmn-Beltrán E. 2020. Miniature patch and slot microstrip arrays for IoT and ISM band applications. *IEEE Access*. 8:102846–102854.
3. Su Z, Klionovski K, Bilal RM, Shamim A. 2018. 3D Printed Near-Isotropic Asymmetric Dipole Antenna-on-Package for IoT Applications. *In: 2018 IEEE Indian Conference on Antennas and Propagation (InCAP). IEEE*; 1–3.
4. Liu Q, Shen J, Yin J, Liu H, Liu Y. 2015. Compact 0.92/2.45-GH dual-band directional circularly polarized microstrip antenna for handheld RFID reader applications. *IEEE Trans Antennas Propag*. 63(9):3849–3856.
5. Falade OP, Rehman MU, Gao Y, Chen X, Parini CG. 2012. Single feed stacked patch circular polarized antenna for triple band GPS receivers. *IEEE Trans Antennas Propag*. 60(10):4479–4484.
6. Liu Q, Liu Y, Wu Y, Li S, Yu C. 2013. Design of a compact wideband circularly polarized microstrip antenna. *Microw Opt Technol Lett*. 55(11):2531–2536.
7. Guo Y-X, Khoo K-W, Ong LC. 2008. Wideband circularly polarized patch antenna using broadband baluns. *IEEE Trans Antennas Propag*. 56(2):319–326.
8. Hsu H-T, Huang T-J. 2014. A 1-by-2 dual-band antenna array for Radio-Frequency Identification (RFID) handheld reader applications. *IEEE Trans Antennas Propag*. 62(10):5260–5267.
9. Chang T-N, Lin J-M. 2011. Serial aperture-coupled dual band circularly polarized antenna. *IEEE Trans Antennas Propag*. 59(6):2419–2423.
10. Caso R, Michel A, Rodriguez-Pino M, Nepa P. 2014. Dual-band UHF-RFID/WLAN circularly polarized antenna for portable RFID readers. *IEEE Trans Antennas Propag*. 62(5):2822–2826.
11. Abdulkawi WM, Sheta A-FA. 2018. Design of Chipless RFID Tag Based on Stepped Impedance Resonator for IoT Applications. *In: 2018 International Conference on Innovation and Intelligence for Informatics, Computing, and Technologies (3ICT). IEEE*; 1–4.
12. Gbadamosi A-Q, Oyedele L, Mahamadu A-M, Kusimo H, Olawale O. 2019. The role of Internet of Things in delivering smart construction. *In:* CIB World Building Congress. 17–21. https://uwe-repository.worktribe.com/output/1492573

13. Sisinni E, Saifullah A, Han S, Jennehag U, Gidlund M. 2018. Industrial internet of things: Challenges, opportunities, and directions. *IEEE Trans Ind Informatics.* 14(11):4724–4734.

14. Xu J, Lu W. 2018. Smart Construction from Head to Toe: A Closed-Loop Lifecycle Management System Based on IoT. *In: Construction Research Congress 2018. New Orleans*; 157–168.

15. Duong TT, Duong TH. 2020, A dual-band MIMO antenna using gradient arcs for construction monitoring and inspection systems based on IIoT. *J Sci Technol Inf Commun.* 1(2):3–8.

16. Banerjee S, Singh A, Dey S, Chattopadhyay S, Mukherjee S, Saha S. 2019. SIW Based Body Wearable Antenna for IoT Applications. *In: 2019 International Conference on Opto-Electronics and Applied Optics (Optronix)*. IEEE; 1–4.

17. Kumar R, Singh J, Sohi BS, Mohali CCG. 2016. Hexagonal shaped body wearable textile antenna on EBG substrate material. *Int J Computer Sci Mob Computing.* 5(6):260–266.

18. Karpagavalli S, Shaaru Nivetha SR, Roland DS. 2016. EBG. Enhancement of Gain in Microstrip Patch Antenna Using EBG Structure for WLAN Application. *In: International Conference on Innovations in Engineering and Technology. Tamilnadu, India*; 429–436.

19. Banu MA, Tamilselvi R, Rajalakshmi M, Lakshmi MP. 2020. IoT-Based Wearable Micro-Strip Patch Antenna with Electromagnetic Band Gap Structure for Gain Enhancement. *In:* Inventive Communication and Computational Technologies. *Springer*; 1379–1396.

20. Lee H, Choi J. 2017. A Polarization Reconfigurable Textile Patch Antenna for Wearable IoT Applications. *In: 2017 International Symposium on Antennas and Propagation (ISAP). IEEE*; 1–2.

21. Azeez HI, Yang H-C, Chen W-S. 2019. Wearable triband E-shaped dipole antenna with low SAR for IoT applications. *Electronics.* 8(6):665.

22. Porter E, Bahrami H, Santorelli A, Gosselin B, Rusch LA, Popović M. 2016. A wearable microwave antenna array for time-domain breast tumor screening. *IEEE Trans Med Imaging.* 35(6):1501–1509.

23. Shrestha S, Agarwal M, Hemati A, Ghane P, Varahramyan K. 2012. Breast tumour detection by flexible wearable antenna system. *Int J Comput Aided Eng Technol.* 4(6):499–516.

24. Guay P, Gorgutsa S, LaRochelle S, Messaddeq Y. 2017. Wearable contactless respiration sensor based on multi-material fibers integrated into textile. *Sensors.* 17(5):1050.

25. Qin W. 2016. Research of antenna for microwave energy transmission system for IoT. *DEStech Trans Eng Technol Res.* ICETA 83–88. DOI: 10.12783/dtetr/iceta2016/6978.

26. Shafique K, Khawaja BA, Khurram MD, et al. 2018. Energy harvesting using a low-cost rectenna for Internet of Things (IoT) applications. *IEEE Access.* 6:30932–30941.

27. Raghavandaar M, Prathiksh M, Revathy S, Habiba HU. 2020. Energy Harvesting Using 2.45 GHz Rectenna for Powering Sensors in IoT Devices. *In: 2020 International Conference on Electronics, Information, and Communication (ICEIC). IEEE*; 1–3.

Index

Note: Locators in *italics* represent figures and **bold** indicate tables in the text.

For Product Safety Concerns and Information please contact our EU
representative GPSR@taylorandfrancis.com
Taylor & Francis Verlag GmbH, Kaufingerstraße 24, 80331 München, Germany

9 780367 702083